Proceedings of the Fourth International Symposium on Selenium in Biology and Medicine
Held July 18–21, 1988, Tübingen, FRG
Sponsored by the International Selenium Organizing Committee in cooperation with the Deutsche Forschungsgemeinschaft

National Committee

Dr. A. Wendel
Program Chairman, University of Konstanz, Konstanz, FRG
Dr. L. Flohé
Grünenthal GmbH, Research Center, Aachen, FRG
Dr. J. Lombeck
University of Düsseldorf, Düsseldorf, FRG
M. Fausel
University of Tübingen, General Secretary, Tübingen, FRG

International Committee

Dr. R.F. Burk
Vanderbilt University, Nashville, TN, USA
Dr. J. Chen
Institute of Nutrition and Food Hygiene, Beijing, People's Republic of China
Dr. G.F. Combs, Jr.
Cornell University, Ithaca, NY, USA
Dr. A.T. Diplock
Guy's Hospital, London, UK
Dr. O.A. Leander
US Department of Agriculture, Human Nutrition Research Center, Beltsville, MD, USA
Dr. J. Spallholz
Texas Tech University, Lubbock, TX, USA
Dr. T.C. Stadtman
National Institutes of Health, Bethesda, MD, USA
Dr. G.L. Xu
University of Xian, People's Republic of China

A. Wendel (Ed.)

Selenium in Biology and Medicine

With 126 Figures and 96 Tables

Springer-Verlag Berlin Heidelberg New York
London Paris Tokyo Hong Kong

Professor Dr. ALBRECHT WENDEL
Universität Konstanz
Fakultät für Biologie
Postfach 5560
7750 Konstanz, FRG

ISBN-13:978-3-642-74423-5 e-ISBN-13:978-3-642-74421-1
DOI: 10.1007/978-3-642-74421-1

Library of Congress Cataloging-in-Publication Data. Selenium in biology and medicine / edited by A. Wendel. p. cm. Includes index. ISBN-13:978-3-642-74423-5 (U.S.: alk. paper) 1. Selenium – Physiological effect. 2. Organoselenium compounds – Physiological effect. 3. Selenium – Health aspects. I. Wendel, A. (Albrecht), 1943– . QP535.S5S442 1989 599′.019214 – dc20

Softcover reprint of the hardcover 1st edition 1989

Typesetting: K+V Fotosatz GmbH, Beerfelden

2131/3145-543210 – Printed on acid-free paper

Preface

In July, 1988, more than 300 scientists from 29 different countries gathered at Tübingen, W. Germany, in order to spend 4 days discussing their favourite trace element, selenium. This meeting continued the good tradition of three previous meetings held in Corvallis/Oregon, 1976, in Lubbock/Texas, 1980, and in Beijing/China, in 1984. Incidentally, the University of Tübingen provided a unique historical background for a Symposium devoted to recent advances in biochemistry, pharmacology, human nutrition and human health; here, the first independent department of Physiological Chemistry in Germany was founded in 1845. Professor Hoppe-Seyler elucidated here the hemin structure and his student Friedrich Miescher discovered the nucleic acids.

This book contains one-half of the oral or poster presentations which were selected before the meeting was held on the basis of a one-page abstract. It is the reader who will have to decide whether this time-consuming policy of quality assessment was warranted or not.

The chapters of this volume reflect the areas of research which are at present either most popular or most under debate. Examples include the diagnosis of the selenium status in man in different countries, the question whether individual or collective saturation leads exist for the trace element, and the usefulness of supplementation against disease or for general improvement of live expectance. Several contributions in this book provide excellent material with respect to this field which allow unequivocal conclusions. For the first time, data are now available on the threshold of chronic toxicity of selenium in man. The interaction of selenium with malignant growth will still remain an area of disagreement; defenders of the pro as well as the contra view are represented in a balanced way in these proceedings. On the other hand, a new field has emerged by the development of new selenium-containing drugs. Especially ebselen, which was represented by a single contribution at the preceding meeting, has attracted widespread interest.

This meeting would have not been possible without generous support from several countries which illustrates a common goal within a growing internationally communicating group of scientists. The local success of the conference was entirely due to my highly motivated co-workers who did a beautiful job. We are looking forward to the next International Symposium on Selenium in Biology and Medicine to be held in 1992 in Nashville, Tennessee.

Konstanz, Spring 1989 A. WENDEL

Contents

Part II Influence of Selenium on the Metabolism of Animals and Man

Part IV Selenium and Human Health

Part V Selenium and Malignant Growth

List of Contributors

You will find the addresses at the beginning of the respective contribution

Part I Biochemistry of Selenoproteins and Selenium Compounds

The Regulation of Glutathione Peroxidase Gene Expression; Implication for Species Difference and the Effect of Dietary Selenium Manipulation

H. TOYODA[1], S. HIMENO[1], and N. IMURA[1,2]

1 Introduction

Glutathione peroxidase (GSH-Px) is considered to be a housekeeping enzyme, functioning as an important antioxidant enzyme which protects cells from oxidative damage. GSH-Px protein contains a selenocysteine (Se-Cys) moiety in its active site, but the mechanism of Se-Cys incorporation into the polypeptide chain of the enzyme protein has long been unresolved. Chambers et al. (1986) discovered that a UGA termiation codon in the open reading frame of murine GSH-Px mRNA was utilized as a codon for Se-Cys. This finding has encouraged the elucidation of the regulatory mechanism of GSH-Px gene expression. Evidence so far obtained has implied a quite complicated regulatory system for the synthesis of this selenoprotein.

In the present study, we aimed to clarify which step in GSH-Px gene expression is most crucial in regulating the activity of the enzyme. We focused on two well-known facts relating to the activity of GSH-Px; one is species difference in GSH-Px activity observed among some rodents and the other is the alteration of the activity by dietary selenium (Se) manipulation. To elucidate which step of the process in GSH-Px gene expression is responsible for the species difference, the copy number of GSH-Px gene and the level of mRNA in tissues of mice, rats and guinea pigs were examined by using molecular biological techniques. On the other hand, it has been reported that progressive Se-deficiency resulted in a decrease in the level of GSH-Px protein parallel to GSH-Px activity (Knight and Sunde 1987; Takahashi et al. 1986). To verify which step of the process in GSH-Px gene expression is most sensitive to the manipulation of tissue Se level, we performed a detailed time-course study concerning the effects of Se-depletion and repletion on the level of GSH-Px mRNA and the enzyme activity in mice.

2 Materials and Methods

Male mice (ICR, 20–25 g), rats (Wistar, 150–200 g) and guinea pigs (Hartley, 180–200 g) were used. Liver, kidney and heart tissues were collected from each

[1] Department of Public Health, School of Pharmaceutical Sciences, Kitasato University, 9-1, Shirokane 5 Chome, Minato-ku, Tokyo 108, Japan
[2] To whom all correspondence should be addressed

species of animals. In the time-course experiment of Se-depletion and repletion, male ICR mice (4-weeks old) were fed a Se-deficient diet based on torula yeast for 6 weeks and then a Se-supplemented diet for 3 weeks. In a Se-supplemented diet, sodium selenite (0.4 ppm as Se) was added to the Se-deficient diet. Mice were killed on days 0, 3, 5, 7, 10, 14, 21, 28, 35 and 42 during the period of Se-depletion, and then on days 3, 5, 7 and 10 after the Se-supplementation. Liver and kidney tissues were collected and frozen at −80 °C for the following measurements.

GSH-Px activity was assayed by the method of Lawrence and Burk (1976) using hydrogen peroxide as a substrate. Se concentration was determined fluorometrically by Watkinson's method (1966).

To determine the copy number of the GSH-Px gene, genomic DNAs were extracted from mouse, rat and guinea pig livers and subjected to Southern blot analysis (Maniatis et al. 1982) after digestion with restriction endonucleases, such as Hind III, EcoR I, BamH I and Bgl II. As a probe, GSH-Px cDNA which has been cloned from mouse liver cDNA library (generous gift from Dr. S. Kuge, School of Medicine, Kitasato University) was used. The detailed condition for the hybridization will be published elsewhere (Toyoda et al. submitted).

To compare the amount of GSH-Px mRNA, total cellular RNA was extracted from tissue homogenate, subjected to Northern blot analysis (Maniatis et al. 1982), using mouse GSH-Px cDNA probe, and the intensity of bands in the autoradiogram was quantified by densitometric scanning. The hybridizing condition for nick-translated probes was the same as that in Southern blot analysis. As a control, the level of superoxide dismutase (SOD) mRNA was also determined by using human SOD cDNA fragment provided by Nippon Kayaku Co. Ltd. as a probe.

3 Results and Discussion

3.1 Species Difference in GSH-Px Activity

Comparison of GSH-Px activity in the liver, kidney and heart among three species of animals revealed that the enzyme activity of guinea pig was extremely low in all the organs examined. However, Se concentrations in guinea pig organs are not so much different from those in other animals. This indicates that a distinctly low activity of this selenoenzyme in guinea pig cannot be explained by the tissue Se level. Therefore, we performed Southern blot and Northern blot analyses to determine which step of the process in GSH-Px gene expression is responsible for the above-mentioned species difference in GSH-Px activity.

First, genomic DNAs extracted from the liver of each animal species were subjected to Southern blot analysis. The resulting autoradiogram showed that DNA extracts obtained from all three species including guinea pig gave a single hybridizing band with mouse GSH-Px cDNA probe in all cases of digestion with different restriction enzymes, indicating that these three species have a single copy of the GSH-Px gene in their genomic DNAs.

Second, the mRNA levels of GSH-Px in the liver, kidney and heart of the three animal species were examined by Northern blot analysis. The intensity of the bands in the autoradiogram was compared under the condition in which the genomic DNAs of each animal species should give the same intensity of signals with the mouse cDNA probe. The amount of mRNA varied to a large extent either between species or between tissues. In tissues of guinea pig, no GSH-Px mRNA was detectable in the autoradiogram even with a longer period of exposure.

Considering the experimental fact described above together with the results of Southern blot analysis, the extremely low activity of GSH-Px in guinea pig tissues should be ascribed to a lack of capacity to transcribe the GSH-Px gene. However, the possibility cannot be excluded that the decreased level of mRNA in guinea pig tissues is a mere reflection of a species-specific degradation rate of mRNA as a post-transcriptional control. Further study such as an in vitro nuclear run-off experiment is needed to clarify this point.

On the other hand, tissue specificity of GSH-Px activity within a species cannot be solely explained by the level of mRNA. Additional factors to those involved in transcription may participate in the regulation of the GSH-Px gene in the respective tissues.

3.2 Effects of Dietary Se Manipulation on the Level of GSH-Px mRNA in Mice

Feeding of mice with a Se-deficient diet for 6 weeks resulted in substantial decrease in GSH-Px activity as well as Se concentration in both the liver and kidney (Table 1). Northern blot analysis for RNA extracts of mouse tissues revealed that the levels of GSH-Px mRNA also decreased remarkably in the liver and kidney after 6 weeks of Se-depletion. Whereas the mRNA levels of SOD or metallothionein-I was not altered by Se-deficiency, suggesting that the depression of the

Table 1. Percent changes in mRNA level and enzyme activity of GSH-Px and Se concentration in the liver and kidney of mice fed a Se-deficient diet

Days after Se-depletion	Liver			Kidney		
	mRNA level	Enzyme activity	Se conc.	mRNA level	Enzyme activity	Se conc.
0	100[a]	100	100	100	100	100
3	83	96	65	90	82	83
7	63	38	43	106	65	70
10	65	22	31	82	41	59
21	41	9	19	54	17	41
28	26	6	16	42	13	36
35	14	4	10	31	7	18
42	5	3	8	23	5	18

[a] Percent of control values obtained from Se-supplemented mice.

Table 2. Percent recovery of mRNA level and enzyme activity of GSH-Px and Se concentration in mouse liver by Se-repletion

Days after Se-repletion	mRNA level	Enzyme activity	Se conc.
0	5[a]	3	7
3	69	16	25
5	84	36	36
7	80	53	58
10	88	73	76

[a] Percent of control values obtained from Se-supplemented mice killed at day 0.

amount of mRNA caused by Se-depletion is specific for GSH-Px. Our results are in good accordance with the reports by Saedi et al. (1988) which also showed an extremely low level of GSH-Px mRNA in the liver of rats fed a Se-deficient diet for about 20 weeks.

The time-course experiment for Se-depletion and repletion revealed a more detailed relationship between the level of GSH-Px mRNA and the activity of GSH-Px in mice. As shown in Table 1, the drop in GSH-Px activity preceded the decrease in the mRNA level in both the liver and kidney, especially 10 days after the start of feeding with the Se-deficient diet. The delay in the decrease of mRNA levels at the early phase of Se-deficiency was more prominent in the kidney than in the liver.

After the dietary Se-repletion (Table 2), both the enzyme activity and Se concentration in the liver increased gradually and reached almost normal values 2 weeks after the change of the diet. In contrast, the recovery of the GSH-Px mRNA level after the diet substitution was markedly rapid, as can be seen in the mRNA level on day 3.

Therefore, the reduced amount of mRNA should be a primary cause for the decreased activity of GSH-Px at the late phase of Se-deficiency, while the decrease in the enzyme activity at the early phase should be ascribed to post-transcriptional factors rather than those involved in transcription. The concept that the amount of mRNA is not the sole factor for determining GSH-Px activity in Se-manipulated mice is strongly supported by the lag between rapid recovery of the mRNA level and the slow rate of increase in the enzyme activity in the liver of Se-repleted mice.

Thus, the results of the present study designed to investigate the regulatory mechanism of GSH-Px gene expression, demonstrated that the species difference observed in rodents can be due to the incapability in gene transcription and that the alteration in GSH-Px activity in mice subjected to dietary Se manipulation is ascribed not only to transcriptional but also to post-transcriptional regulation.

Acknowledgments. We are grateful to Dr. Shusuke Kuge for mouse liver cDNA library, to Dr. Akio Nomoto for synthetic oligo-nucleotides for the cloning of mouse GSH-Px cDNA and to Nippon Kayaku Co. Ltd. for human SOD cDNA fragment. This work was supported in part by a research grant awarded by the Ministry of Education, Science and Culture, Japan (No. 6378019).

References

Chambers I, Frampton J, Goldfarb P, Affara N, McBain W, Harrison PR (1986) The structure of the mouse glutathione peroxidase gene: the selenocysteine in the active site is encoded by the "termination" codon, TGA. EMBO J 5:1221–1227

Knight SAB, Sunde RA (1987) The effect of progressive selenium deficiency on anti-glutathione peroxidase antibody reactive protein in rat liver. J Nutr 117:732–738

Lawrence RA, Burk RF (1976) Glutathione peroxidase activity in selenium-deficient rat liver. Biochem Biophys Res Commun 71:952–958

Maniatis T, Fritsch EF, Sambrook J (1982) Molecular cloning: a laboratory manual. Cold Spring Harbor Laboratory, Cold Spring Harbor, New York

Saedi MS, Smith CG, Frampton J, Chambers I, Harrison PR, Sunde RA (1988) Effect of selenium status on mRNA levels for glutathione peroxidase in rat liver. Biochem Biophys Res Commun 153:855–861

Takahashi K, Newburger PE, Cohen HJ (1986) Glutathione peroxidase protein: absence in selenium deficiency states and correlation with enzymatic activity. J Clin Invest 77:1402–1404

Watkinson JH (1966) Fluorometric determination of selenium in biological material with 2,3-diaminonaphthalen. Anal Chem 38:92–97

Regulation of Expression of Glutathione Peroxidase by Selenium

R.A. SUNDE[1,2], M.S. SAEDI[1], S.A.B. KNIGHT[1], C.G. SMITH[1], and J.K. EVENSON[1]

1 Introduction

Selenium was shown in 1957 to be an essential dietary nutrient for the rat (Schwarz and Foltz 1957). In the same year, Mills (1957) reported the discovery of an enzyme, glutathione peroxidase (GSH-Px, GSH:H_2O_2 oxidoreductase, EC 1.11.1.9), that was important for protection of erythrocytes against hemolysis. These two discoveries, plus the apparent interaction of sulfur compounds with Se and GSH-Px, led J.T. Rotruck and W.G. Hoekstra at Wisconsin in 1971 to identify GSH-Px as a Se-dependent enzyme (Rotruck et al. 1972, 1973). The background to this discovery was described succinctly at the Second Selenium in Biology and Medicine Symposium (Rotruck 1981).

2 Effect of Selenium Status on GSH-Px Activity

I was lucky to have chosen to do a senior thesis project in Prof. Hoekstra's laboratory with D.G. Hafeman in the fall of 1971. Following Rotruck's discovery that GSH-Px was a selenoenzyme, we supplemented weanling rats with 0, 0.05, 0.1, 0.5, 1.0, and 5.0 ppm Se (as Na_2SeO_3) in a 30% torula yeast-based diet (containing 0.008 ppm Se and supplemented with 50 IU of vitamin E and 0.3% D,L-methionine to prevent liver necrosis and to promote adequate growth) (Hoekstra et al. 1973; Hafeman et al. 1974). Erythrocyte GSH-Px activity decreased exponentially over the first 66 days with a half-life of 30 days when the rats were fed the Se-deficient diet (Fig. 1A). Supplementation with 0.05 ppm Se slowed the decline in erythrocyte GSH peroxidase ($t_{1/2} = 45$ days) over the first 45 days of supplementation; supplementation with 1.0 or 5.0 ppm Se elicited a rapid rise in erythrocyte GSH-Px. After 45 days, erythrocyte GSH-Px levels began increasing in all rats supplemented with Se, whereas the Se-deficient rats continued to have erythrocyte GSH-Px activities less than 20% of the initial level. Liver GSH-Px activity responded to Se deficiency in a similar fashion, but the decrease was more rapid in liver ($t_{1/2} = 4$ days) than in erythrocytes, and liver GSH-

[1] Department of Biochemistry and Department of Nutrition and Food Science, University of Arizona, 309 Shantz Building, Tucson, AZ 85721, USA
[2] To whom all correspondence should be addressed

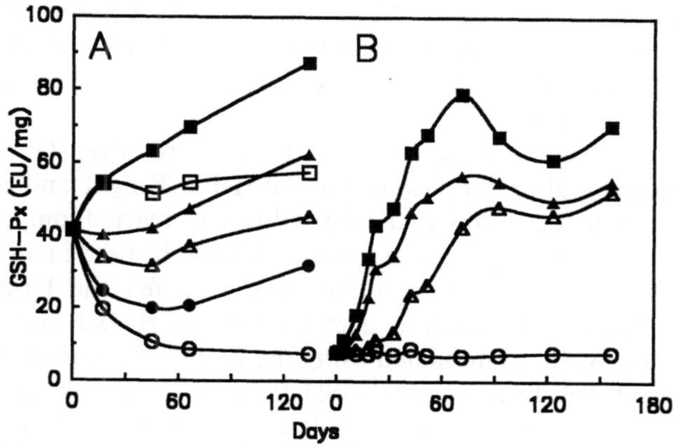

Fig. 1A, B. Erythrocyte GSH-Px activity. **A** Mean GSH-Px activities of weanling rats that were supplemented with 0 (○), 0.05 (●), 0.1 (△), 0.5 (▲), 1 (□) and 5 ppm Se (■). **B** Mean GSH-Px activities of rats fed a Se-deficient diet for 9 months that were resupplemented with 0 (○), 0.1 (△), 0.5 (▲), and 5 ppm Se (■) for the indicated days. GSH-Px activities are expressed in direct-assay units and replotted from Hafeman et al. (1974)

Px activity in rats supplemented with 1.0 or 5.0 ppm Se did not increase above that observed in rats fed 0.5 ppm Se (Hoekstra et al. 1973; Hafeman et al. 1974).

When long-term Se-deficient rats were resupplemented with 0.1, 0.5, or 5.0 ppm Se (Hoekstra et al. 1973; Hafeman et al. 1974), erythrocyte GSH-Px levels were restored to levels at or above the initial levels (Fig. 1 B). GSH-Px activities in rats resupplemented with 0.5 or 5.0 ppm Se increased significantly within 4–10 days, whereas the increase in GSH-Px activity in rats supplemented with 0.1 ppm Se was delayed over 20 days as compared to the Se-deficient rats. This suggests that with 0.1 ppm Se, other tissues had a higher priority for Se than did erythrocytes in Se-deficient rats.

These experiments were the first Se depletion/repletion studies that monitored GSH-Px activity. The studies clearly demonstrated that GSH-Px activity declines in the Se-deficient animal, and that GSH-Px activity can be restored by dietary Se repletion. These reports were quickly followed by other experiments which further illustrated that GSH-Px activity is regulated by the dietary Se content in animals and man (cf. National Research Council 1983). Rotruck et al. (1973) also reported that addition of Se to Se-deficient hemolysates would not restore GSH-Px activity, and that dialysis of Se-adequate hemolysates would not result in loss of GSH-Px activity. This indicates that more than in vitro addition of selenite is necessary for activity and that Se is bound tightly to GSH-Px.

3 Effect of Selenium Status on GSH-Px Protein

It is now known that Se in GSH-Px is localized at the active site of the enzyme (Epp et al. 1983) and that loss or derivation of Se results in complete loss of enzyme activity. Thus, in hindsight, it is not surprising that changes in nutritional

Se status would alter the cellular Se concentration, which in turn would result in loss or restoration of enzyme activity. At Arizona we next became interested in whether or not changes in Se status would affect the level of the protein portion of GSH-Px.

We injected rabbits with purified rat liver GSH-Px in order to produce polyclonal antibodies against GSH-Px (Knight and Sunde 1987). The resulting antisera coprecipitated with GSH-Px, reacted only with GSH-Px during immunoblotting of liver cytosol, and could be used in an ELISA to quantitate GSH-Px protein. When weanling rats were fed the basal diet supplemented with 0.2 ppm Se for 28 days, liver cytosolic GSH-Px activity increased 66% and GSH-Px protein, as measured using the ELISA, increased 50%. In rats fed the Se-deficient diet, liver cytosolic GSH-Px activity decreased to undetectable levels by day 21 with an exponential half-life of 2.8 days, and GSH-Px protein also decreased exponentially with a half-life of 5.2 days (Knight and Sunde 1987). Thus, Se deficiency decreases both GSH-Px protein and activity.

To further study the effect of dietary Se on GSH-Px expression, Se-deficient rats were resupplemented with 0.1 or 0.5 ppm Se for 14 days (Knight and Sunde 1988). With 0.1 ppm Se, neither GSH-Px activity nor protein was increased significantly until 14 days after the start of repletion. With 0.5 ppm Se, both liver GSH-Px activity and protein were significantly increased 1 day after the start of Se repletion, and both parameters increased in a hyperbolic manner such that by day 7 for GSH-Px activity and day 5 for GSH-Px protein, the values were no longer significantly different from Se-adequate rats. By interpolation, it took 6.6 and 3.8 days of resupplementation with 0.5 ppm Se to raise GSH-Px activity and protein, respectively, to 50% of the levels found in Se-adequate rats (Knight and Sunde 1988). Additional short-term injection studies revealed that the increase in detectable GSH-Px activity and protein was delayed 12 h in Se-deficient rats in spite of an increase in liver Se levels within 1 h of injection (Knight and Sunde 1988). This indicates that more than just an increase in cellular Se is necessary for increases in GSH-Px protein as well as activity.

4 Effect of Selenium Status on GSH-Px mRNA

I. Chambers in P.R. Harrison's laboratory in Glasgow reported in 1986 that a gene that they had cloned and sequenced was the gene for murine GSH-Px (Chambers et al. 1986). Unknowingly, they had solved a basic question of Se biochemistry by showing that the codon for the selenocysteine ([Se]Cys) moiety of GSH-Px is the nonsense codon, "UGA", and that this codon lies in the middle of the open reading frame of the first exon. This cloned GSH-Px gene also provided us with a probe for further characterization of the expression of GSH-Px.

To determine whether or not Se status controls the level of liver mRNA for GSH-Px, we isolated total cellular RNA from the livers of mature Se-adequate rats (liver GSH-Px: 0.615 EU mg^{-1} protein) or from long-term (>135 days) Se-deficient rats (liver GSH-Px: 0.025 EU mg^{-1}) using the guanidine isothiocyanate method (Saedi et al. 1988). Total polyadenylated RNA (poly A^{+} RNA) was also

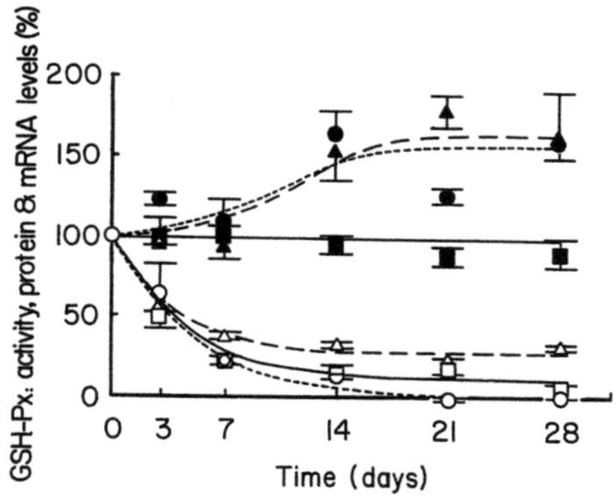

Fig. 2. Liver GSH-Px activity, protein, and mRNA. Liver from rats fed Se-deficient (○, △, □) or Se-adequate (●, ▲, ■) diets were assayed for GSH-Px activity (○, ●), GSH-Px protein (△, ▲) and GSH-Px mRNA (□, ■). The 100% value represents the mean at day 0 for GSH-Px activity (0.66 EU mg^{-1} protein, expressed in coupled-assay units, Knight and Sunde 1987), for GSH-Px protein ($A_{405} = 0.447$ for 1.5 µl liver cytosol), and for mRNA (30 mV as measured by densitometry). Error *bars* represent the SEM for three samples. Over the first 7 days of Se deficiency, GSH-Px activity, GSH-Px protein, and GSH-Px mRNA decreased exponentially with half-lives of 3.3, 5.0 and 3.2 days, respectively

isolated from one (±Se) set of total RNA using oligo-(d)T-cellulose chromatography. Sets of total RNA and poly A^+ RNA were subjected to RNA blot-hybridization using the 0.7 kb EcoR I fragment of the murine GSH-Px gene, kindly provided by P. R. Harrison and co-workers (Chambers et al. 1986). Autoradiography showed that total RNA from Se-adequate liver contained only a 13 S species that hybridized with the [^{32}P]-labeled 0.7 kb probe; 13 S is the same size as reported for murine GSH-Px mRNA (Chambers et al. 1986). The total RNA isolated from Se-deficient liver contained only 7–10% of the level of GSH-Px mRNA found in Se-adequate rat liver, and poly A^+ RNA from Se-deficient rat liver contained only 17% of that from Se-adequate rat liver. Control RNA blot-hybridization of the same RNA samples using a 0.4 kb probe from chicken *β*-actin showed that actin mRNA levels were unchanged by Se deficiency (Saedi et al. 1988).

To further examine the time course of this decrease in GSH-Px mRNA levels during progressive Se deficiency, weanling rats were fed a Se-deficient diet or supplemented with 0.2 ppm Se, and three rats per group were killed 0, 3, 7, 14, 21, and 28 days after the start of the experiment. Liver cytosol was assayed for GSH-Px activity and protein, and total cellular RNA was prepared and subjected to RNA blot-hybridization analysis to determine the effect of progressive Se deficiency on GSH-Px expression. As shown in Fig. 2, GSH-Px activity and GSH-Px protein increased, and GSH-Px mRNA levels were unchanged during the 28-day Se-supplementation period. In Se-deficient rats, GSH-Px activity and GSH-Px protein decreased exponentially ($t_{1/2} = 3.3$ and 5.0 days, respectively), clearly indicating that these rats were becoming Se-deficient. GSH-Px mRNA levels in this experiment also decreased exponentially (Fig. 2) to a level 6±2% of that observed in the Se-adequate rats. The half-life of the decrease in GSH-Px mRNA was 3.2

days. RNA blot-hybridization using the 0.4 kb β-actin probe again showed that β-actin levels were unchanged by Se deficiency, indicating that the effect on GSH-Px mRNA was not a general effect of Se deficiency on all mRNA species. Thus, Se status regulates the level of GSH-Px mRNA as well as GSH-Px activity and protein.

5 Discussion

These studies, begun in Prof. Hoekstra's laboratory at Wisconsin and continued at the University of Arizona, have shown that Se status regulates the expression of GSH-Px in the rat whether measured as GSH-Px activity, as antibody-reactive GSH-Px protein, or as GSH-Px mRNA. During Se depletion in the young, rapidly growing rat, GSH-Px activity and GSH-Px mRNA levels fell with a half-life of 3 days. The longer apparent half-life for GSH-Px protein may have been due to the persistence of inactive enzyme after the loss of activity, but it also may have been due to an artifact caused by low levels of contaminating antibodies in the anti-GSH-Px IgG fraction which elicited an additional, constant increment of reactivity in the ELISA. The 3-day half-life for GSH-Px activity is likely to be an overestimate for the half-life of individual GSH-Px molecules, because some Se is undoubtedly recycled as GSH-Px turns over. The 3-day half-life also implies that GSH-Px turnover is similar to the turnover of many other cytosolic liver enzymes (Waterlow et al. 1978).

The coordinated decrease in GSH-Px activity and mRNA levels as Se deficiency progresses thus suggests that changes in liver Se concentration may be the factor that regulates both GSH-Px enzyme and mRNA. With the discovery that the [Se]Cys moiety of GSH-Px is encoded by UGA (Chambers et al. 1986) and with our demonstration that serine is the precursor of the [Se]Cys (Sunde and Evenson 1987), we suggested that Se is incorporated into GSH-Px in a cotranslational process: our hypothesis is that Se incorporation is mediated by a tRNA that is specific for UGA, that is acylated by serine, and that the serine is coverted to [Se]Cys while attached to the tRNA (Sunde and Evenson 1987). Leinfelder et al. (1988) have recently discovered just such a tRNA in bacteria, thus strongly implicating such a process in higher animals. A low cellular Se concentration might block translation of GSH-Px at the UGA codon, thus leading to the observed fall in GSH-Px protein as well as activity. An even more involved process, however, would be necessary for Se regulation of the level of GSH-Px mRNA.

The discovery that GSH-Px is a Se-dependent enzyme led to the widespread use of tissue GSH-Px activity as an indicator of Se status. The results presented here indicate that the reason that GSH-Px activity proved to be an exceptionally good index of biochemical status, as compared to other nutrient-dependent enzymes or proteins (such as pyridoxal phosphate-dependent transaminases or serum iron-binding proteins), was because the expression of GSH-Px is linked directly to the Se status of the animal. Elucidation of the mechanism used to regulate GSH-Px mRNA will help to further characterize the relation between measured GSH-Px activity and biochemical Se status.

Acknowledgments. This work was supported in part by the University of Arizona Experiment Station, and by grants from the National Institutes of Health (DK 32942), the National Cancer Institute (CA 45164), and the International Life Sciences Institute-Nutrition Foundation.

References

Chambers I, Frampton J, Goldfarb P, Affara N, McBain W, Harrison PR (1986) The structure of the mouse glutathione peroxidase gene: the selenocysteine in the active site is encoded by the 'termination' codon, TGA. EMBO J 5:1221–1227

Epp O, Ladenstein R, Wendel A (1983) The refined structure of the selenoenzyme glutathione peroxidase at 0.2 nm resolution. Eur J Biochem 133:51–69

Hafeman DG, Sunde RA, Hoekstra WG (1974) Effect of dietary selenium on erythrocyte and liver glutathione peroxidase in the rat. J Nutr 104:580–587

Hoekstra WG, Hafeman DG, Oh SH, Sunde RA, Ganther HE (1973) Effect of dietary selenium on liver and erythrocyte glutathione peroxidase in the rat. Fed Proc 32:885

Knight SAB, Sunde RA (1987) The effect of progressive selenium deficiency on anti-glutathione peroxidase antibody reactive protein in rat liver. J Nutr 117:732–738

Knight SAB, Sunde RA (1988) Effect of selenium repletion on glutathione peroxidase protein level in rat liver. J Nutr 118:853–858

Leinfelder W, Zehelein E, Mandrand-Berthelot M-A, Bock A (1988) Gene for a novel tRNA species that accepts L-serine and cotranslationally inserts selenocysteine. Nature (London) 331:723–725

Mills GC (1957) Hemoglobin catabolism. I. Glutathione peroxidase, an erythrocyte enzyme which protects hemoglobin from oxidative breakdown. J Biol Chem 229:189–197

National Research Council (1983) Selenium in nutrition. National Academy Press, Washington DC

Rotruck JT (1981) Discovery of the role of selenium in glutathione peroxidase. In: Spallholz JE, Martin JL, Ganther HE (eds) Selenium in biology and medicine. AVI Press, Westport CN, pp 10–16

Rotruck JT, Hoekstra WG, Pope AL, Ganther H, Swanson A, Hafeman D (1972) Relationship of selenium to GSH peroxidase. Fed Proc 31:691

Rotruck JT, Pope AL, Ganther HE, Swanson AB, Hafeman DG, Hoekstra WG (1973) Selenium: biochemical role as a component of glutathione peroxidase. Science 179:588–590

Saedi MS, Smith CG, Frampton J, Chambers I, Harrison PR, Sunde RA (1988) Effect of selenium status on mRNA levels for glutathione peroxidase in rat liver. Biochem Biophys Res Commun 153:855–861

Schwarz K, Foltz CM (1957) Selenium as an integral part of factor 3 against dietary necrotic liver degeneration. J Am Chem Soc 79:3292–3293

Sunde RA, Evenson JK (1987) Serine incorporation into the selenocysteine moiety of glutathione peroxidase. J Biol Chem 262:933–937

Waterlow JC, Garlick PJ, Millward DJ (1978) Protein turnover in mammalian tissues and in the whole body. North-Holland, Amsterdam New York Oxford, pp 481–528

Studies on New Selenoproteins and Specific Selenium Target Tissues

D. Behne[1,3], H. Hilmert[1], S. Scheid[1], H. Gessner[1], A. Kyriakopoulos[1], and W. Elger[2]

1 Introduction

Due to the early findings that Se deficiency or a combined low Se and low vitamin E status leads to diseases in certain tissues such as the liver, the skeletal muscle and the heart muscle, research has naturally concentrated on the role of Se in the affected tissues. Since the element was then identified as the essential component of glutathione peroxidase, the effects of Se have been investigated mainly with regard to the functions of this enzyme. Through our studies on Se metabolism we became interested in the following two questions: First, is Se of importance also in other parts of the organism? Second, are there other biologically important Se compounds besides glutathione peroxidase? The studies which are presented here deal with these two aspects of Se research.

2 Specific Selenium Target Tissues

In our investigations a great deal of information was obtained by comparing the distribution and the metabolism of Se in rats with a normal and a very low Se status. In experiments on Se in the reproductive organs we found in this way that regulation mechanisms exist with the help of which the organism strives to maintain the Se content in the testes during periods of insufficient Se intake (Behne et al. 1982). The results of some of our investigations carried out later showed, however, that the testis is not the only organ which in Se deficiency is preferentially supplied with the element (Behne and Höfer-Bosse 1984).

This also became evident in a long-term depletion experiment in which rats were fed a Se-deficient diet for about 8 months (Behne unpublished data). In Table 1 the element contents in some tissues and in the blood are shown for the deficient animals and for Se-supplemented controls. After the long depletion period the main Se pools, liver and muscle, and also the blood fractions had lost most of their Se and only contained between 0.2 and 5% of the normal Se level. In the pituitary, the testes and the adrenals, however, the decreases were comparatively small. This could not have been due solely to a greatly reduced turnover

[1] Hahn-Meitner-Institut Berlin, Postfach 390128, 1000 Berlin 39, FRG
[2] Schering AG, 1000 Berlin 65, FRG
[3] To whom correspondence should be addressed

Table 1. Effects of a low selenium diet on the selenium content in tissues of male rats after a feeding period of 8 months

Tissue	Selenium content[a]		Se−/Se+ (%)
	Se+	Se−	
Pituitary	1960 ± 300	680 ± 30	34.7
Testes	6660 ± 130	1300 ± 150	19.5
Adrenals	2250 ± 190	430 ± 20	19.1
Muscle	440 ± 50	24 ± 5	5.4
Erythrocytes	1560 ± 130	27 ± 6	1.7
Liver	2610 ± 170	29 ± 3	1.1
Plasma	4980 ± 480	9 ± 1	0.2

[a] The values for the Se content (μg Se kg^{-1} dry weight, means ± SD for four animals) were taken from an investigation (Behne, unpublished data) on the changes in the tissue Se levels in rats which from the age of 3 weeks onwards were fed either a low Se diet with less than 15 μg Se kg^{-1} (Se−) or the same diet with 250 μg Se kg^{-1} added as sodium selenite (Se+).

Table 2. Effects of the selenium status on the selenium retention in rats[a]

Tissue	RF	Tissue	RF
Brain	51.9	Heart	1.9
Thyroid	46.9	Muscle	1.8
Pituitary	42.0	Plasma	1.7
Testes	22.5	Liver	0.9
Adrenals	21.9	Erythrocytes	0.6
Ovaries	22.6		
Corpora lutea	42.7		

[a] A small dose of selenite labeled with ^{75}Se was injected into rats which had been kept on either a Se-deficient or a Se-adequate diet. Six weeks after the administration of the tracer the tissue ^{75}Se activities were determined. For each tissue the retention factor, RF, was calculated as the ratio of the specific ^{75}Se (percent of dose g^{-1} tissue) in the deficient rats to that in the control animals. The data (Behne et al. 1988) represent the means ± SE for five male animals in each group. Corpora lutea and ovaries were taken from similarly treated females.

of the element and thus a slower loss than in the other tissues as alone by the growth of the organs during the depletion period the Se levels would already have been reduced to a much greater extent. It is rather here a case of reutilization of the metabolized element which in Se deficiency is excreted to a much lesser extent and instead is transported back to certain priority target tissues (Behne and Höfer-Bosse 1984).

As the next step we then carried out a tracer experiment in which the retention of the element was investigated in Se-deficient rats and adequately supplied control animals (Behne et al. 1988). From the specific ^{75}Se activities found in the tissues 6 weeks after the application of the tracer, retention factors were calculated by relating the value in the deficient rats to the corresponding value in the control animals (Table 2). On the left-hand side of Table 2 the tissues are shown in which the highest retention factors were found. They include the brain, thyroid, pituitary, testes, adrenals, ovaries and corpora lutea; that is, besides the brain, all organs with reproductive or endocrine functions. In these compartments the depleted animals retained about 20 to 50 times more of the Se dose than the controls. On the right-hand side the tissues with the lowest retention factors are listed, namely the heart, the skeletal muscle, the liver and the blood fractions. The retention factors of all the other tissues were between those two extremes.

By comparing these findings with the data for the Se contents in Table 1, it can be seen that the small Se dose administered was not used to refill the compartments which were most severely Se-depleted but that, on the contrary, the retention was highest in the tissues with the smallest element losses. This similarity in the distribution pattern indicates that the differences in the ^{75}Se retention are likewise due to the hierarchy in the Se supply in Se deficiency. Regulatory mechanisms apparently exist which ensure that during periods of insufficient Se intake the content of the element is maintained above all in the brain, the reproductive and the endocrine organs. As it may be assumed that Se is preferentially supplied to the sites in which it is most needed, the findings indicate important functions of the element in these organs.

The differences in the Se retention between the tissues might also explain the fact that in Se deficiency or a combined low Se and low vitamin E state lesions are found first in the liver, the heart and the skeletal muscle. The occurrence of lesions during the first stages of Se depletion does not therefore suggest that the tissues affected are the main sites of action of the element but indicates that these tissues are only of secondary importance with regard to the maintenance of the Se level.

3 New Selenoproteins

In the experiments which have been discussed so far, Se metabolism was investigated without consideration of the chemical forms of the element in the tissues. In one of our earlier experiments we had calculated from the data for Se content and glutathione peroxidase activity that more than 60% of the Se in the rat is not bound to the enzyme (Behne and Wolters 1983), and in the tissues of several species other Se-containing proteins have been detected as has been reported in a survey (Whanger 1987) and some more recent publications (Hawkes et al. 1985; Danielson and Medina 1986; Calvin et al. 1987). It was therefore of interest to find out to what extent other selenoproteins besides glutathione peroxidase are present in the organism and should be included in the studies on the element.

For these investigations the tissue selenoproteins were labeled in vivo by injecting Se-deficient rats with ^{75}Se-selenite several times during a period of 5 weeks. The animals were then killed, the proteins separated using SDS-polyacrylamide gel electrophoresis and the ^{75}Se-containing compounds identified by means of autoradiography (Behne et al. 1988). Figure 1 shows an example of ^{75}Se-containing proteins or protein subunits found in the brain and in the thyroid. In the brain 11 labeled proteins were detected in this way. In the thyroid there was an additional protein which is marked here as No. 7. Several tests strongly suggest that

Fig. 1. Autoradiogram of ^{75}Se-labeled selenoproteins in rat tissue homogenates (Behne et al. 1988). The Se-containing proteins or protein subunits, separated by SDS-polyacrylamide gel electrophoresis, were numbered in order of their relative molecular weights (see Table 3)

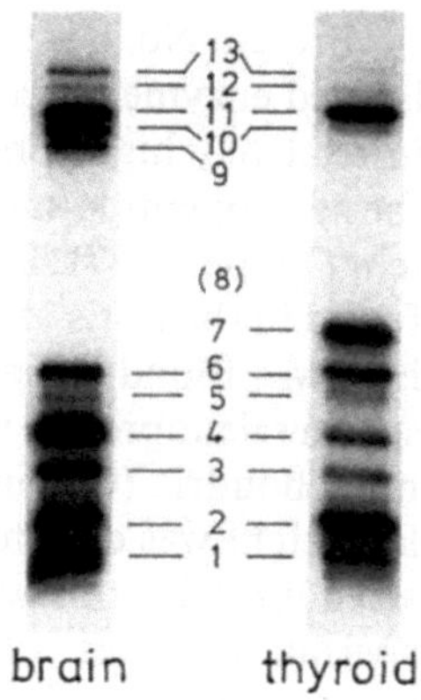

Table 3. Se-containing proteins or protein subunits in rat tissues[a]

Protein	M_r	Protein	M_r
1 +++	12100 ± 500	8 +	33300 ± 1300
2 +++	15600 ± 300	9 ++	55500 ± 1300
3 +++	18000 ± 300	10 +++	59900 ± 2200
4 +++	19700 ± 400	11 +++	64900 ± 1200
5 +	22200 ± 600	12 +	70100 ± 1300
6 +++	23700 ± 700	13 ++	75400 ± 2300
7 +	27800 ± 400		

[a] The ^{75}Se-labeled proteins in homogenates of adrenals, brain, corpora lutea, heart, kidneys, liver, lungs, muscle, ovaries, pituitary, prostate, spleen, testes and thyroid were separated by SDS-polyacrylamide gel electrophoresis and identified by autoradiography (Behne et al. 1988). Their relative molecular weights M_r are the mean ± SD of five determinations. The selenoproteins were numbered in the order of their increasing molecular weights. The crosses indicate that the protein was detected either in a few (+), nearly all (++) or all (+++) of the tissues investigated.

these compounds are genuine selenoproteins. The boiling of the homogenates with SDS and reducing agents, which would have removed adsorbed or loosely bound Se, did not change the ^{75}Se activity of the proteins. Nonspecific labeling by only randomly replacing the chemically similar sulfur atoms in the compounds can also be ruled out, since in the main tissue proteins, which are present in high concentrations and therefore contain relatively large amounts of sulfur, no ^{75}Se could be detected. The results of some further experiments which indicate that the metabolism of these Se-containing compounds is regulated, will be discussed later.

By investigating a large number of rat tissues a total of 13 Se-containing proteins have been detected. Their relative molecular weights are shown in Table 3. Most of the selenoproteins were found in all of the tissues. With the proteins No. 9 and No. 13, which were found in nearly all of the tissues, it can be assumed that they too occur in all tissues but in some were not sufficiently labeled to be detectable.

Protein No. 6 is the glutathione peroxidase subunit. It was identified using the labeled enzyme from the erythrocytes as a marker. Protein No. 9 was also present in relatively high concentrations in the plasma and is most probably identical to the selenopeptide subunit of a protein which is thought to be a Se-transport protein (Tappel 1987). Protein No. 8 was a weakly labeled compound which was only found in the testis and in the spermatozoa. The main spermatozoa protein, which here was found to have a molecular weight of 19700, was described as a specific sperm selenoprotein with structural functions in the outer membrane of the sperm mitochondria (Calvin et al. 1981). However, protein No. 4 with the same molecular weight was detected in all the tissues investigated. If it is identical to the sperm selenoprotein, this would suggest that it has not only a specific function in the spermatozoa but is also active in other cells. Protein No. 7 is a special case as, apart from a weakly labeled band in the kidney and in the liver, it was only present in the thyroid.

After these selenoproteins had been detected it was of interest to obtain information on their tissue levels and the Se distribution among these compounds in the normal state. The direct analysis of the Se concentration in these proteins was not yet possible as in the separation methods with sufficiently high resolution only very small amounts of the proteins could be separated and their Se concentrations were below the limit of detection. We could, however, measure the ^{75}Se tracer quantitatively. We therefore carried out an experiment in which we tried to match the distribution of the tracer with that of the "native" Se. For this purpose we repleted the Se pools in Se-deficient rats by feeding a Se-adequate diet and injected the tracer at regular intervals during the repletion period. We then determined the tracer activity in the tissues and compared it with the content of the native Se which was analyzed in animals with a normal Se status. The highly significant correlation ($R = 0.98$), which was found between the two sets of values, indicates that under these conditions the tracer is distributed in the same way as the native Se and the ^{75}Se activity therefore reflects the Se content in the normal state.

We then separated the selenoproteins in the tissues of the repleted animals and measured the tracer activities in the different fractions. Table 4 shows as an example the relative distribution of the main selenoproteins in the adrenal cortex, the brain and the thyroid in the normal state. In most of the tissues about 90% of the Se was present in seven main proteins including the glutathione peroxidase. These are the proteins Nos. 1, 2, 3, 4, 6, and 10+11. Only in the thyroid does protein No. 7 constitute an additional main Se compound.

After having found these new selenoproteins, we were interested in the question of whether, as with the priority supply of Se to certain tissues, there is also a priority supply of the element at a molecular level. To answer this question we compared the Se incorporation into the proteins from a small dose of Se given either to Se-deficient rats or to the animals which were repleted during the experiment. In all tissues of the deficient animals a much higher percentage of the ^{75}Se activity was retained in other selenoproteins than in the glutathione peroxidase. In the adrenal cortex, for instance, the proportion of the Se dose incorporated into the glutathione peroxidase was only 8% in the deficient animals compared with 44% in the animals with a normal Se status. In the liver this difference in the

Table 4. Selenium distribution among selenoproteins in rat tissues[a]

Protein	Relative protein ^{75}Se activity (%)		
	Adrenal cortex	Brain	Thyroid
1	5	15	7
2	14	13	16
3	3	9	7
4	11	15	7
6 (GSH-Px)	44	15	23
7	–	–	13
10+11	15	20	11

[a] Se-deficient male rats were fed a Se-adequate diet for 4 weeks. ^{75}Se-selenite was administered repeatedly during repletion. The ^{75}Se activity in the main labeled proteins was measured and expressed as a percentage of the total protein-bound tissue ^{75}Se (Behne et al. 1988).

retention in the glutathione peroxidase was even greater with 6% in the deficient and 52% in the repleted animals.

4 Conclusions

The hierarchy in the long-term distribution of Se found for the tissues and for the selenoproteins is summarized schematically in Fig. 2. It shows that with an insufficient Se intake the priority pathways are to the brain, the reproductive organs and the endocrine organs, and within these tissues to selenoproteins other than glutathione peroxidase. The maintenance of the Se level in tissues such as the liver, the heart and the skeletal muscle is last in this hierarchy, and in these tissues, too, the supply to other selenoproteins has priority over that to the glutathione peroxidase.

So far nothing is known about the biological significance of the selenoproteins found here. Because of the homeostatic regulation of the tissue levels of

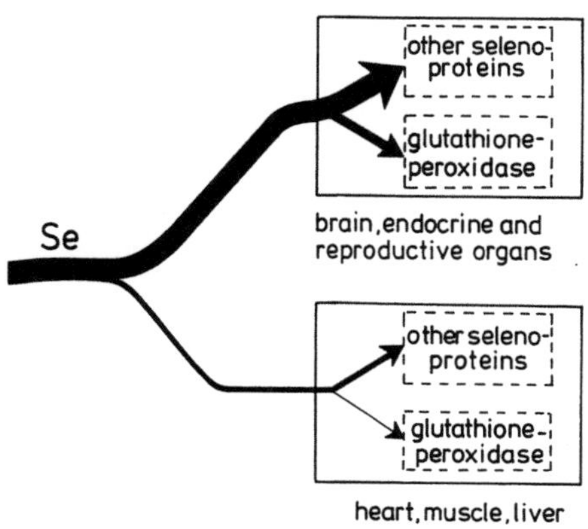

Fig. 2. Supply of Se to body compartments of the rat. The width of the *arrow* represents the ratio of the Se retention in Se-depleted to that in Se-adequate rats and thus indicates the hierarchy in the supply in the long-term distribution of the element during inadequate Se intake (Behne et al. 1988)

these Se compounds it can be assumed that most of the metabolic disturbances which were observed during the first stages of Se depletion are due to the decreases in the glutathione peroxidase activity. However, from the existence of the regulatory mechanisms it can also be concluded that the newly found selenoproteins have important biological functions, especially in the specific target tissues, and it is to be expected that these findings will open up new areas in selenium research in the fields of biochemistry, biology and medicine.

Acknowledgment. The studies described here were supported in part by the Deutsche Forschungsgemeinschaft.

References

Behne D, Höfer-Bosse T (1984) Effects of a low selenium status on the distribution and retention of selenium in the rat. J Nutr 114:1289–1296

Behne D, Wolters W (1983) Distribution of selenium and glutathione peroxidase in the rat. J Nutr 113:456–461

Behne D, Höfer T, Berswordt-Wallrabe R von, Elger W (1982) Selenium in the testis of the rat: studies on its regulation and its importance for the organism. J Nutr 112: 1682–1687

Behne D, Hilmert H, Scheid S, Gessner H, Elger W (1988) Evidence for specific selenium target tissues and new biologically important selenoproteins. Biochim Biophys Acta 966:12–21

Calvin HI, Cooper GW, Wallace E (1981) Evidence that selenium in rat sperm is associated with a cysteine-rich structural protein of the mitochondrial capsules. Gamete Res 4:139–149

Calvin HI, Grosshans K, Musicant-Shikora SR, Turner SI (1987) A developmental study of rat sperm and testis selenoproteins. J Reprod Fertil 81:1–11

Danielson KG, Medina D (1986) Distribution of selenoproteins in mouse mammary epithelial cells in vitro and in vivo. Cancer Res 46:4582–4589

Hawkes WC, Wilhelmsen EC, Tappel AL (1985) Abundance and tissue distribution of selenocysteine-containing proteins in the rat. J Inorg Biochem 23:77–92

Tappel AL (1987) Glutathione peroxidase and other selenoproteins. In: Combs GF, Levander OA, Spallholz JE, Oldfield JE (eds) Selenium in biology and medicine, part A. AVI Press, New York, pp 122–132

Whanger PD (1987) Evidence for non-glutathione peroxidase selenium-containing proteins in mammalian tissues. In: Combs GF, Levander OA, Spallholz JE, Oldfield JE (eds) Selenium in biology and medicine, part A. AVI Press, New York, pp 133–146

Selenium-Containing Proteins in Plasma of Higher Primates

P. D. WHANGER[1,2], J. T. DEAGEN[1], J. A. BUTLER[1], and M. A. BEILSTEIN[1]

Only one major selenium-containing peak was found when animal plasma was subjected to gel filtration (Sephadex G-150) (Beilstein and Whanger 1983). The selenium peak eluted slightly ahead of the glutathione peroxidase (GPX) activity peak, suggesting that the majority of the plasma selenium may not be associated with GPX. In subsequent work (Beilstein and Whanger 1988), less than 10% of the selenium in plasma was calculated to be associated with GPX. Surprisingly, when plasma from humans was subjected to gel filtration, two major selenium peaks were obtained but neither one of them cochromatographed with GPX activity (Deagen et al. 1988). Because of this difference in the gel filtration patterns of plasma between animals and humans, further work was undertaken to investigate the reasons for this disagreement.

The chemical forms of selenium in the two peaks were investigated in plasma from a person who took 200 μg selenium daily as selenium-enriched yeast for about 3 months. The plasma was subjected to gel filtration and the first peak was further chromatographed on a column of DEAE Sephacel and another gel filtration column yielding a 30-fold purification. The second selenium peak was chromatographed on a column of reactive blue 2-Sepharose, yielding a 40-fold purification.

After these proteins were acid hydrolyzed and chromatographed on Dionex DC6A resins (Beilstein and Whanger 1986), the eluted fractions were monitored for selenium with the atomic absorption spectrophotometer with a Zeeman background corrector. The form of selenium in the first peak is selenocysteine as judged by elution at the exact location as the standard selenocysteine, coelution of selenium and radioactivity upon rechromatography with radioactive selenocysteine, and coelution of the carboxymethyl derivatives of standard ^{75}Se-selenocysteine and the plasma selenoprotein hydrolysate (Deagen et al. 1988). It is suspected that this peak is selenoprotein P as was first described (Burk and Gregory 1982) and recently purified from rat plasma (Yang et al. 1987) and later described in monkeys (Motsenbocker and Tappel 1982; Beilstein et al. 1984).

Selenium analysis of the fractions from the second selenium peak revealed this element to be present at the exact location of selenomethionine (SEM). Based on the fact that this peak was retained on reactive blue 2-Sepharose, which has been taken as evidence for albumin (Gordon et al. 1987), it is suspected that it is this protein. The presence of selenocysteine in the first peak but the presence of SEM

[1] Department of Agricultural Chemistry, Oregon State University, Corvallis, OR 97331-6502, USA

[2] To whom correspondence should be addressed

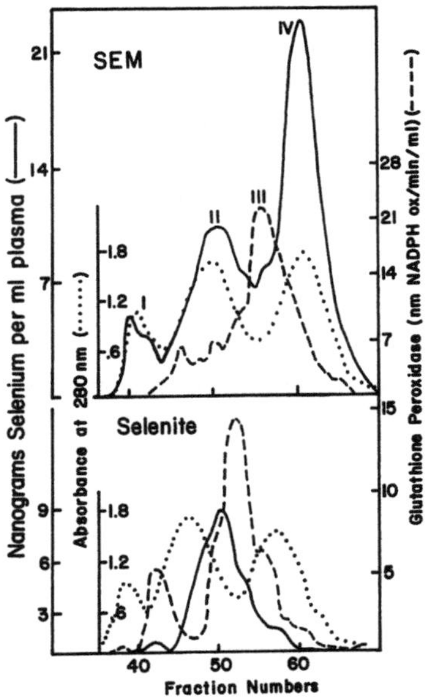

Fig. 1. Gel filtration of plasma from monkeys given selenite or selenomethionine. The plasma was diluted with ammonium acetate buffer (1:3) and chromatographed on a column (2×120 cm) of Sephadex G-150 at 4 degrees C with 0.33 M ammonium acetate buffer, pH 6.8, at a flow rate of 10 ml h^{-1}. About 5 ml were collected per fraction

in the second peak is analogous to the situation when SEM is administered to rats. Upon gel filtration of erythrocyte lysates, the form of selenium in the first peak, GPX, has been shown to be selenocysteine but the form in the second peak, the hemoglobin fraction, has been identified as SEM (Beilstein and Whanger 1986).

Further work was done with rhesus monkeys fed a commercial chow and given selenium as either selenite or SEM in the drinking water. At 1 month and later the plasma selenium levels were significantly higher in the monkeys given SEM than in those given selenite (Butler and Whanger 1988; Butler et al. 1988). Upon gel filtration one main selenium peak was obtained for monkeys receiving selenite (Fig. 1, bottom) but the main GPX peak eluted slightly after the selenium peak. In contrast to the pattern for selenite, the majority of the selenium in plasma from monkeys given SEM accumulated in a peak which was absent (peak IV) in the plasma of monkeys given selenite (Fig. 1, top). Two main selenium peaks were obtained but neither one cochromatographed with GPX activity. The greatest increase of selenium occurred in the peak labeled IV, with a smaller increase in peak II. There was no increase in the selenium content of the GPX peak (labeled III) since the lowest point (fraction 55) between peaks II and IV (6.5 ng Se ml^{-1}) was no higher than the values for those given selenite (7.5 ng Se ml^{-1}, fraction 55, Fig. 1, bottom) (Butler et al. 1988).

The plasma GPX from humans has been purified to homogeneity in our laboratory (Broderick et al. 1987) and shown to exhibit 28 units (μM NADPH ox/min) per mg protein, which is similar to the value of 25 units found by another research group (Takahashi et al. 1987). Using 28 as the value, this calculates to be 5.93 units per mg selenium on the basis of 4 g-atoms of selenium per mole and a molecular weight of 92000. This was used as the standard to calculate the per-

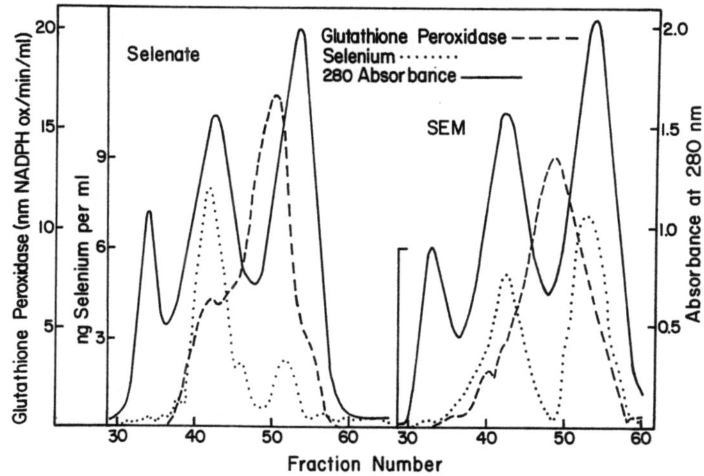

Fig. 2. Gel filtration of plasma from women taking selenium as either selenate or selenomethionine. The chromatography procedures are the same as indicated for Fig. 1

centage of Se associated with GPX in the plasma of these monkeys from the GPX activity and selenium content. The percentage of selenium associated with plasma GPX ranged from 31 to 41 in the monkeys given selenite. In contrast, this percentage was significantly lower in plasma from the monkeys given SEM, ranging between 10 and 19.

The percentage of selenium associated with each of the plasma peaks in monkeys given SEM was calculated from the areas under each curve (Lawrence and Burk 1978). This percentage was multiplied by the plasma concentration of selenium to obtain the amount of selenium in each peak. The plasma selenium content in each of the four peaks indicated that peak IV contained the most selenium at all times during SEM supplementation but the selenium in GPX remained fairly constant throughout the entire study (Butler et al. 1988).

Our laboratory is currently conducting a cooperative study with the personnel at the University of Otago, Dunedin, New Zealand. Women took tablets containing selenium (200 µg daily) as either selenate or SEM. The gel filtration patterns of plasma after 6 months of supplementation (Fig. 2) are very similar to those obtained for the monkeys (Fig. 1). The major peak of selenium eluted ahead of the GPX peak, with a much smaller peak eluting later in plasma of women taking selenate. In contrast, there were two major selenium-containing peaks in plasma from women taking SEM in which the second one contained more selenium than the first one. The GPX activity did not cochromatograph with either of the two selenium-containing peaks in plasma from either group of women.

In conclusion, it is felt that rhesus monkeys are good model animals in the study of human selenium metabolism. Apparently, the two selenium-containing peaks in human plasma are a reflection of SEM as the major form of selenium in food. Until analytical methods are developed for determining the forms in food, it is suggested that the ratio of selenium in the two peaks in plasma be used as a crude measurement of the forms of selenium consumed in food. In preliminary work, the relative amounts of selenium in these two peaks have been shown to vary with the quantitative intake of SEM. The form of selenium in mammalian proteins is suggested to be used as an initial measure of a "true" selenoprotein. If present as selenocysteine, it is probably a selenoprotein, but if present as SEM it is probably not such a protein. The percentage of selenium associated with plasma GPX was lower when monkeys were given SEM rather than selenite and initial studies indicate that this is also true in humans.

References

Beilstein MA, Whanger PD (1983) Distribution of selenium and glutathione peroxidase in blood fractions from humans, rhesus and squirrel monkeys, rats and sheep. J Nutr 113:2138–2146

Beilstein MA, Whanger PD (1986) Chemical forms of selenium in rat tissues after administration of selenite or selenomethionine. J Nutr 116:1711–1719

Beilstein MA, Whanger PD (1988) Glutathione peroxidase activity and chemical forms of selenium in tissues of rats given selenite or selenomethionine. J Inorg Biochem 33:31–46

Beilstein MA, Butler JA, Whanger PD (1984) Metabolism of ^{75}Se-selenite by rhesus monkeys. J Nutr 114:1501–1509

Broderick DJ, Deagen JT, Whanger PD (1987) Properties of glutathione peroxidase isolated from human plasma. J Inorg Biochem 28:299–308

Burk RF, Gregory PE (1982) Some characteristics of ^{75}Se-P, a selenoprotein found in rat liver and plasma and comparison of it with selenoglutathione peroxidase. Arch Biochem Biophys 213:73–80

Butler JA, Whanger PD (1988) Selenite and selenomethionine metabolism in the rhesus monkey (*Macaca mulatta*). FASEB J 2:A1439 (Abstr)

Butler JA, Whanger PD, Kaneps AJ, Patton NM (1989) Metabolism of selenite and selenomethionine in the rhesus monkey (*Macaca mulatta*). J Nutr (in press)

Deagen JT, Beilstein MA, Whanger PD (1988) Plasma selenoproteins in human blood. FASEB J 2:A1440 (Abstr)

Gordon DT, Leinart AS, Cousins RJ (1987) Portal copper transport in rats by albumin. Am Physiol Soc 252:E327–E333

Lawrence RA, Burk RF (1978) Species, tissue and subcellular distribution of non Se-dependent glutathione peroxidase activity. J Nutr 108:211–215

Motsenbocker MS, Tappel AL (1982) Selenocysteine-containing proteins from rat and monkey plasma. Biochem Biophys Acta 704:253–260

Takahashi K, Avissar N, Whitin J, Cohen HJ (1987) Purification and characterization of human plasma glutathione peroxidase: a selenoglycoprotein distinct from the human cellular enzyme. Arch Biochem Biophys 256:677–686

Yang J-G, Morrison-Plummer J, Burk RF (1987) Purification and quantitation of a rat plasma selenoprotein distinct from glutathione peroxidase using monoclonal antibodies. J Biol Chem 262:13372–13376

Involvement of a Selenoprotein in Glycine, Sarcosine, and Betaine Reduction by *Eubacterium acidaminophilum*

W. Freudenberg[1], K. Hormann, M. Rieth[1], and J.R. Andreesen[1,2]

Eubacterium acidaminophilum is a versatile anaerobic bacterium fermenting amino acids and related compounds (Zindel et al. 1988). Glycine degradation to NH_3, CO_2, and acetate requires the presence of 1 μM selenite (optimum) as does the reduction of betaine, sarcosine, or creatine to acetate, methylamines and urea (Fig. 1) irrespective of NAD(P)H, H_2, formate, or e.g., valine being the reductant. However, growth on serine is independent of selenite addition as that on alanine, aspartate, valine, leucine, and malate in the presence of an H_2-scavenging organism. The involvement of selenium in glycine metabolism is well documented for a variety of anaerobic bacteria such as *Clostridium sticklandii, C. lentoputrescens* (Turner and Stadtman 1973; Stadtman 1978), *C. sporogenes* (Costilow 1977), the purinolytic species *C. purinolyticum, C. acidiurici,* and *C. cylindrosporum* (Dürre and Andreesen 1982, 1983), *Peptostreptococcus magnus, P. glycinophilus* (Dürre et al. 1983), and *C. histolyticum* (Lebertz and Andreesen 1988). Betaine reduction has so far only been studied with *C. sporogenes* (Naumann et al. 1983). The observed formation of trimethylamine was dependent on the presence of 1 μM selenite.

In *C. sticklandii* ^{75}Se-selenite is incorporated into the selenoprotein A of the glycine reductase complex (Turner and Stadtman 1973; Cone et al. 1976, 1977; Tanaka and Stadtman 1979; Sliwkowski and Stadtman 1987, 1988), into formate dehydrogenase (Stadtman 1978), and into seleno-tRNAs specific for L-prolyl-, L-valyl-, and L-glutamyl-tRNA (Chen and Stadtman 1980; Chen et al. 1982; Ching et al. 1985). In contrast, cells of *E. acidaminophilum* labeled by ^{75}Se-selenite resulted only in three radioactive protein peaks of about 13, 26, and 52 kD size if crude extracts were separated by gel filtration or SDS-PAGE (Rieth 1987). All three protein fractions reacted positively using a stain specific for glycoproteins and complemented glycine reductase activity using the *C. cochlearium* system of Stadtman (1978). Under harsh denaturation conditions, the amount of radioactively labeled protein of about 13 kD increased by a subsequent decrease in the content of the 52 and 26 kD proteins (Rieth 1987). This can be taken as indication of the existence of one selenoprotein occurring as a mono-, di-, and tetramer.

The dimeric form of the selenoprotein prevailed in glycine-grown cells and was, therefore, purified to homogeneity by using established procedures such as chromatography on Whatman DE 52 and gel filtration by Sephadex G-100 (Rieth 1987). The homogeneous protein contained 0.33 g-atom of selenium. The UV ab-

[1] Institut für Mikrobiologie der Universität Göttingen, Grisebachstr. 8, 3400 Göttingen, FRG
[2] To whom correspondence should be addressed

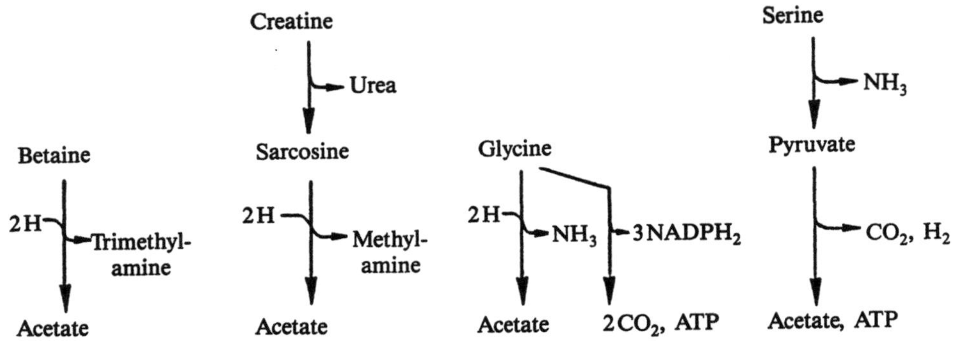

Fig. 1. Characteristic metabolic reactions of *Eubacterium acidaminophilum*

sorption spectrum in the 250–277 nm region and the fluorescence spectra were closely related to those of *C. sticklandii* as published by Cone et al. (1976, 1977). The glycoprotein of *E. acidaminophilum* contained as sugars mannose, glucose, galactose, arabinose, and some unidentified sugars, but not galactosamine or glucosamine. Therefore, the sugar moiety could not be separated by treatment with commercially available endoglycosidases. The glycoprotein nature might be responsible for the tendency of the selenoprotein to aggregate and to give sometimes divergent molecular weight estimations. On the other hand, it might serve as an anchoring device to the cytoplasmic membrane.

Antibodies could be raised against the purified selenoprotein involved in glycine reductase. Immunological cross-reactions were observed with extracts of *Clostridium sticklandii, C. sporogenes, Eubacterium angustum, Peptostreptococcus barnesae, P. glycinophilus,* and *P. variabilis*. Western blots with crude extracts of *E. acidaminophilum* using FITC-labeled antibodies demonstrated a cross-reaction also with the selenoproteins of 13 and 52 kD besides the most prominent band at 26 kD. This also leads to different aggregation sizes of the same protein. Blot analysis of extracts of *E. acidaminophilum* revealed that both a positive cross-reaction and about the same labeling pattern were observed in cells grown on sarcosine, creatine, and betaine, but no reaction was found in serine-grown cells in the absence of selenite. This result correlated well to the ability of *E. acidaminophilum* to exhibit no requirement for selenium compounds during growth on serine and points to an inductive formation of the selenoprotein. Formate dehydrogenase and hydrogenase activities did not alter significantly in cells grown in the presence or absence of selenite (Zindel et al. 1988). Therefore, these two enzymes do not seem to be selenoproteins in this microorganism. In addition, no cochromatography of ^{75}Se-radioactivity was found with respect to both activities. Thus, the selenoprotein involved in the glycine, sarcosine, and betaine reductase reaction is the only selenoprotein formed by *E. acidaminophilum* in prominent amounts.

The three reductase reactions are catalyzed by different terminal reductases as evidenced by their induction spectrum. Both betaine and sarcosine reductase activities (by NADPH or dithioerythreitol) are only formed after growth on their respective substrates (sarcosine reductase by creatine also), whereas glycine reduc-

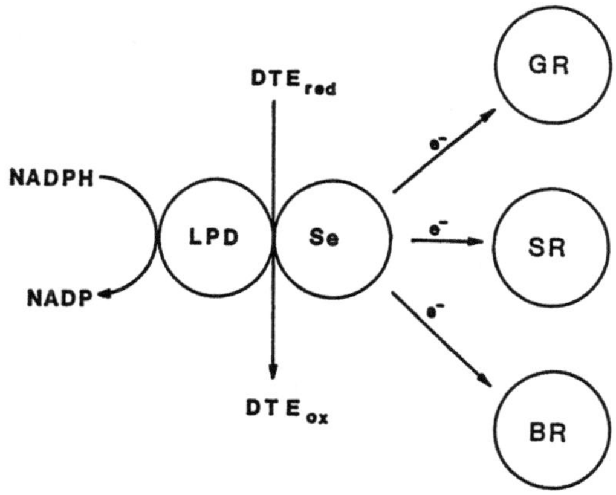

Fig. 2. Proposed electron flow towards the terminal reductases. *LPD* lipoamide dehydrogenase; *Se* selenoprotein; *GR* glycine reductase; *SR* sarcosine reductase; *BR* betaine reductase; *DTE* dithioerythreitol

tase is present in all these cells, however, at a decreased activity (15–60%) (Hormann 1988), which was somewhat higher than the 5% level observed in cells grown in serine media supplemented with selenite (Zindel et al. 1988).

Glycine, sarcosine, and betaine grown cells contained a lipoamide dehydrogenase which exhibited an atypically low molecular weight, reacted preferentially with NADPH instead of NADH, and was part of the glycine decarboxylase complex (Freudenberg et al. 1989). Using antibodies and immunogold techniques, the lipoamide dehydrogenase was nearly exclusively located at the cytoplasmic membrane. Double-labeling studies with immunogold of different sizes indicated that the lipoamide dehydrogenase was surrounded by particles representing the selenoprotein, which in addition was found in the cytoplasm by about 50%. This close association of lipoamide dehydrogenase and selenoprotein should serve for a direct coupling of an electron flow from glycine oxidation by glycine decarboxylase towards the selenoprotein of glycine reductase. The enzymes methylene-FH_4 dehydrogenase and formate dehydrogenase prefer as the lipoamide dehydrogenase and the glycine reductase NADP and NADPH as electron carrier, respectively (Zindel et al. 1988). This unusual preference might further facilitate the electron flow towards one of the three terminal reductases. Since both lipoamide dehydrogenase and the selenoprotein are specifically induced during growth on glycine, sarcosine, and betaine, it seems reasonable to propose that these two electron-transferring proteins generally constitute the first part in the electron transfer chain to one of the three different terminal reductases. The latter are induced by their specific substrate (Fig. 2).

References

Chen CS, Stadtman TC (1980) Selenium-containing tRNAs from *Clostridium sticklandii*: cochromatography of one species with L-prolyl-tRNA. Proc Natl Acad Sci USA 77:1403–1407

Chen CS, Wen TN, Tuan HM (1982) Selenium-containing tRNAs from *Clostridium sticklandii*. Cochromatography of seleno-tRNA I with L-valyl-tRNA. Biochim Biophys Acta 699:92–97

Ching WM, Alzner-De Weerd B, Stadtman TC (1985) A selenium-containing nucleoside at the first position of the anticodon in seleno-tRNAGlu from *Clostridium sticklandii*. Proc Natl Acad Sci USA 82:347–350

Cone JE, Martin del Rio R, Davis JN, Stadtman TC (1976) Chemical characterization of the selenoprotein component of clostridial glycine reductase: identification of selenocysteine as the organoselenium moiety. Proc Natl Acad Sci USA 73:2659–2663

Cone JE, Martin del Rio R, Stadtman TC (1977) Clostridial glycine reductase complex. Purification and characterization of the selenoprotein component. J Biol Chem 252:5337–5344

Costilow RN (1977) Selenium requirement for the growth of *Clostridium sporogenes* with glycine as the oxidant in Stickland reaction. J Bacteriol 131:366–368

Dürre P, Andreesen JR (1982) Selenium-dependent growth and glycine fermentation by *Clostridium purinolyticum*. J Gen Microbiol 128:1457–1466

Dürre P, Andreesen JR (1983) Purine and glycine metabolism by purinolytic clostridia. J Bacteriol 154:192–199

Dürre P, Spahr R, Andreesen JR (1983) Glycine fermentation via glycine reductase in *Peptococcus glycinophilus* and *Peptococcus magnus*. Arch Microbiol 134:127–135

Freudenberg W, Dietrichs D, Lebertz H, Andreesen JR (1989) Isolation of an atypically small lipoamide dehydrogenase involved in the glycine decarboxylase complex from *Eubacterium acidaminophilum*. J Bacteriol 171:1346–1354

Hormann K (1988) Sarkosin- und Betainstoffwechsel von *Eubacterium acidaminophilum*. Dipl thesis, Univ Göttingen

Lebertz H, Andreesen JR (1988) Glycine fermentation by *Clostridium histolyticum*. Arch Microbiol 150:11–14

Naumann E, Hippe H, Gottschalk G (1983) Betaine: new oxidant in the Stickland reaction and methanogenesis from betaine and L-alanine by a *Clostridium sporogenes-Methanosarcina barkeri* co-culture. Appl Environ Microbiol 45:474–483

Rieth M (1987) Untersuchungen zur selenabhängigen Glycinreduktase aus *Eubacterium acidaminophilum*. PhD thesis, Univ Göttingen

Sliwkowski MX, Stadtman TC (1987) Purification and immunological studies of selenoprotein A of the clostridial glycine reductase complex. J Biol Chem 262:4899–4904

Sliwkowski MX, Stadtman TC (1988) Selenoprotein A of the clostridial glycine reductase complex: purification and amino acid sequence of the selenocysteine-containing peptide. Proc Natl Acad Sci USA 85:368–371

Stadtman TC (1978) Selenium-dependent clostridial glycine reductase. Methods Enzymol 53:373–382

Tanaka H, Stadtman TC (1979) Selenium-dependent clostridial glycine reductase. Purification and characterization of the two membrane-associated protein components. J Biol Chem 254:447–452

Turner DC, Stadtman TC (1973) Purification of protein components of the clostridial glycine reductase system and characterization of protein A as a selenoprotein. Arch Biochem Biophys 154:366–381

Zindel U, Freudenberg W, Rieth M, Andreesen JR, Schnell J, Widdel F (1988) *Eubacterium acidaminophilum* sp. nov., a versatile amino acid-degrading anaerobe producing or utilizing H_2 or formate. Description and enzymatic studies. Arch Microbiol 150:254–266

Dependence of Mouse Liver Phospholipid Hydroperoxide Glutathione Peroxidase on Dietary Selenium

F. WEITZEL[1], F. URSINI[2], and A. WENDEL[1,3]

1 Introduction

In addition to the selenoenzyme glutathione peroxidase (GSH-Px), another enzyme catalyzing the reduction of hydroperoxides by oxidation of GSH has been recently discovered (Ursini et al. 1982) and named "peroxidation inhibiting protein" and later substituted by the name phospholipid hydroperoxide glutathione peroxidase (PHGPx). This newly described enzyme is membrane-associated and occurs ubiquitously in mammalian organs (Ursini and Bindoli 1987). Selenocysteine has been identified in this protein in a stoichiometry of 1 g-atom Se per subunit. Basic differences exist in the substrate specificity of the two GSH-peroxidases. While the "classical" GSH-Px does not reduce membrane-bound, esterified lipid hydroperoxides unless phospholipase activity released the free fatty acid hydroperoxides (Grossmann and Wendel 1983), the PHGPx acts directly upon complex lipid hydroperoxides even in the heterogeneous phase, i.e., in the form of micelles (Ursini and Bindoli 1987). On the other hand, the enzymatic rate of PHGPx towards H_2O_2 is poor compared to GSH-Px.

In spite of growing enzymological knowledge, no information is available on the dependence of the PHGPx on dietary selenium. We therefore established an assay system based on the different substrate specificities of the two enzymes and applied it to the time course of the activities during selenium deficiency in the mouse liver. Since GSH transferase isoenzymes exhibit also a so-called non-Se-dependent GSH-Px activity and were observed to react to prolonged, severe Se deficiency by dramatic increases in activity (Reiter and Wendel 1984), we included this parameter into our study.

2 Animals and Methods

Male NMRI mice were kept from weanling on a torula yeast-based diet containing less than 12 ppb selenium. Controls were raised on a diet supplemented with 500 μg Se kg^{-1} (0.5 ppm) as described (Reiter and Wendel 1984). Soybean phos-

[1] Fakultät für Biologie, Universität Konstanz, Postfach 5560, 7750 Konstanz, FRG
[2] Department of Biological Chemistry, University of Padova, 35100 Padova, Italy
[3] To whom correspondence should be addressed

phatidylcholine hydroperoxide was enzymatically synthesized as described elsewhere (Ursini et al. 1983). GSH-Px and GSH-S-transferase were measured according to Reiter and Wendel (1984). PHGPx activity was assayed as follows.

Hamoglobin-free perfused mouse liver was homogenized in Tris-HCl buffer (1+4, w/w), pH = 7.6 containing 0.3 M KCl and 0.1% Triton X-100 (peroxide-free). A total volume of 1.0 ml containing 0.1 M Tris-HCl buffer, pH = 7.6, 5 mM EDTA, 1 mM azide, 3 mM GSH, 0.1% Triton X-100 (peroxide-free), 0.1 mM NADPH and 0.6 units GSSG-reductase (Boehringer, Mannheim) was mixed with 10 µl liver homogenate and pre-incubated for 10 min at 37 °C. The reaction was started by addition of 10 µl phosphatidylcholine hydroperoxide in methanolic solution resulting in a final concentration of 57 µM. NADPH consumption was recorded at 340 nm in a Shimadzu UV-300 dual wavelength spectrophotometer against a blank, which contained 10 µl methanol and lacked either GSH or hydroperoxide.

3 Results and Discussion

The assay system resulted in a linear time course of the reaction until the substrate was consumed. Linearity was given until 110 µM hydroperoxide was exceeded. Another linear reaction was restored upon a second addition of substrate. The rate was proportional to the amount of homogenate added up to 40 µl, i.e., to about 1.6 mg hepatic protein. A constant blank reaction due to non-GSH-specific oxidation of NADPH was observed and could be subtracted from the enzymatic rate. When pure bovine GSH-Px was assayed in this system, essentially no reaction was observed. This indicates that the substrate preparation contained no free fatty acid hydroperoxides. Based on these finding, we consider this assay as suitable for determining PHGPx in biological material in the presence of GSH-Px.

In Se-deficient mice, liver GSH-Px activity dropped to 10% within 19 days and reached undetectable levels after 130 days of dietary Se deficiency (Fig. 1). In contrast, PHGPx was only partially lost and still amounted to 44% of control activity after 130 days (Fig. 2). When GSH-Px activity was measured with cumene hydroperoxide as substrate, a similar time course was observed as with the phosphatidylcholine hydroperoxide-dependent activity (Fig. 3). However, the absolute rate was about two orders of magnitude higher. Since the maximum rates of PHGPx with phosphatidyl choline hydroperoxide and cumene hydroperoxide are in the same range (Ursini and Bindoli 1987), this observation indicates that a cumene hydroperoxide-sensitive, non-Se-dependent GSH-Px activity, as observed earlier (Cykrit et al. 1982), obscures the decrease in PHGPx activity. On the other hand, GSH-transferase activity did not change significantly within the first 20 days of dietary Se deficiency. A more than fivefold increase of this enzyme activity was observed after 130 days (Fig. 4).

Therefore, it seems that this transferase increase is related to the induction of another isoenzyme without peroxidase activity. These results demonstrate primarily that mouse liver PHGPx is subject to an activity decrease in Se deficiency. The dependence on dietary Se of this enzyme is, however, entirely different from the

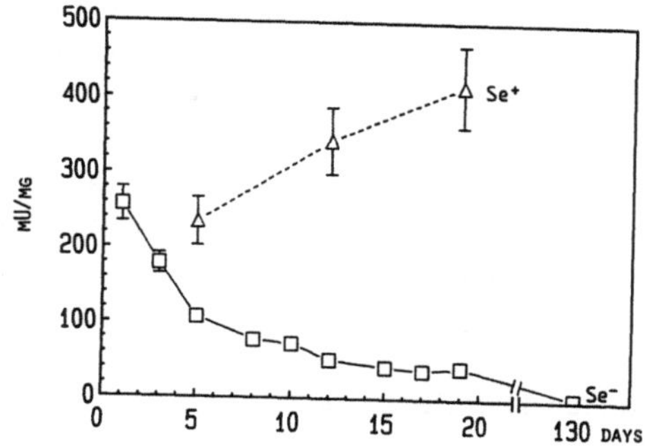

Fig. 1. Time course of the specific activity of hepatic selenium-dependent glutathione peroxidase assayed with H_2O_2 in selenium-deficient mice (*squares*) and Se-supplemented controls (*triangles*)

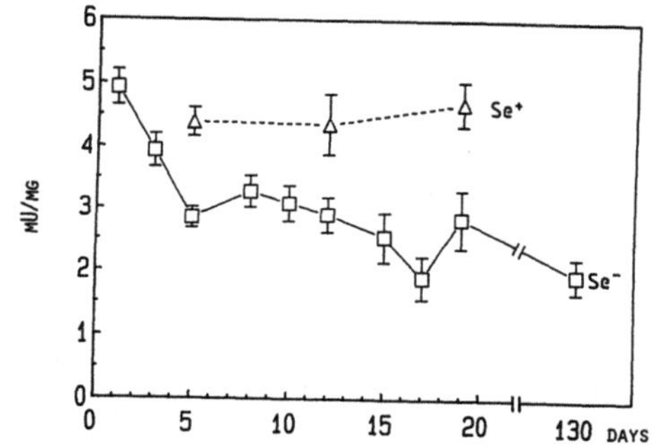

Fig. 2. Time course of the specific activity of hepatic phospholipid hydroperoxide glutathione peroxidase assayed with phosphatidylcholine hydroperoxide during dietary Se deficiency in mice

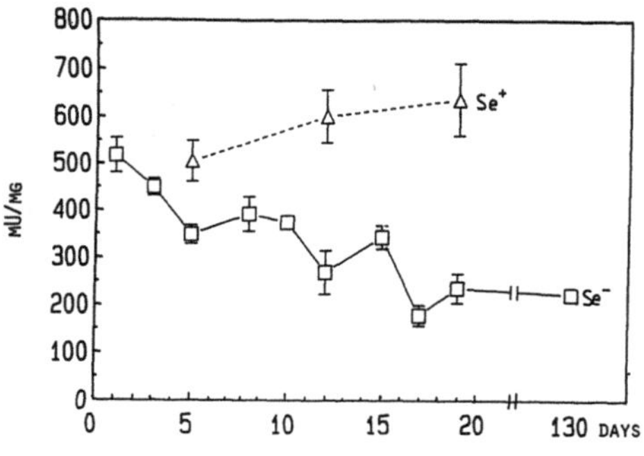

Fig. 3. Time course of the specific activity of hepatic glutathione peroxidase activity assayed with cumene hydroperoxide during dietary Se deficiency in mice

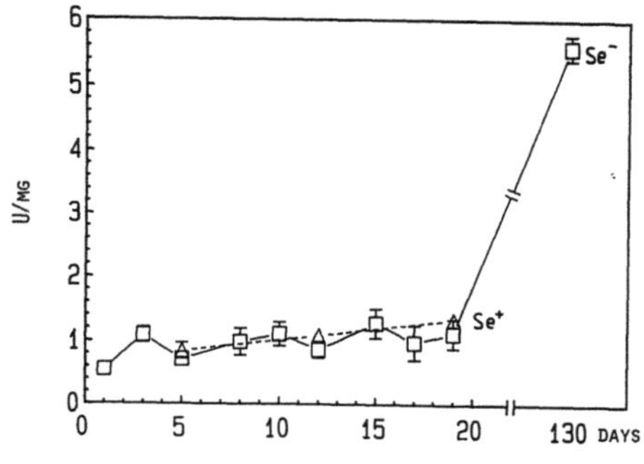

Fig. 4. Time course of the specific activity of the hepatic glutathione transferase activity assayed with chlorodinitrobenzene during dietary Se deficiency in mice

one of the "classical" GSH-Px. It is obvious that PHGPx has a much more efficient capacity of utilizing Se in the deficient animals. Whether this phenomenon is due to a decreased catabolism of this selenoprotein or an enhanced re-utilization cannot be decided at present.

The question arises as to the nature of the residual PHGPx activity after 130 days. A methodological pitfall due to an NADPH oxidase seems unlikely to us for two reasons: firstly, this residual activity could only be measured in the presence of GSH. Secondly, in some other organs measured such as heart, lungs and kidneys, markedly different residual activities were found in 130-day-deficient mice. Therefore, we cannot decide at present whether a yet undefined non-Se-dependent PHGPx activity may be present, or whether the very low amount of 10 ppb Se in our diet is still sufficient to keep PHGPx activity on a substantial level.

Acknowledgment. Supported by the Deutsche Forschungsgemeinschaft, Grant We 686/8-3

References

Cykrit P, Feuerstein S, Wendel A (1982) Biochem Pharmacol 31:2873–2877
Grossmann A, Wendel A (1983) Eur J Biochem 135:549–552
Reiter R, Wendel A (1984) Biochem Pharmacol 33:1923–1928
Ursini F, Bindoli A (1987) Chem Phys Lipids 44:255–276
Ursini F, Maiorino M, Valente M, Ferri L, Gregolin C (1982) Biochim Biophys Acta 710:197–211
Ursini F, Bonaldo L, Maiorino M, Gregolin C (1983) J Chromatogr 270:301–308

Interaction of Heparin with a Selenoprotein in Human Plasma

B. ÅKESSON[1,2] and B. MÅRTENSSON[1]

Summary

A major selenium-containing peak at gel filtration (Sephadex G-150) of human serum had a K_{av} of 0.19, but using heparinized plasma this peak had a K_{av} of 0.01. This indicated an interaction between heparin and a major selenium-containing protein, which was not identical to glutathione peroxidase. The effect of heparin was eliminated by the addition of protamine. Approximately 50% of selenium in plasma from healthy subjects was retained on a heparin-Sepharose column, and this protein could be eluted by a gradient of heparin or salt. The data explain some discrepancies on selenium distribution in plasma as due to the use of different anticoagulants.

1 Introduction

The concentration of selenium in plasma has been widely used as a marker of selenium status in epidemiological and clinical studies. Less attention has been paid to the form of selenium occurring in plasma and how different forms may reflect selenium status.

Glutathione peroxidase occurs in human blood plasma, but there are conflicting estimates of the proportion of plasma selenium that is localized in the enzyme. Recent gel filtration experiments of EDTA-containing plasma suggested that less than 10% of total selenium was located in the enzyme (Borglund et al. 1988). We also found that the major part of plasma selenium was located in a protein with an apparent molecular weight of approx. 174 kD, which could essentially be separated from glutathione peroxidase. Previously, another pattern of selenium distribution in human plasma with two large peaks was reported (Beilstein and Whanger 1983), where the major peak, close to the void volume upon gel filtration on Sephadex G-150, would have a much larger calculated molecular weight than that found by us (Borglund et al. 1988). Data presented in this communication indicate that the anticoagulant used can affect the distribution of plasma selenium studied by gel filtration.

[1] Department of Clinical Chemistry, University Hospital, University of Lund, S-221 85 Lund, Sweden

[2] To whom correspondence should be addressed

2 Materials and Methods

Blood was collected from healthy adult subjects using either heparin or EDTA as anticoagulant. Serum/plasma was separated by gel filtration using a 2.6×90 cm column packed with Sephadex G-150 as described previously (Borglund et al. 1988). The elution buffer was 0.01 M Tris-HCl (pH 7.8), 0.04 M ammonium acetate. Samples of plasma/serum were also separated on a Superose 6 column as described elsewhere (Åkesson and Mårtensson 1988).

Serum was also chromatographed on a heparin-Sepharose column essentially according to McKay and Laurell (1980). The eluate from different columns was analyzed for absorbance at 280 nm, glutathione peroxidase activity and selenium (Borglund et al. 1988).

3 Results and Discussion

When serum was separated by gel filtration on Sephadex G-150, a major selenium-containing fraction migrated close to immunoglobulin G with a K_{av} of 0.19 (Fig. 1). EDTA-containing plasma gave the same pattern as serum. Gel filtration of heparinized plasma instead gave a major selenium-containing peak close to the void volume (K_{av} at the peak approx. 0.01), but the distribution of A_{280} and glutathione peroxidase activity was the same in serum and heparinized plasma. The recovery of selenium from the columns was 70–85%. The findings indicate that heparin interacted with a major selenoprotein, not identical to

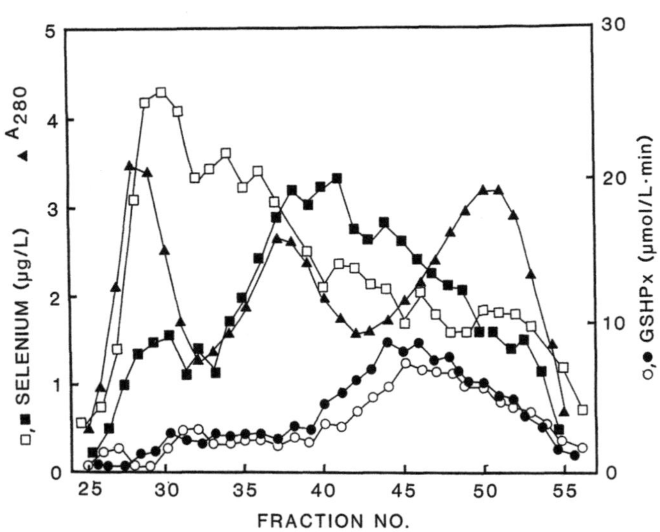

Fig. 1. Distribution of selenium (■, □), A_{280} (▲) and glutathione peroxidase activity (●, ○) after gel filtration of serum (●, ■) or heparin-containing plasma (○, □) on a Sephadex G-150 column (Åkesson and Mårtensson 1988)

Table 1. Selenium-containing peaks after chromatography of serum/plasma on Superose 6

	Peak A	Peak B
Serum		+
Heparinized plasma	+	+
Serum + heparin	+	+
Heparinized plasma + protamine		+

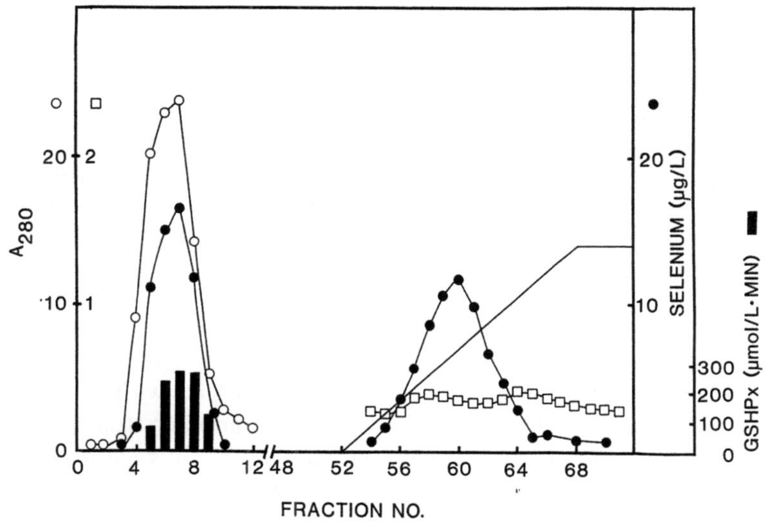

Fig. 2. Separation of selenium-containing proteins on a heparin-Sepharose column. Twenty ml serum was diluted with an equal volume of 0.05 M Tris-HCl (pH 7.4), 0.1 M ammonium acetate, 1 mM EDTA and applied to a 1.6×19 cm column of heparin-Sepharose. After washing with buffer, elution was continued with a gradient of heparin, 0 to 500 IU ml^{-1} in the same buffer, and then with heparin (500 IU ml^{-1}). The flow rate was 20 ml min^{-1}, and 10-ml fractions were collected. *Symbols:* Absorbance at 280 nm (○, □; note the difference in scales); selenium (●); glutathione peroxidase activity (■)

glutathione peroxidase. There was no indication that heparin was contaminated with selenium, since total selenium concentrations in serum and heparinized plasma were not significantly different.

A more rapid evaluation of factors affecting the chromatographic behaviour of selenoproteins was possible using a Superose 6 column. When serum was chromatographed, one major selenium peak with a K_{av} of 0.25 was observed (peak B), but heparinized plasma gave rise to two peaks (K_{av} 0.19 and 0.25). Two peaks were also observed after chromatography of serum to which heparin (30–300 IU ml^{-1}) had been added (Table 1). The addition of protamine (0.2–0.6 mg ml^{-1}) to heparin-containing plasma changed the chromatographic pattern from two peaks to one selenium-containing peak. These experiments show that heparin reversibly interacted with a major selenoprotein in human plasma.

Table 2. Plasma proteins binding to heparin-Sepharose (McKay and Laurell 1980)

Complement	Cell surface	Protease inhibitors	Others
Properdin	Protein HC	Inter-α-trypsin inhibitor	β-Lipoprotein
C1q	Fibronectin	Antithrombin III	β_2-glycoprotein 1
Factor B		C3b inactivator	C-reactive protein
β1H			
C2			

It was also investigated whether this selenoprotein could interact with heparin-Sepharose. Twenty ml of serum (corresponding to 1.7 μg selenium) was applied to a heparin-Sepharose column, and 0.6 μg selenium was eluted with unretained proteins. A heparin gradient was then applied, which eluted another 0.7 μg selenium with the selenium peak, corresponding to a heparin concentration of 250 IU ml^{-1} (Fig. 2). Glutathione peroxidase activity was only detected among unretained proteins. Thus, approximately half of the selenium in plasma resided in a protein(s) with affinity for heparin.

Glutathione peroxidase is the most well-known selenoprotein in mammals. Evidence for other selenoproteins has been obtained in animals (Whanger 1987), but only limited information is available from humans. Previously, a major part of selenium in human plasma was found in proteins with an apparent molecular weight of 174 kD (Borglund et al. 1988), but another study indicated that most of plasma selenium was located in protein(s) that eluted close to the void volume on a Sephadex G-150 column (Beilstein and Whanger 1983). This pattern was similar to that found here after chromatography of heparinized plasma (Fig. 1), indicating that heparin had been used as anticoagulant in the latter study.

The interaction of heparin with the human plasma selenoprotein has not been described previously. Herrman (1977) demonstrated that what is now called selenoprotein P (Burk and Gregory 1982; Motsenbocker and Tappel 1982) from heparinized rat plasma had a lower pI determined from isoelectric focussing than that in serum. He also suggested that 2 M NaCl did not dissociate the heparin-protein complex, which differs from our experience with the human plasma selenoprotein. McKay and Laurell (1980) reported that at least 17 plasma proteins bind to heparin-Sepharose (Table 2). Whether the heparin-binding selenoprotein belongs to any of these groups is unknown.

Thus, in conclusion, this study demonstrates an interaction between heparin and a major selenoprotein in human plasma. The choice of anticoagulant is therefore important in studies of selenium distribution in plasma. The interaction could be reversed by protamine, and the protein could be enriched by chromatography on heparin-Sepharose. The structure and function of the protein and also the physiological relevance of its interaction with heparin are unknown.

Acknowledgment. The study was supported by the Swedish Medical Research Council (project 3968) and the Påhlsson Foundation.

References

Åkesson B, Mårtensson B (1988) Heparin interacts with a selenoprotein in human plasma. J Inorg Biochem 33:257–261
Beilstein MA, Whanger PD (1983) Distribution of selenium and glutathione peroxidase in blood fractions from humans, rhesus and squirrel monkeys, rats and sheep. J Nutr 113:2138–2146
Borglund M, Åkesson A, Åkesson B (1988) Distribution of selenium and glutathione peroxidase in plasma, compared in healthy subjects and rheumatoid arthritis patients. Scand J Clin Lab Invest 48:27–32
Burk RF, Gregory PE (1982) Some characteristics of ^{75}Se-P, a selenoprotein found in rat liver and plasma, and comparison of it with selenoglutathione peroxidase. Arch Biochem Biophys 213:73–80
Herrman JL (1977) The properties of a rat serum protein labelled by the injection of sodium selenite. Biochim Biophys Acta 500:61–70
McKay EJ, Laurell CB (1980) The interaction of heparin with plasma proteins. J Lab Clin Med 95:69–80
Motsenbocker MA, Tappel AL (1982) Selenocysteine-containing proteins from rat and monkey plasma. Biochim Biophys Acta 704:253–260
Whanger PD (1987) Evidence for non-glutathione peroxidase selenium-containing proteins in mammalian tissues. In: Combs Jr GF, Spallholz JE, Levander OA, Oldfield JE (eds) Selenium in biology and medicine. Proc 3rd Int Symp Selenium Biol Med. Van Nostrand Reinhold, New York, pp 133–146

Selenium-Independent Glutathione Peroxidase in Malaria Parasites

A. JUNG[1,2], B. FRITSCH[1], A. DIECKMANN[1], B. BLEIHOLDER[1], and R. OTCHWEMAH[1]

1 Introduction

Inherited and/or acquired alterations of red blood cell metabolism, like glucose-6-phosphate dehydrogenase- or vitamin E deficiency, which influence peroxide metabolism, inhibit the regular intracellular development of the malaria parasite. Many antimalarials are peroxides (e.g., t-butyle-hydroperoxide/t-BHP, qinghaosu/QHS) or radical generators (e.g., primaquine, isouramil). Reactive oxygen species are thought to belong to the main mediators in the unspecific immune reactions against intracellular protozoan parasites (reviewed by Clark et al. 1986; Roth et al. 1986). It is still unknown as to whether the parasite, the host erythrocyte, or both are responsible for the increased susceptibility to oxidative damage and whether the parasite participates in the redox cycle capacity of the host cell. The susceptibility of the normal red blood cell, the parasitized erythrocyte or the parasite to oxidative damage depends on the individual antioxidant capacity. The relation of *O*xidative *S*tress *V*ersus *A*ntioxidant *C*apacity (OSVAC) is very decisive for either the survival or non-survival of malaria parasites or infected erythrocytes. An increase in the oxidant stress and/or a decrease in the antioxidant capacity (e.g., inhibition of the redox cycle enzymes, vitamin E or vitamin C deficiency) disturbs the critical OSVAC balance within the infected cell and might thus lead to a more or less selective destruction of the parasite or the host RBC. Because the parasitized red blood cell is a multicompartment system, it is very difficult to distinguish between the metabolic activities of host and guest cells. In the course of its intracellular multiplication, the parasite degrades up to 75% of host's cell stroma and as such the metabolic capacity of the residual erythrocytes becomes more and more labile. The aim of these investigations was to determine whether malaria parasites depend on the host cell's redox cycle and whether OSVAC of host and guest cells are of different qualities.

2 Materials and Methods

NMRI mice were fed for 8 to 12 weeks with a selenium-deficient diet (Se^-). The diet of control animals (Se^+) was supplemented with 0.33 ppm selenium.

[1] Physiologisch-Chemisches Institut, Universität Tübingen, Hoppe-Seyler-Str. 4, 7400 Tübingen, FRG
[2] To whom correspondence should be addressed

Glutathione peroxidase activity of (Se^-) RBC was below 10% of (Se^+) control cells. Se^-- and Se^+-mice were infected with *Plasmodium vinckei* and chloroquine-sensitive or -resistant *P. berghei* (NK65). Parasitemia was determined in Giemsa-stained smears.

P. vinckei infected RBC were separated into different stages of blood schizogony by a modified Percoll-TM gradient centrifugation (Rüssmann et al. 1981). Free parasites (*P. vinckei, P. berghei,* both strains) were prepared by a modification of our cross-linking method (Heidrich et al. 1979).

For enzyme assays, RBC and parasites were lysed in 10 mM 2-(N-morpholino)-propane sulfonic acid buffer, pH 7.4; 5 mM EDTA; 10 mM $MgCl_2$; 2 mM dithiothreitol; 0.5% Triton-X-100. The lysates were centrifuged for 30 min at 100000×g (r = min). The following enzymes were assayed according to described procedures: the parasitic marker enzyme, NADP-dependent glutamate dehydrogenase/GDH (EC 1.4.1.3) (Picard-Maureau et al. 1975); glutathione synthetase/GS (EC 6.3.2.3) (Wendel et al. 1975); glutathione reductase/GR (EC 1.6.4.21) (Benöhr et al. 1974); glutathione peroxidase/GSH-Px (EC 1.11.1.9) with the substrates t-butyle hydroperoxide, cumene hydroperoxide and H_2O_2 (Cycrit et al. 1982; Wendel 1981).

3 Results and Discussion

Selenium deficiency of the host had no influence on the development of the malaria parasites; the growth rate of malaria parasites (*P. vinckei* and *P. berghei*) was the same in Se^+ and Se^- mice (Fig. 1). In *P. vinckei*-infected Se^+ and Se^- RBC from the Percoll gradient, we found a correlation to the intraerythrocytic growth rate of the parasite and parallel to the parasitic marker enzyme (GDH), the following results:

1. Increasing activities of the enzymes GS and GR (Fig. 2). This indicates that these enzymes are synthesized by the parasite. There were no differences between Se^+ and Se^- cells.
2. In parasitized Se^+ erythrocytes, decreasing GSH-Px activities with H_2O_2 as substrate, slightly increasing activities with t-BHP and steeply increasing activities with cumene hydroperoxide as substrates (Fig. 3).
3. In parasitized Se^- erythrocytes, however, low GSH-Px activities with H_2O_2 as substrate in all stages of parasite development and increasing activities with both organic hydroperoxides as substrates (Fig. 3).

With free parasites (*P. vinckei, P. berghei*), isolated from Se^- and Se^+ RBC, GSH-Px activity was minimal with the substrate H_2O_2 but very high with cumene hydroperoxide (Fig. 4).

4 Conclusion

The development of malaria parasites is not inhibited by selenium deficiency or by low GSH-Px activity of the host RBC. This confirms the findings that Se defi-

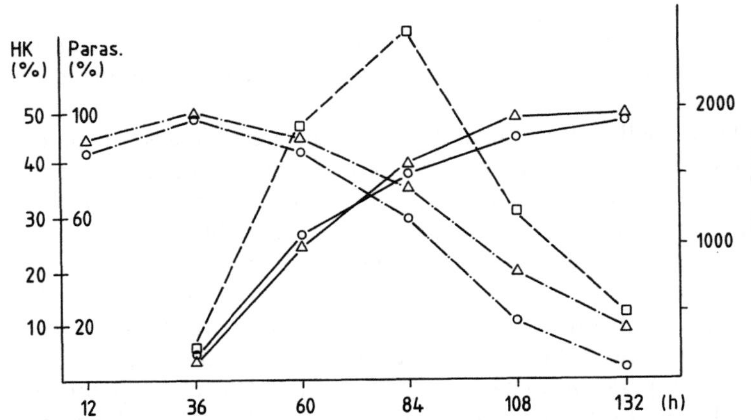

Fig. 1. Parasitemia (*Paras*) and Hematocrit (*Hk*) in selenium-deficient (Se^-) and selenium-supplemented (Se^+) NMRI mice parasitized with *Plasmodium vinckei*. ○—○ Parasitemia in Se^- mice; △—△ parasitemia in Se^+ mice; ○ – · – ○ hematocrit in Se^- mice; △ – · – △ hematocrit in Se^+ mice; □ – – □ product of hematocrit and parasitemia (*Hk*×*Paras*) for Se^- and Se^+ mice

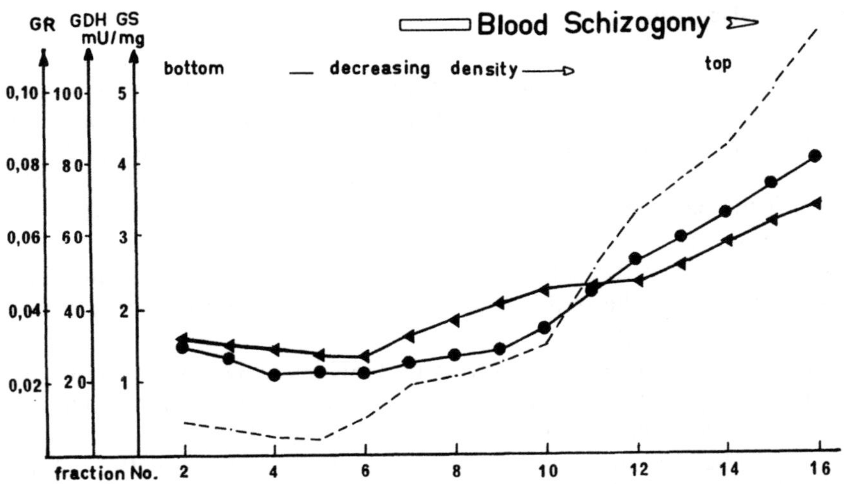

Fig. 2. Glutathione synthetase and glutathione reductase in *P. vinckei* parasitized erythrocytes (Se^+/Se^-) separated into different stages of blood schizogony by Percoll density gradient centrifugation. ● – – ● Glutathione synthetase (*GS*); ◄ – – ◄ glutathione reductase (*GR*); – – – – glutamate dehydrogenase (*GDH*)

ciency does not influence the antimalaria activity of Qinghaosu (Levander et al. 1986). Malaria parasites possess their own GSH-Px with defined and different molecular properties compared to the host cell enzyme and do not depend on GSH-Px of the host cell. The substrate specificities of the parasitic GSH-Px are similar to that of a selenium-independent GSH-Px described elsewhere (Cycrit et al. 1982). The antioxidant capacity of the host erythrocyte decreases continuously

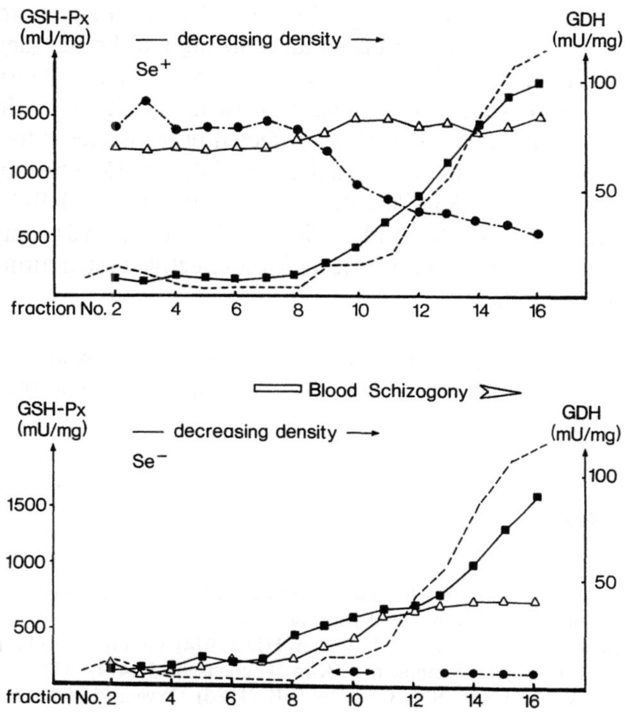

Fig. 3. Glutathione peroxidase activities with different substrates in *P. vinckei* parasitized erythrocytes of selenium-supplemented (*Se*$^+$) and selenium-deficient (*Se*) mice. The cells were separated into different stages of blood schizogony by Percoll density gradient centrifugation. The substrates were: ●–●–●–● H_2O_2; ■–■ cumene hydroperoxide; △—△ t-butyle hydroperoxide; – – – – glutamate dehydrogenase (*GDH*)

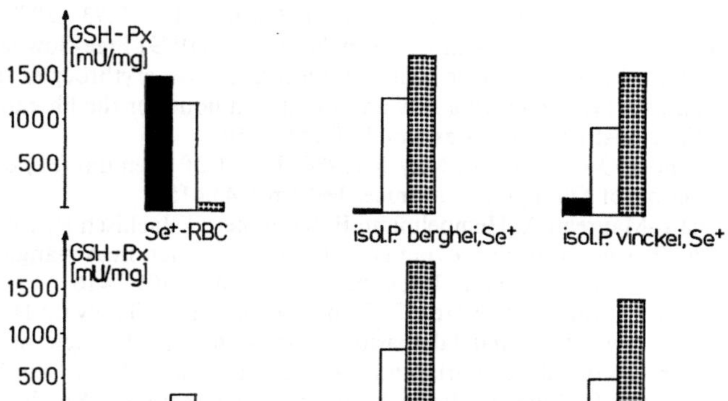

Fig. 4. Specific activity and hydroperoxide specificity of glutathione peroxidase in Se$^+$ (*above*) and Se$^-$ (*below*) uninfected mouse erythrocytes and in isolated *P. vinckei* and *P. berghei* from Se$^+$- and Se$^-$-erythrocytes. ■ H_2O_2; ░ cumene hydroperoxide; □ t-butyle hydroperoxide

in the course of the intracellular development of the parasite, because of the degradation of hemoglobin and erythrocytic enzymes which are, for example, involved in the protection of the red cell against oxidative damage. GSH was found to be markedly increased in rodent erythrocytes infected with malaria parasites (Picard-Maureau et al. 1975; Stocker et al. 1985). The existence of a parasitic glutathione cycle and the high GSH concentration reveal that the antioxidant capacity of plasmodia is remarkable high, although they have no catalase (Picard-Maureau et al. 1975) and a GSH-Px which is unable to detoxify H_2O_2. This conclusion is supported by the observations that the antioxidant vitamins E and C are accumulated in malaria-infected erythrocytes (Stocker personal communication); and that the antimalaria activity of Qinghaosu is reversed by vitamin E (Ager et al. 1986). The differences between host RBC and guest parasite in their quality of OSVAC and in the capabilities to resist oxidative damage by distinct peroxides are good prospects for a selective destruction of the parasite by suitable compounds.

Acknowledgement. This work was supported by the World Health Organization and by the Münchner Apparatebau KG, FRG; A. D. was a scholar of the Studienstiftung des deutschen Volkes.

References

Ager AL, Levander OA, May R, Morris VC (1986) Effect of vitamin E on the anti-malaria action of Qinghaosu. Fed Proc 46:1163

Benöhr H-Ch, Tigges FJ, Walter HD (1974) Glutathionreduktase-Aktivität in menschlichen Erythrozyten-Membranen. Klin Wochenschr 52:474–477

Clark IA, Hunt NH, Cowden WB (1986) Oxygen-derived free radicals in pathogenesis of parasitic diseases. Adv Parasitol 25:1–44

Cycrit P, Feuerstein S, Wendel A (1982) Selenium- and non-selenium-dependent glutathione peroxidase in mouse liver. Biochem Pharmacol 31:2873–2877

Heidrich H-G, Rüssmann L, Bayer B, Jung A (1979) Free-flow electrophoresis for the separation of malaria-infected and uninfected mouse erythrocytes and for the isolation of free parasites (*P. vinckei*): a new and rapid technique for the liberation of malaria parasites from their host cells. Z Parasitenkd 58:151–159

Levander OA, Ager AL, May R (1986) Effect of selenium (Se) deficiency on the anti-malaria action of Quinghaosu in mice. Fed Proc 45:415

Picard-Maureau A, Hempelmann E, Krämmer G, Jackisch R, Jung A (1975) Glutathion Status in *Plasmodium vinckei* parasitierten Erythrocyten in Abhängigkeit vom intraerythrocytären Entwicklungsstadium. Tropenmed Parasitol 26:405–416

Roth E Jr, Raventos-Suarez C, Gilber H, Stump D, Tanowitz H, Rowin KS, Nagel RL (1984) Oxidative stress and falciparum malaria: a critical review of the evidence. In: Eaton JW, Brewer GJ (eds) Malaria and the red cell. Alan R Riss, New York, pp 13–22

Rüssmann L, Jung A, Heidrich H-G (1981) The use of Percoll gradients, elutriator rotor and Mithramycin staining for the isolation and identification of intraerythrocytic stages of *Plasmodia berghei.* Z Parasitenkd 66:273–278

Stocker R, Hunt NH, Weidemann MJ, Lewis-Huges PH, Clark IA (1985) Oxidative stress and protective mechanisms in erythrocytes in relation to *Plasmodium vinckei* load. Proc Natl Acad Sci USA 82:548–551

Wendel A (1981) Glutathione peroxidase. Methods Enzymol 77:325–333

Wendel A, Gumboldt G, Hahn R (1975) Aktivitätsbestimmung und Normalwerte von L-Glutamat-L-Cystein-τ-Ligase (EC 6.3.2.2) in menschlichen Erythrocyten; Glutathionbiosynthese V. Z Klin Chem Klin Biochem 13:157–162

Synthesis of a Selenium Analogue of *Neurospora crassa* Metallothionein and Its Properties

H. Tanaka[1,2], T. Oikawa[1], M. Sugimoto[1], N. Esaki[1], and K. Soda[1]

1 Introduction

Selenium is located between sulfur and tellurium in the periodic table and has both metallic and non-metallic properties. It resembles sulfur in chemical properties, but generally, the reactivity of selenium compounds with metals is higher than that of sulfur counterparts. There have been many reports on the biological interaction between selenium and a number of heavy metals (Kar et al. 1960; Parizek and Ostadalova 1967; Ganther et al. 1972; Naganuma et al. 1983).

On the other hand, metallothioneins (MTs) are cysteine-rich proteins with low molecular weights and bound with various kinds of metal ions such as copper and cadmium. They have been demonstrated in a variety of eukaryotic and prokaryotic organisms. The biological functions of MTs have been proposed to participate in heavy metal storage and detoxification. All cysteinyl residues are involved in metal binding. The *Neurospora crassa* copper-MT is the smallest MT isolated so far; it consists of only 25 amino acids, including 7 cysteinyl residues to which 6 Cu(I) ions are bound (Lerch 1980). The MT shows a high sequence homology to the mammalian MTs; in particular, the seven cysteinyl residues are found in the same positions as the first seven cysteinyl residues of the amino terminal region of the mammalian MTs. Therefore, the *N. crassa* MT is the simplest system to study the structure-function relationship of MTs. We are interested in examining whether the structure and properties of the *N. crassa* MT are influenced by substituting selenium atoms for the sulfur ones. Thus, we have studied the synthetic method for a selenium-containing MT (selenometallothionein, SeMT), in which all the cysteinyl residues are replaced by selenocysteine residues, and its properties in comparison with those of the MT synthesized in a similar manner.

2 Synthesis of SeMT and MT

SeMT and MT were synthesized with an Applied Biosystems peptide synthesizer (430A) according to the method of Merrifield. The protected amino acid

[1] Laboratory of Microbial Biochemistry, Institute for Chemical Research, Kyōto University, Uji, Kyoto-Fu 611, Japan
[2] To whom correspondence should be addressed

derivatives except for selenocysteine derivatives were purchased from Applied Biosystems. *Se*-(*p*-methylbenzyl)-*N*-*t*-butyroxycarbonyl(BOC)-selenocysteine was synthesized as follows from selenocysteine, which was prepared from disodium diselenide and 3-chloroalanine as described previously (Chocat et al. 1985). Selenocystine (2.84 g, 8.5 mmol) was dissolved in 200 ml ethanol water (1 : 1, v/v), to which was added 58 ml (410 mmol) triethylamine in an N_2-flushed, stoppered, 1000-ml, three-necked flask. To this solution was added dropwise 2 g (53 mmol) sodium borohydride. After 1 h, α-bromo-*p*-xylene (4.5 g, 24 mmol) was added, and the mixture was stirred at 40 °C for 14 h in an oil bath. The precipitated white crystals were collected and washed with water and ethanol thoroughly (yield 84%). The *Se*-(*p*-methylbenzyl)-selenocysteine (4.3 g) obtained and triethylamine (15 ml, 108 mmol) were added to 11.3 ml water. To the resulting solution, *S*-*t*-Boc-4,6-dimethyl-2-thiopyrimidine (4.81 g, 20 mmol) dissolved in 9.3 ml dioxane was added, and then the mixture was incubated at room temperature for 65 h under stirring. After completion of the reaction, 50 ml water was added to the reaction mixture. After 5 min, the solution became turbid and unreacted carbonate was extracted twice with 50 ml ethyl acetate. The aqueous layer was cooled to 0 °C, and adjusted to pH 2.3 by addition of 5% $KHSO_4$ solution cooled to 0 °C. An oily precipitate appeared which was extracted with 100 ml ethyl acetate and twice with 50 ml ethyl acetate. The combined ethyl acetate layer was dried over anhydrous sodium sulfate at 4 °C. After the ethyl acetate was evaporated in vacuo, the resulting solid was dissolved in 3 ml ethyl acetate, and 1 ml petroleum ether was added. These solvents were evaporated in vacuo again, and the residue was cooled at 4 °C for 24 h. The resulting solid was broken into pieces and refluxed in 40 ml petroleum ether for 1 h. Cream-colored, micropowdered crystals of *Se*-(*p*-methylbenzyl)-*N*-*t*-Boc-selenocysteine were obtained (yield, 94%). The product was confirmed as the desired one by elemental analysis, and mass and ^{1}H-NMR spectra.

The average coupling efficiency of SeMT and MT was over 99%. The *Se*-*p*-methyl benzyl group was confirmed to be stable during peptide synthesis. The removal of the side chain protecting groups and the cleavage of the synthesized peptides from the resin were performed with anhydrous HF. Purification of the crude SeMT and MT was carried out with a reverse-phase HPLC column (4.6×150 mm for analysis and 10×250 mm for preparation, Sinwa kako, Japan). The chromatography was carried out with a linear gradient between 0 and 50% acetonitrile in 0.1% trifluoroacetic acid at room temperature with a flow rate of 0.8 ml min^{-1}. Retention times of SeMT and MT were 21 and 20 min, respectively. Table 1 shows the amino acid compositions of the purified SeMT and MT. The composition of the synthesized MT was identical with that of *N. crassa* MT (Lerch 1980). In the case of SeMT, the number of other amino acid residues than selenocysteine accorded with that of *N. crassa* MT. Seven residues of selenocysteine were observed as *Se*-carboxyamidomethyl-selenocysteine in the reduced peptide alkylated with iodoacetamide. Unless alkylated, the selenium moiety was decomposed during acid hydrolysis of the peptide, and was detected neither in hydrolysis of the native peptide nor in the peptide oxidized with performic acid. Thus, both synthesized peptides were identified as the desired SeMT and MT.

Table 1. Amino acid analysis of SeMT and MT

Amino acid	Number of residues		
	Syn. MT	Syn. SeMT	*N. crassa* MT[a]
Asx	2.7	2.4	3
Ser	7.2	6.6	7
Gly	6.2	6.0	6
Ala	1.0	1.0	1
Cys	6.7	–	7
SeCys	–	7.1	–
Lys	0.8	0.7	1

[a] K. Lerch (1980)

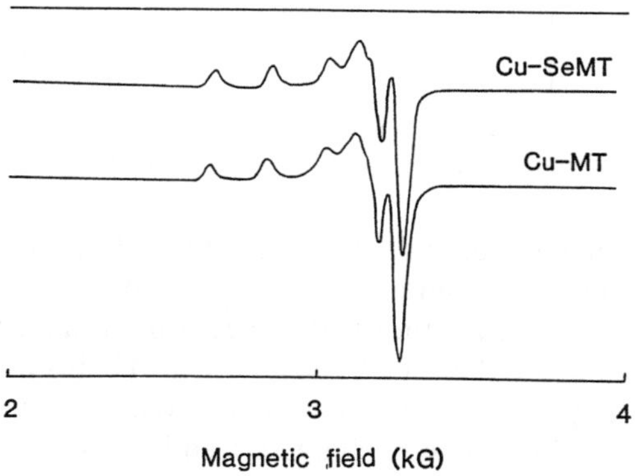

Fig. 1. ESR spectra of SeMT and MT

3 Properties of Cu-SeMT

Cu-bound SeMT was prepared as follows in a nitrogen-purged glove box (oxygen concentration, less than 10 ppm). After the synthesized apo-SeMT (800 µg, 312 nmol) was dissolved in 20 mM Tris-HCl, pH 8.6, and reduced with 2-mercaptoethanol, the solution was incubated with $CuSO_4$ at 37 °C for 30 min. The Cu-bound SeMT was separated from excess reagents with an Asahipak GS-220 HPLC column equilibrated with 20 mM Tris-HCl buffer, pH 8.6. Cu-bound MT was also prepared in the same way. The copper contents were analyzed with a Shimadzu AA-670G atomic absorption spectrophotometer. The Cu-SeMT and Cu-MT were confirmed to contain 3 g-atoms of copper per mole of the peptides. These values differed from that of the native *N. crassa* MT, but accorded with those of cadmium, mercury and zinc-substituted MT (Lerch 1980). So, we have examined the valency of the copper bound to the peptide. Figure 1 shows the ESR spectra. Both spectra gave pictures of typical divalent copper. This fact is also supported by the following evidence. First, these signals disappeared by addition of reducing agents such as cysteine and ascorbic acid. Second, the spectra were identical with that of *N. crassa* copper bound to the oxidized MT, which is

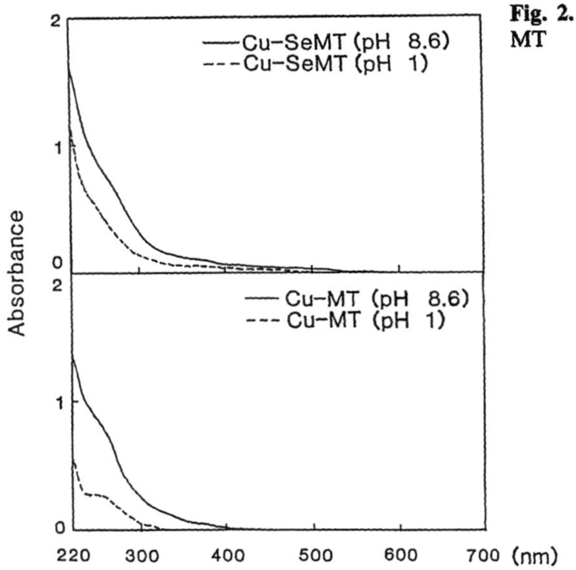

Fig. 2. Absorption spectra of SeMT and MT

known to contain divalent copper. Therefore, both apothioneins were found to be capable of binding the divalent copper ions.

We have studied the spectrophotometric properties of Cu-SeMT in order to compare them with those of Cu-MT. Both absorption spectra were rather similar (Fig. 2). No features were observed in the visible region. The spectra of Cu-SeMT and Cu-MT were characterized by a very broad peak with a shoulder around 270 and 260 nm, respectively. These are compatible with the lack of aromatic amino acid residues. Figure 2 also shows the spectra at pH 1 by addition of hydrogen chloride. We observed a difference between two thioneins; the binding of the Cu-Se bond is stronger than that of the Cu-S bond. We are currently studying the binding of Se-MT to various metal ions and its configuration with EXAFS and ESR.

References

Chocat P, Esaki N, Tanaka H, Soda K (1985) Anal Biochem 148:485

Ganther HE, Goudie C, Sunde ML, Kopecky MJ, Wagner P, Hoekstra WG (1972) Science 175:1122

Kar AB, Das RP, Mukenji B (1960) Proc Natl Inst Sci India Part B 26:40

Lerch K (1980) Nature (London) 284:368

Naganuma A, Tanaka T, Maeda K, Matsuda R, Tabata J, Imura N (1983) Toxicology 29:77

Parizek J, Ostadalova I (1967) Experientia 23:142

Reactions of the Thioredoxin System with Selenium

A. HOLMGREN[1,2] and S. KUMAR[1]

1 Introduction

Thioredoxin is a small (M_r = 12000) ubiquitous redox protein with the conserved active site structure: -Trp-Cys-Gly-Pro-Cys- (Holmgren 1985). The oxidized form (trx-S_2) contains a disulfide bridge which is reduced by NADPH and thioredoxin reductase (T.R.) (Reaction 1).

$$\text{Trx-S}_2 + \text{NADPH} + \text{H}^+ \xrightarrow{\text{T.R.}} \text{Trx(SH)}_2 + \text{NADP}^+ \quad (1)$$

$$\text{Trx-(SH)}_2 + \text{protein-S}_2 \xrightarrow{\text{spontaneous}} \text{Trx-S}_2 + \text{protein-(SH)}_2 \quad (2)$$

$$\text{Net: NADPH} + \text{H}^+ + \text{protein-S}_2 \xrightarrow{\text{T.R.+Trx}} \text{NADP}^+ + \text{protein-(SH)}_2 \ . \quad (3)$$

The reduced form of thioredoxin [trx-$(SH)_2$] is a powerful protein disulfide oxidoreductase (Reaction 2). The thioredoxin system, a name used collectively for NADPH, thioredoxin and thioredoxin reductase is thus an NADPH-dependent protein disulfide reducing system (Reaction 3).

Thioredoxins have been characterized in a wide variety of prokaryotic and eukaryotic cells (Holmgren 1985; Gleason and Holmgren 1988), and generally show about 50% amino acid sequence homology to *Escherichia coli* thioredoxin; mammalian thioredoxins about 30%. *E. coli* thioredoxin has 108 amino acids and a known amino acid sequence and a three-dimensional structure for Trx-S_2 to 2.8 Å resolution. The redox-active disulfide is located in a unique protrusion of the molecule.

Functions of thioredoxin involve, e.g. acting as hydrogen donor for ribonucleotide reductase that synthesizes the precursors for DNA replication. In addition, thioredoxin can regulate the activity of many systems via thiol redox control (Holmgren 1985; Holmgren 1986). This process involves changes in the activity of an enzyme or receptor via dithiol-disulfide interchange reactions (Holmgren 1981).

In view of the known reactivity of sulfhydryl groups with selenium compounds (Spallholz et al. 1981) the interaction with the thioredoxin system has been investigated in the present study. We have discovered that selenite is a substrate for the thioredoxin system. Under aerobic conditions a large nonstoichiometric ox-

[1] Department of Physiological Chemistry, Karolinska Institutet, Box 60400, 10401 Stockholm, Sweden
[2] To whom correspondence should be addressed

idation of NADPH is given by physiologically relevant concentrations of selenite. The results have implications for the effects of selenite on cell growth and for the metabolic reduction of selenite resulting in the incorporation of selenocysteine residues in proteins.

2 Materials and Methods

Na_2SeO_3, Na_2SeO_4 and DTNB [5,5′-dithiobis-(2-nitrobenzoic acid)] were from Sigma. NADPH was from Boehringer and bovine insulin from Nordisk Insulin A/S, Gentofte, Denmark.

2.1 Enzyme Preparations

Thioredoxin from *E. coli* (Holmgren and Reichard 1967) and calf liver (Engström et al. 1974) were homogeneous preparations. Thioredoxin reductase from *E. coli* was from IMCO Corporation, S-113 30 Stockholm, Sweden. Thioredoxin reductase from calf thymus was purified to homogeneity essentially as described for the rat liver enzyme (Luthman and Holmgren 1982). The enzyme ($M_r = 116000$) was stored at 8 mg ml^{-1} in 50 mM Tris-Cl, 1 mM EDTA, pH 7.5 at −70 °C. Dilutions of the enzyme for the experiments were made to 1.0 mg ml^{-1} (8.6 μM) in the same buffer.

2.2 Spectrophotometric Measurements

The activity with selenite was determined in semimicro-quarts cuvettes at 25 °C using a Zeiss PMQ3 spectrophotometer equipped with an automatic sample exchanger and a recorder. Measurements were done at 25 °C in air or anaerobically in carefully evacuated and N_2-equilibrated special cuvettes. The reaction was started by injecting the enzyme through a rubber septum with a gas-tight syringe.

2.3 Enzymatic Assays

Measurements of selenite reduction were performed in 50 mM Tris-Cl, 1 mM EDTA, pH 7.5 generally with 200 μM NADPH. Control cuvettes used as reference cells contained either no enzyme or no selenite and were automatically subtracted by the spectrophotometer. The oxidation of NADPH was followed at 340 nm using an $E_{340} = 6.2 \times mM^{-1} \times cm^{-1}$.

Assays of the disulfide reduction activity of the thioredoxin system were based on reduction of insulin as described previously (Method 4; Luthman and Holmgren 1982). Incubations in a final volume of 100 μl contained 80 mM HEPES buffer, pH 7.6, 3 mM EDTA, 0.7 mM NADPH, 330 μM insulin and the indicated amounts of thioredoxin, selenite and thioredoxin reductase. After incubation for 30 min at 37 °C (in air) the reaction was stopped by 0.50 ml 6 M guanidine-HCl, 0.20 M Tris-Cl, pH 8.0, 0.2 mg ml^{-1} DTNB and the absorbance at 412 nm was measured to determine SH groups ($E_{412} = 13.6\ mM^{-1} \times cm^{-1}$). Blanks contained no thioredoxin and thioredoxin reductase.

3 Results and Discussion

3.1 Thioredoxin Reductase from Calf Thymus

Oxidation of NADPH in cuvettes containing only 10 µM selenite and calf thymus thioredoxin reductase was noted. With 50 nM enzyme an extensive oxidation of NADPH was observed at A_{340} nm (Fig. 1). No oxidation of NADPH by selenite was recorded in the absence of the enzyme. Low concentrations of selenite (5 – 10 µM) gave a pronounced lag phase of the reaction for unknown reasons. This was also observed at early time points with higher concentrations (40 µM) of selenite. The oxidation of NADPH was nonstoichiometric and 5 nmol selenite caused the oxidation of more than 60 nmol NADPH (Fig. 1).

Use of sodium selenate in place of selenite with calf thymus thioredoxin reductase failed to give any oxidation of NADPH. This demonstrated that the reaction was specific for Se in the +4 oxidation state.

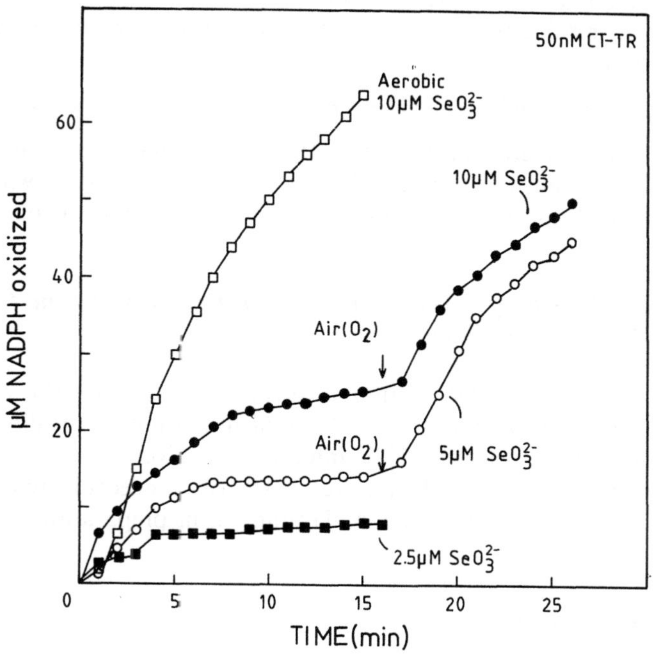

Fig. 1. Selenite-dependent oxidation of NADPH catalyzed by calf thymus thioredoxin reductase (*CT-TR*). Cuvettes contained 0.50 ml of 50 mM Tris-Cl, 2 mM EDTA, pH 7.5, 200 µM NADPH and □–□, 10 µM selenite in air, 5×10^{-8} M CT-TR. ●–● 10 µM selenite anaerobically followed by admission of air; the same at ○–○ 5 µM selenite or ■–■ 2.5 µM selenite

3.2 *Reactions with E. coli Thioredoxin Reductase and Thioredoxin*

Thioredoxin reductase from *E. coli* was also tested for its reactivity with selenite. No NADPH oxidation was observed and selenite is thus not a substrate for this enzyme. However, after addition of 3 μM Trx-S_2 a fast reaction was observed. Thus, also *E. coli* Trx-$(SH)_2$ catalyzes reduction of selenite. Also calf thymus thioredoxin (3 μM) stimulated the reduction of selenite when added to 100 nM calf thymus thioredoxin reductase (data not shown). Selenite is thus reduced by both the calf thymus thioredoxin reductase and thioredoxin.

3.3 *Apparent K_m Values for Selenite*

We measured the maximal rate of NADPH oxidation per min at different concentrations to obtain the apparent K_m values for selenite. For *E. coli* thioredoxin and thioredoxin reductase the apparent K_m value was about 5 μM. A higher apparent K_m value of about 20 μM was obtained with calf thymus thioredoxin reductase.

3.4 *Reduction of Selenite Anaerobically*

As seen from Fig. 1 anaerobic reduction of selenite with calf thymus thioredoxin reductase stopped after oxidation of about 3 mol NADPH per mole of selenite. Admission of air resulted in a continued reaction consistent with redox cycling. The result suggests the following stoichiometry for the anaerobic reaction (Reaction 4):

$$SeO_3^{2-} + 3\ NADPH + 3\ H^+ \xrightarrow{CT\text{-}TR} Se^{2-} + 3\ NADP^+ + 3\ H_2O\ . \qquad (4)$$

Thus, anaerobically selenite (Se^{2-}) is probably generated. This is known to be highly sensitive to autoxidation (Ganther 1971). Also with L-selenocysteine, continued nonstoichiometric oxidation of NADPH was observed (data not shown).

3.5 *Effects of Selenite on Disulfide Reduction by the Calf Thymus Thioredoxin System*

Selenite inhibited insulin disulfide reduction by the calf thymus thioredoxin system as shown in Table 1. Total inhibition of insulin reduction was obtained with high selenite; 50% inhibition is given below 10 μM, demonstrating the powerful action of selenite. Also, ribonucleotide reduction was inhibited by selenite (S. Kumar and A. Holmgren, in preparation).

4 Conclusions

We have discovered that selenite is reduced by NADPH and thioredoxin reductase and thioredoxin. This novel dithiol-dependent reaction has been characterized. The anaerobic reduction of selenite oxidizes 3 mol NADPH and gives rise to

Table 1. Effect of sodium selenite on insulin disulfide reduction catalyzed by the mammalian thioredoxin system

Sodium selenite (μM)	μM SH groups generated per 30 min	
	CT-TR (10 nM) CT-Trx (3 μM)	CT-TR (100 nM) CT-Trx (0.3 μM)
0	583	300
10	76	100
40	52	100
200	17	54
400	10	12

selenide. Aerobic redox cycling of an intermediate leads to a large nonstoichiometric oxidation of NADPH. These results may explain the effects of selenite on cell growth, in particular, the inhibition of tumor cell growth (Shamberger 1983).

Acknowledgements. This investigation was supported by grants from the Swedish Cancer Society (961) and the Swedish Medical Research Council (13X-3529). S.K. was supported by a fellowship from the Swedish Cancer Society (961-B89-03V).

References

Engström N, Holmgren A, Larsson A, Söderhäll S (1974) Isolation and characterization of calf liver thioredoxin. J Biol Chem 249:205–210

Ganther HE (1971) Reduction of the selenotrisulfide derivative of glutathione to a persulfide analog by glutathione reductase. Biochemistry 10:4089–4098

Gleason FK, Holmgren A (1988) Thioredoxin and related proteins in procaryotes. FEMS Microbiol Rev 54:271–298

Holmgren A (1981) Thioredoxin: structure and functions. Trends Biochem Sci 6:26–29

Holmgren A (1985) Thioredoxin. Annu Rev Biochem 54:237–271

Holmgren A (1986) Thioredoxin and glutaredoxin systems – an overview. In: Holmgren A, Brändén C-I, Jörnvall H, Sjöberg B-M (eds) Thioredoxin and glutaredoxin systems: structure and function. Raven Press, New York, pp 1–9

Holmgren A, Reichard P (1967) Thioredoxin 2: cleavage with cyanogen bromide. Eur J Biochem 2:187–196

Luthman M, Holmgren A (1982) Rat liver thioredoxin and thioredoxin reductase: purification and characterization. Biochemistry 21:6628–6633

Shamberger RJ (1983) Biochemistry of selenium. Plenum, New York

Spallholz JE, Martin JL, Ganther HE (1981) (eds) Selenium in biology and medicine. AVI Press, Westport, CT

Selenite-Induced Variation of Glutathione Peroxidase Activity in Mammalian Cells and Its Consequences for Cellular Radiation Resistance

B.E. SANDSTRÖM[1,4], J. CARLSSON[2], and S.L. MARKLUND[3]

1 Introduction

The current work was initiated upon the finding of a positive correlation ($0.05 < P < 0.1$) between the GSH peroxidase activities of seven human cell lines and their resistance against γ-irradiation (Marklund et al. 1984), as indicated by quasithreshold doses (D_q) and extrapolation numbers (n). The correlation between GSH peroxidase and D_q was particularly striking (Fig. 1). There was no correlation between GSH peroxidase and the general radiosensitivity (D_0) of the cell lines. GSH peroxidase forms the primary intracellular enzymic defence against oxygen radicals (abundantly formed when water solutions are irradiated under oxic conditions) together with catalase, CuZn superoxide dismutase and Mn superoxide dismutase. No correlation with any survival parameter could be found for the latter enzymes. D_0, D_q and n are parameters derived from the survival curve (Fig. 2) and are different measures of the size of the so-called shoulder

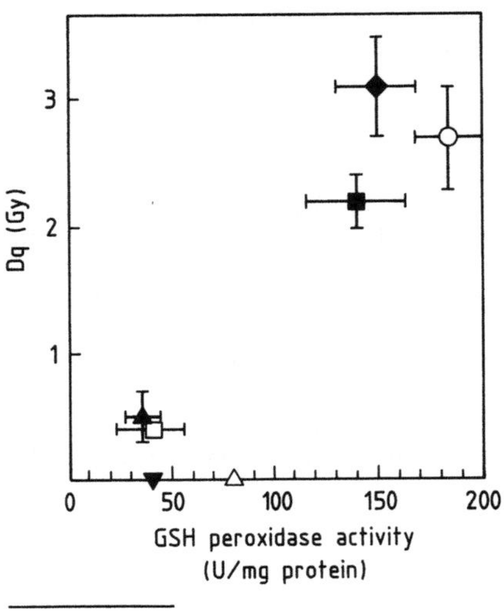

Fig. 1. Relation between GSH peroxidase activity of seven human cell lines and their radiation sensitivity as indicated by quasithreshold doses (D_q). Each parameter is the mean of at least two separate determinations. *Error bars* show maximal error (max. value minus min. value)

[1] Division of Cell- and Microbiology, Swedish Defence Research Establishment, 90182 Umeå, Sweden
[2] Department of Radiation Sciences, University of Uppsala, Uppsala, Sweden
[3] Department of Clinical Chemistry, Umeå University Hospital, Umeå, Sweden
[4] To whom correspondence should be addressed

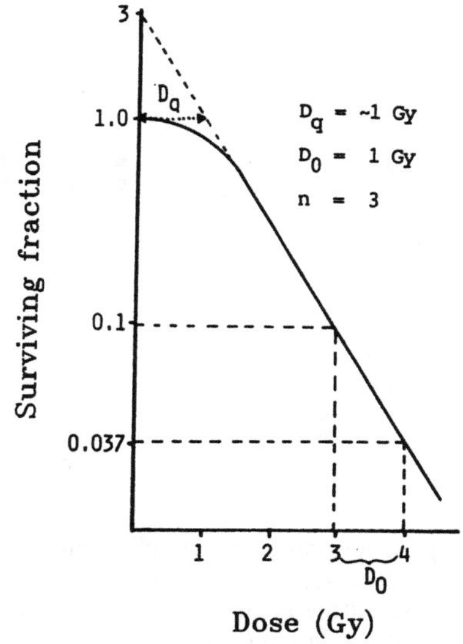

Fig. 2. Generalized survival curve for radiation-induced cell death in mammalian cells (Adapted from Coggle 1983)

of the survival curve. The size of the shoulder is thought to be the result of repair of radiation damage (Coggle 1983). There are mechanisms by which GSH peroxidase might contribute to this repair. Alkyl radicals are formed by ionizing radiation in target molecules. In the presence of oxygen, peroxy-radicals will be instantly formed, which subsequently form hydroperoxides by hydrogen abstraction. Such ionizing radiation-induced hydroperoxides have been found to be substrates for GSH peroxidase (Tan et al. 1987), which by reduction will form less toxic and probably more easily repaired alcohols.

Our approach to control the validity of the found correlation was to vary the GSH peroxidase activities of different cell lines and to examine whether increased or decreased enzyme activities would lead to correlated alterations in radiation tolerance. We subsequently developed a procedure for the reproducible modification of GSH peroxidase activities of different cell lines (Sandström et al. 1987). This was accomplished by supplementing the culture medium with 100 nM selenite and reducing the fetal calf serum content from 10 to 2%.

2 Effects of Selenite on GSH Peroxidase Activity and Cell Growth

We studied the GSH peroxidase activities of five different cell lines: P31 (Fig. 3), P7 (both human mesotheliomas), HT29 (human colon carcinoma; Fig. 4), U-118 MG (human glioma) and N-18 (mouse neuroblastoma; Fig. 5), after selenite supplementation with 10–5000 nM to medium containing 2% serum. In comparison, the effects of addition of 100 nM selenite to medium containing 10% serum were also studied (Table 1). When cultured in 2% serum, all cell lines increased their GSH peroxidase activities in response to increasing selenite concentrations. The human cell lines needed only a 10 nM addition of selenite to reach a plateau

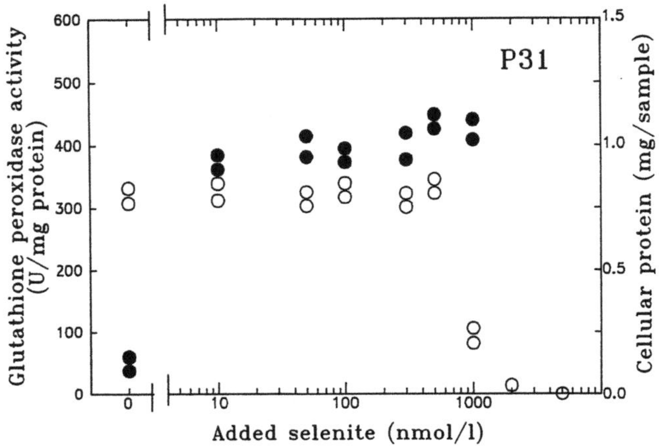

Fig. 3. Response of P31 cells, cultured in medium containing 2% fetal calf serum, to 11 days of selenite supplementation with various amounts of selenite. The basal selenium concentration in the culture medium was 6 nM (from added serum). *Filled circles* show GSH peroxidase activities; *open circles* show cellular protein levels

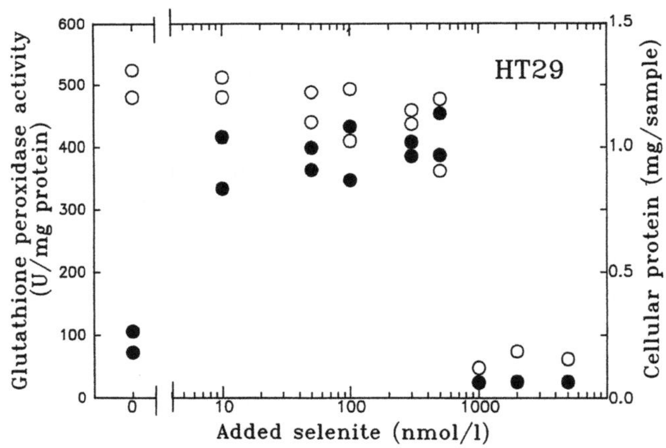

Fig. 4. Response of HT29 cells to 11 days of selenite supplementation. Conditions as in Fig. 3

in GSH peroxidase activity, extending to 300–1000 nM, where the growth-retarding effects of the selenite became apparent and the GSH peroxidase activity also fell (cf. Figs. 3 and 4). N-18 (Fig. 5) showed a different response to supplementation with selenite and increased its GSH peroxidase activity exponentially proportional until effects of toxicity set in at 500–1000 nM selenite. For P31, P7 and HT29 cells a higher GSH peroxidase level was obtained in 10% serum than could be achieved with any selenite concentration in 2% serum. U-118 MG cells cultured in 10% serum were aberrant and did not respond with a significant increase in GSH peroxidase upon addition of 100 nM selenite. Interestingly, the GSH peroxidase activity of U-118 MG cells was higher when the cells were cultured in 2% serum.

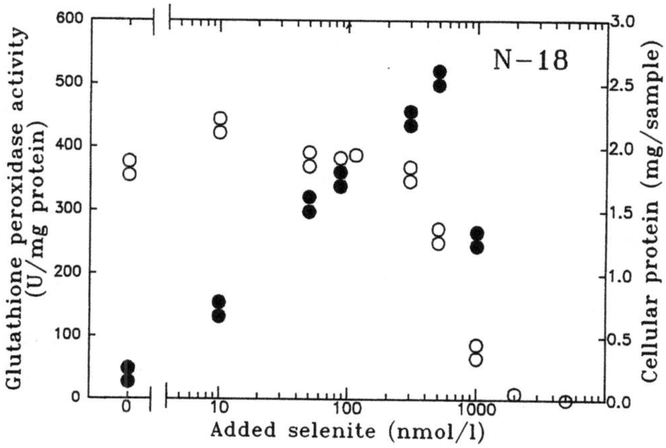

Fig. 5. Response of N-18 cells to 11 days of selenite supplementation. Conditions as in Fig. 3

Table 1. Effect of treatment with 100 nM selenite for 11 days

Cell line	Fetal calf serum added to culture medium (%)	GSH peroxidase activity in control cultures ($U\ mg^{-1}$ protein)	GSH peroxidase activity after 11 days of 100 nM Se treatment ($U\ mg^{-1}$ protein)	Relative change in GSH peroxidase activity
N-18	2	38 ± 1[a]	356 ± 6[a]	9.4[b]
	10	202 ± 4	360 ± 3	1.8
P31	2	49 ± 1	384 ± 1	7.9
	10	256 ± 1	476 ± 23	1.9
P7	2	22 ± 8	98 ± 9	4.6
	10	142 ± 22	202 ± 21	1.4
HT29	2	89 ± 17	388 ± 40	4.4
	10	362 ± 21	492 ± 18	1.4
SW1116	2	59 ± 1	185 ± 45	3.1
	10	114 ± 22	152 ± 7	1.3
P27	2	84 ± 8	211 ± 34	2.5
	10	164 ± 14	228 ± 2	1.4
U-118 MG	2	250 ± 28	454 ± 29	1.8
	10	118 ± 29	143 ± 1	1.2
U-2 OS	2	80 ± 5	130 ± 15	1.6
	10	112 ± 9	114 ± 1	1.0
U-393 OS	2	90 ± 15	123 ± 4	1.4
	10	81 ± 7	91 ± 1	1.1

[a] Mean ± maximal variation of at least two duplicate enzyme determinations
[b] The mean of the GSH peroxidase activity in the selenite-supplemented cultures divided by the mean of the GSH peroxidase activity in the control cultures

Four other human cell lines, P27 (mesothelioma), U-2 OS, U-393 OS (osteosarcomas) and SW1116 (colon carcinoma), were investigated using a reduced test protocol (Table 1). These cells were cultured in 2 and 10% serum, either unsupplemented or supplemented with 100 nM selenite for 11 days. Both osteosarcoma cell lines (U-2 OS and U-393 OS) showed weak increases in enzyme activity when selenite was added to the medium and may become interesting as negative controls of this type of selenite supplementation. They could prove effective in distinguishing between selenium-induced and GSH peroxidase-induced effects.

The cell growth was assessed from the content of cellular protein in the cultures. We found no general selenite-induced stimulation of the cell growth when the cells were cultured in 2% serum (cf. Figs. 3–5). The small amount of selenium (around 6 nM) in our medium containing 2% serum, in combination with other complex components of the serum, thus seems to have been sufficient to allow maximal growth under these conditions. The growth was, as expected, in general better when the cells were cultured in 10% serum.

3 Effects of GSH Peroxidase Activity on Radiation Resistance

We chose three of the cell lines (Figs. 3–5) that increased their GSH peroxidase activities fourfold or more to investigate whether or not a low or high GSH peroxidase activity would change the radiation response of the cells (Sandström et al. 1989). Control and selenite-supplemented cells were exposed to different doses of γ-radiation and the surviving fractions were estimated by clonogenic survival or $[^3H]$thymidine uptake. The number of DNA strand breaks in cells irradiated under repairing, as well as non-repairing, conditions were analyzed with a slight modification of the recently introduced DNA precipitation technique (Olive 1988) using double-labelled cells. Cells, from cultures parallel to the ones irradiated, were harvested for each experiment and GSH peroxidase activity was analyzed. The activities of catalase, CuZn superoxide dismutase, Mn superoxide dismutase and non-selenium-dependent GSH peroxidase were also analyzed, but only on samples.

Our selenite supplementation procedure resulted in large differences in GSH peroxidase activity between supplemented and non-supplemented cells: about 40-fold for N-18, about 5-fold for P31 and about 2-fold for HT29 (Table 2). The other studied enzymes, CuZn superoxide dismutase, Mn superoxide dismutase and catalase, were not significantly influenced by the selenite treatment (Table 2), although there was slightly more catalase in selenite-supplemented N-18 cells. The increases, however, was almost negligible compared with the increase in GSH peroxidase activity. The GSH peroxidase reaction is also catalyzed by GSH transferases (Prohaska and Ganther 1977). The GSH transferases, however, cannot, unlike GSH peroxidase, react with H_2O_2. To establish the contribution of GSH transferases to the total GSH peroxidase activity of our cell lines, cell homogenates were analyzed with cumene hydroperoxide and H_2O_2. The difference between the results should represent GSH transferase-dependent GSH peroxidase activity. There appeared to be some non-selenium-dependent GSH peroxidase activity in the control homogenates (Table 2). In selenite-supplemented cultures, the very high background selenium-dependent GSH peroxidase activity precluded estimation of non-selenium-dependent GSH peroxidase activity with reasonable precision.

Table 2. Effect of selenite supplementation on activities of antioxidant enzymes

Cell line		GSH peroxidase (U mg^{-1} protein)	Non-Se-GSH peroxidase (U mg^{-1} protein)	CuZn superoxide dismutase (U mg^{-1} protein)	Mn superoxide dismutase (U mg^{-1} protein)	Catalase (U mg^{-1} protein)
HT29	−Se	164±69[a]	17±2[b]	88±3[b]	6.0±0.1[b]	4.2±0.6[b]
	+Se	376±125	n.d.[c]	89±6	7.0±0.4	3.2±0.6
P31	−Se	69±21	25±1	154±6	5.2±0.2	0.5±0.0
	+Se	379±68	n.d.[c]	156±3	6.0±0.7	0.6[d]
N-18	−Se	7±3	16±2	237±21	4.0±0.2	1.2±0.1
	+Se	293±7	n.d.[c]	186±12	3.5±0.0	2.9±0.1

[a] Mean ± SEM of all experiments (5 per cell line)
[b] Mean ± maximal variation of 2 samples
[c] Not detectable due to low precision of assay
[d] Single determination

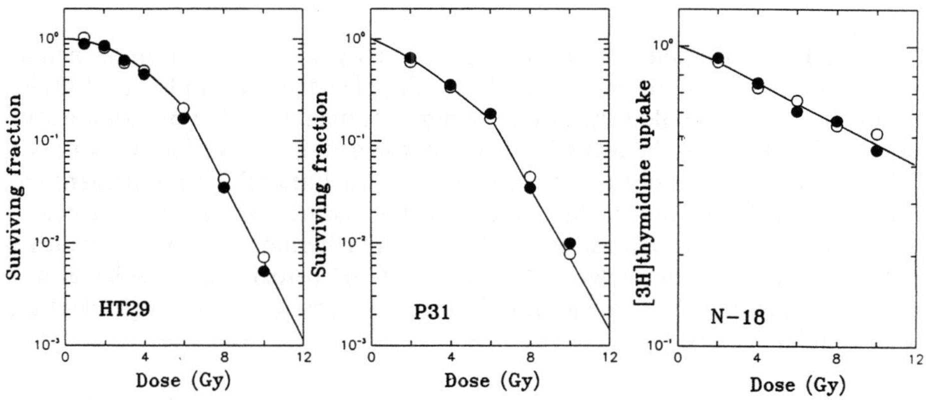

Fig. 6. Clonogenic survival of HT29 and P31 cells and change in [^{3}H]thymidine uptake of N-18 cells exposed to different doses of ^{137}Cs γ-rays. Each *point* corresponds to the mean value of control cells (*open*) and selenite-supplemented cells (*filled*) of two to three experiments per cell line

There were no changes between control and selenite-supplemented cells in the surviving fractions after irradiation (Fig. 6), in spite of the large differences in GSH peroxidase activity (Table 2). We analyzed the specific effect of the selenite supplementation on the induction of DNA strand breaks by mixing ^{3}H-labelled selenite-supplemented cells with ^{14}C-labelled control cells and vice versa. We monitored the number of DNA breaks in the three cell lines before and after irradiation, both at 0° and 37 °C, but found no significant differences between selenite-supplemented and control cells (Fig. 7).

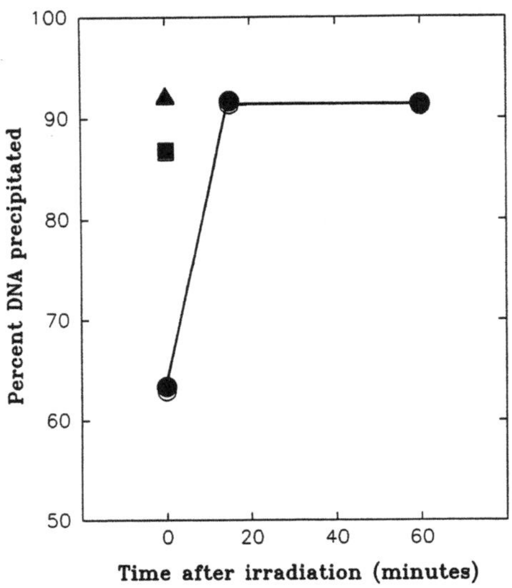

Fig. 7. DNA damage induced by ionizing radiation and measured by the DNA precipitation assay. Data in triplicates from two experiments. HT29 cells were irradiated with 10 Gy at 37 °C (*squares*) and 0 °C and allowed to repair at 37 °C for the indicated times (*circles*). Unirradiated cells (*triangles*) are shown for comparison. *Filled symbols* show selenite-supplemented cells and *unfilled* (mostly hidden) symbols show the response of unsupplemented cells. SEM less than size of symbols

4 Conclusions

There is a large variation in GSH peroxidase activity in response to supplementation with selenite of different cell lines. Supplementation with 50 – 100 nM selenite for 1 week should generally be suitable for maximal GSH peroxidase synthesis. It is possible to increase GSH peroxidase activity severalfold with no, or at least, small changes in the activities of the other intracellular antioxidant enzymes, which should gain further studies in elucidating the role of GSH peroxidase in the protection against oxidative stress. Furthermore, the results indicate that the previously found relation between the GSH peroxidase activities of different cell lines and their radiation resistance was fortuitous, and dependent on other cellular factors.

Acknowledgements. This investigation was supported by the Swedish Natural Science Research Council Grant No. 9204, the Swedish Medical Research Council Grant No. 04761 and the Swedish Defence Research Establishment, Umeå. We gratefully acknowledge the skilful technical assistance of Mrs. Els-Marie Öhman.

References

Coggle JE (1983) Biological effects of radiation, 2nd edn. Taylor & Francis, New York
Marklund SL, Westman NG, Roos G, Carlsson J (1984) Radiation resistance and the CuZn superoxide dismutase, Mn superoxide dismutase, catalase, and glutathione peroxidase activities of seven human cell lines. Radiat Res 100:115 – 123
Olive PL (1988) The DNA precipitation assay: a rapid and simple method for detecting DNA damage in mammalian cells. Environ Mol Mutagen 11:487 – 495

Prohaska JR, Ganther HE (1977) Glutathione peroxidase activity of glutathione-S-transferases purified from rat liver. Biochem Biophys Res Commun 76:437–445
Sandström BE, Carlsson J, Marklund SL (1987) Variations among cultured cells in glutathione peroxidase activity in response to selenite supplementation. Biochim Biophys Acta 929:148–153
Sandström BE, Carlsson J, Marklund SL (1989) Selenite-induced variation in glutathione peroxidase activity of three mammalian cell lines: no effect on radiation-induced cell killing or DNA strand breakage. Radiat Res 117:318–325
Tan KH, Meyer DJ, Coles B, Gillies N, Ketterer B (1987) Detoxication of peroxidized DNA by glutathione transferases. Biochem Soc Trans 15:628–629

Effects of Selenium Supplementation on Platelet Function as Assessed by Platelet Aggregation and Glutathione Peroxidase Activity

H. W. VAN DER TORRE[1,2], J. VEENSTRA[1], H. VAN DE POL[1], H. VAN STEENBRUGGE[1], S. PELUPESSY[1], G. SCHAAFSMA[1], and Th. OCKHUIZEN[1]

1 Introduction

Increased platelet aggregation and an altered arachidonic acid metabolism was reported in selenium-deficient animals (Schoene et al. 1986; Bryant et al. 1983). The selenium-dependent enzyme glutathione peroxidase (GSH-Px) may influence the platelet metabolism by regulating the hydroperoxide level within the cell (Schoene 1985). The activity of the GSH-Px can be markedly influenced by changes in dietary selenium intake (Levander 1983). The purpose of this pilot study was twofold, firstly to establish the effect of a selenium supplementation on platelet GSH-Px activity and secondly to gain an insight into the influence of elevated selenium intake on the platelet aggregation and the production of arachidonic acid metabolites in man.

2 Methods

The subjects were four healthy male volunteers (20–30 years of age). The study consisted of a supplementation period of 4 weeks and a post-dosing period of 1 week. The supplements were given daily as two tablets of Se-rich yeast and provided 200 μg of selenium per day per subject. The subjects consumed their habitual diet and were encouraged to maintain their usual daily routines. Fasting blood samples were taken at weekly intervals. For the analyses of selenium in plasma and red blood cells (RBC) and GSH-Px activities in plasma, RBC and platelets, 14 ml blood (EDTA; anticoagulant) was sampled and separated by centrifugation in platelet-rich plasma (PRP), platelet-poor plasma (PPP) and RBC. The platelets for the GSH-Px assay were prepared from PRP by gel filtration on Sepharose 2B, and suspended in 0.32 M sucrose. GSH-Px activities were assayed by the method of Pleban et al. (1982); selenium was measured by the fluorimetric method of Koh and Benson (1983); and protein was measured by the method of Lowry et al. (1951). For the platelet aggregation study, 10 ml blood was anticoagulated with 0.1 ml trisodiumcitrate 32% (w/v). PRP was diluted with PPP to give a standard-

[1] Department of Nutrition, TNO-CIVO Toxicology and Nutrition Institute, P.O. Box 360, 3700 AJ Zeist, The Netherlands
[2] To whom correspondence should be addressed

ized PRP with a final platelet count of 220×10^6 ml^{-1}. Platelet aggregation, induced with collagen, was studied in a Chrono-log aggregometer for 5 min. The reaction was stopped by adding cold NaCl 0.9% (w/v). The extracts of the stimulated platelets were assayed for 12-hydroxy eicosatetraenoic acid (12-HETE) and 12-hydroxy heptadecatrienoic acid (HHT) by HPLC and for thromboxane B_2 (TxB_2) by enzyme immunoassay.

To evaluate the effects, the variables in question were regressed on time in each individual, the significance of the average regression coefficients were examined with the student's t-test.

3 Results and Discussion

The plasma and RBC selenium concentrations increased significantly during the supplementation period from 92 to 107 ng ml^{-1} ($p < 0.002$) and from 106 to 120 ng ml^{-1} ($p < 0.001$) respectively. The GSH-Px activities in plasma and RBC did not show any significant change. Probably the activities had reached a plateau at the habitual selenium intake level of 40–80 µg day^{-1} (Staarink 1987). In contrast, the platelet GSH-Px activity increased rapidly and significantly during the supplementation period as shown in Fig. 1. Thus, it appeared to be possible to influence platelet GSH-Px activity by increasing the selenium intake. Table 1 shows the results of the platelet aggregation study. Aggregation after induction with 0.4 µg ml^{-1} collagen was decreased at the end of the dosing period (t = 4).

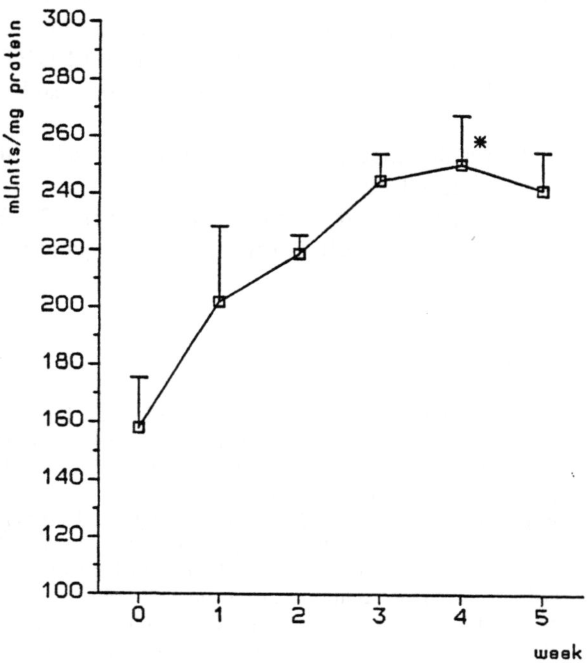

Fig. 1. GSH-Px activities in platelets during daily supplementation with 200 µg selenium as selenium-rich yeast for 4 weeks and 1 week post-dosing. Each *point* represents the mean ± SEM of four subjects. The difference between week 0 and week 4 is statistically significant ($p < 0.002$)

Table 1. Effect of selenium supplementation on platelet aggregation and arachidonic acid metabolites. Each point represents the mean ± SEM of four subjects

	t = 0	t = 4	p-Value
Platelet aggregation (collagen 0.4 μg ml^{-1})			
Maximum aggregation (%)	39.5 ± 9.5	24.2 ± 11.6	< 0.05
Thromboxane B_2 (ng ml^{-1})	318 ± 26	264 ± 29	< 0.005
Platelet aggregation (collagen 0.7 μg ml^{-1})			
Maximum aggregation (%)	62.6 ± 2.6	52.8 ± 2.6	n.s.
Thromboxane B_2 (ng ml^{-1})	717 ± 25	456 ± 46	< 0.01
Platelet aggregation (collagen 16 μg mg^{-1})			
Maximum aggregation (%)	81.1 ± 3.7	79.4 ± 3.7	n.s.
HHT (peak area)	340 ± 94	407 ± 36	n.s.
12-HETE (peak area)	149 ± 40	209 ± 30	< 0.05

Likewise, we found significant decreases in TxB_2 production after induction of aggregation with both 0.4 and 0.7 μg ml^{-1} collagen. Stimulation by using an overdosis of collagen (16 μg ml^{-1}) showed that selenium supplementation resulted in the formation of significantly increased amounts of 12-HETE.

Several laboratories have demonstrated a relationship between arachidonic acid metabolism and GSH-Px activity (Schoene et al. 1986). It is hypothesized that the activity of two dioxygenases, cyclooxygenase and lipoxygenase, is regulated by the peroxide tone within the cell. The GSH-Px can modulate the peroxide tone by reducing the hydroperoxides (Schoene 1985). The increased GSH-Px activity and the decreased TxB_2 production found in our study are in good agreement with the experiment of Guidi et al. (1984), who suggested that in normal platelets, GSH-Px activity controls the amount of TxB_2 formation.

Bryant et al. (1983) indicated a role for platelet GSH-Px in the reduction of the lipoxygenase-derived 12-HPETE into 12-HETE. We found an increase in 12-HETE formation which could be the result of an increased activity of GSH-Px.

The results suggest that an increase in selenium intake causes an effect on platelet metabolism.

The effects found in both TxB_2- and in 12-HETE production suggest that selenium exerts it is effect on the cyclooxygenase as well as on the lipoxygenase pathway.

References

Bryant RW, Simon TC, Bailey JM (1983) Biochem Biophys Res Commun 117:183–189
Guidi G, Schiavon R, Biasioli A, Perona G (1984) J Lab Clin Med 104:574–582
Koh TS, Benson TH (1983) J Assoc Off Anal Chem 66:918–926
Levander OA (1983) Fed Proc 42:1721–1725
Lowry OH, Roseborough NJ, Farr AC, Randall RJ (1951) J Biol Chem 193:265–275
Pleban PA, Munyani A, Beachum J (1982) Clin Chem 28:311–316
Schoene NW (1985) In: Lands WEM (ed) Biochemistry of arachidonic acid metabolism. Nijhoff Boston/Dordrecht/Lancaster, pp 151–160
Schoene NW, Morris VC, Levander OA (1986) Nutr Res 6:75–83
Staarink T, Hakkenbrak P (1987) Het contaminantenboekje 1987. Staatsuitgeverij, Den Haag

Role of Glutathione Peroxidase and α-Tocopherol in Protecting Rat Erythrocytes Against *tert*-Butyl Hydroperoxide-Induced Hemolysis in Vitro

K. YASUMOTO[1,2], C.H. KIM[1], and T. SUZUKI[1]

1 Introduction

Simultaneous induction of vitamin E deficiency exacerbates most of the Se-deficiency disorders in animals (Thompson and Scott 1969, 1970). These two nutrients have sparing effects upon the dietary requirement for both (Bieri 1961; Hamilton and Tappel 1963), and hence are considered to have overlapping, complementary functions in the antioxidative defense mechanisms in the cells of different mammalian organs. The quantitative significance remains yet unclear, however, regarding the protection afforded by glutathione peroxidase (GSH-Px) compared with that afforded by vitamin E (Tappel 1980; Levander 1986). This chapter presents three lines of experimental evidence pointing to the predominant function of GSH-Px in antioxidant defense mechanisms.

2 Materials and Methods

Animals and Diets. Male weanling rats of Wistar strain were fed a torula yeast-based diet which was prepared as described previously (Kim et al. 1989). Analysis showed the basal diet to contain Se at a barely measurable level (<0.01 ppm). The Se-adequate diet was prepared by supplementing the basal diet with 0.1 ppm Se as sodium selenite. After 3 months feeding on the experimental diets, Se content in the erythrocytes from the Se-deficient rats decreased to about 5% of that observed for the Se-adequate counterparts, and GSH-Px activities to a barely detectable level in the Se-deficient erythrocytes.

t-BuOOH-Induced Hemolysis. A peroxide-induced hemolysis test was conducted according to the procedure summarized in Fig. 1. After prescribed periods of incubation with *tert*-butyl hydroperoxide (t-BuOOH) the erythrocyte suspensions were centrifuged and the absorbance of the supernatants was measured at 540 nm.

Phospholipid Analysis. Phospholipids extracted according to the method of Bligh and Dyer (1959) were chromatographed on a Silica Gel G thin-layer plate and individual components were quantitated densitometrically (Rouser et al. 1970).

SDS-Polyacrylamide Gel Electrophoresis. Erythrocytes' ghosts were subjected to electrophoresis on SDS-polyacrylamide gel with a linear gradient gel concentration from 4 to 20%. Protein bands were visualized by the silver staining method (Merril et al. 1981).

[1] The Research Institute for Food Science, Kyōto University, Gokasho, Uji, Kyōto 611, Japan
[2] To whom correspondence should be addressed

Whole blood

├ erythrocyte isolation

Erythrocytes

├ suspended in the basal medium

│ basal medium: erythrocyte 10% as packed cells, 1 mM KCN, 1 mM NaN_3, in isotonic phosphate buffer (pH 7.4)

├ added 2 mM GSH or not

├ added 0.6 mM *tert*-BuOOH

Incubation at 37 °C

│

Hemolysis estimation

Fig. 1. Procedure for determining susceptibility of rat erythrocytes to t-BuOOH

3 Results and Discussion

3.1 t-BuOOH-Induced Hemolysis Test

Effect of GSH Addition on Hemolysis. Effects of GSH addition on the time course of t-BuOOH-induced hemolysis differed by the Se status of erythrocytes (Fig. 2). Without addition of GSH, hemolysis continued with time and reached almost 100% in 2 h after incubation of both Se-adequate and Se-deficient erythrocytes. Addition of 2 mM GSH suppressed remarkably the hemolysis of Se-adequate erythrocytes. Even after 6 h of incubation, the rate of hemolysis was lower than 20%. Therefore, observed susceptibility of erythrocytes to t-BuOOH-induced hemolysis in the presence of added GSH is attributable to Se deficiency.

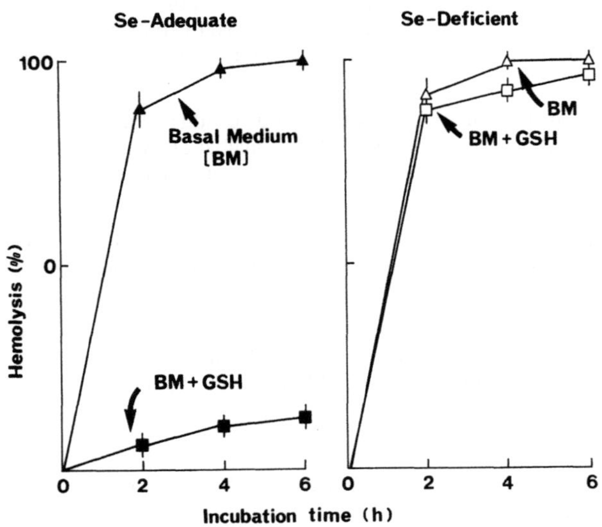

Fig. 2. Effects of GSH on t-BuOOH-induced hemolysis in Se-adequate and Se-deficient erythrocytes

Effect of GSSG, NADPH, and Glucose Addition on Hemolysis. The effects of GSSG, NADPH, and glucose addition on hemolysis vary according to the Se status. GSSG, either in the presence or absence of NADPH in the incubation medium, failed to suppress the hemolysis of both Se-adequate and Se-deficient erythrocytes. GSH of endogenous and exogenous origins was not effective by itself and its protective effect on hemolysis was entirely dependent on dietary Se. This dependence suggests that reducing equivalents of externally added GSH is transported into the erythrocyte possibly via an hitherto undescribed system so that it can collaborate with the GSH-Px present in the cytosol.

3.2 Hemolysis of α-Tocopherol-Decreased Erythrocytes

Depletion of α-Tocopherol by Preincubation with AAPH. In the second series of experiments, the t-BuOOH-induced hemolysis was followed by α-tocopherol-decreased erythrocytes. The tocopherol level of the erythrocytes was decreased by preincubation with 2,2′-azobis(2-amidinopropane)dihydrochloride (AAPH) under the condition described previously (Kim et al. 1989). This azo compound generates a water-soluble alkylperoxy radical in the incubation medium and selectively decomposes α-tocopherol in erythrocytes (Yamamoto et al. 1985). Irrespective of Se status, AAPH decreased α-tocopherol levels to about 40% of the native levels after 6 h of incubation. Compared with the reported values (Bieri and Poukka 1970; Tanaka and Mino 1986), the levels of α-tocopherol in the AAPH-treated erythrocytes are considered to be sufficiently low enough to allow us to assess the contribution of α-tocopherol in protecting erythrocytes against hemolysis.

t-BuOOH-Induced Hemolysis of AAPH-Treated Erythrocytes. The hemolysis test revealed that irrespective of the Se status and AAPH pretreatment, over 90% of the erythrocytes were hemolyzed within 2 h after the addition of t-BuOOH. GSH suppressed remarkably the hemolysis of Se-adequate erythrocytes pretreated with and without AAPH. Less than 20% of hemolysis occurred even after 6 h of incubation with t-BuOOH. On the other hand, t-BuOOH-induced hemolysis proceeded rapidly with Se-deficient erythrocytes even in the presence of added GSH. These results led us to propose that a reasonable working hypothesis is to picture GSH-Px as the most predominant defense against peroxides. The antioxidant function of α-tocopherol appears to be hitherto overestimated.

3.3 Cellular Integrity of Se-Deficient Erythrocytes

Finally, Se-deficient and Se-adequate erythrocytes were analyzed for membrane phospholipids and cytoskeletal proteins. A significant proportion of the erythrocytes from Se-deficient rats was found to contain Heinz bodies and to exhibit an echinocyte conformation. Pyruvate kinase activity, which is known to decline with the age of the erythrocytes (Seaman et al. 1980), became significantly high after long-term feeding of rats on a Se-deficient diet.

Compositional Changes of Membrane Phospholipids of Erythrocytes Incubated with t-BuOOH. The above described facts suggested that Se deficiency produces a gross morphological change and increased hemolysis of erythrocytes. Hence, the experiments were designed to determine whether dietary Se protected against membrane damage as well as against hemolysis on exposure to t-BuOOH in vitro. Obtained results showed that there were no significant differences in phospholipid composition between Se-adequate and Se-deficient erythrocytes collected after 3 months feeding. On the other hand, Se-adequate erythrocytes incubated without addition of GSH and Se-deficient erythrocytes without regards to GSH addition in the incubation medium hemolyzed readily. These hemolysates revealed a decrease of phospholipids, especially of phosphatidylethanolamine, phosphatidylserine and phosphatidylinositol after an incubation period of 4 h. The content of phosphatidylcholine did not particularly decrease except for erythrocytes from Se-deficient rats incubated with GSH.

Alteration of Fatty Acid Composition of Erythrocyte Membrane Phospholipid by the Incubation with t-BuOOH. The fatty acid composition of the total phospholipid fraction was similar between Se-adequate and Se-deficient erythrocytes before incubation. However, after incubation with t-BuOOH, with or without GSH, remarkable differences developed between these erythrocytes. The Se-adequate erythrocytes without GSH addition and Se-deficient erythrocytes, irrespective of GSH addition, showed remarkable decreases in polyunsaturated fatty acids; arachidonic, docosapolyenonic acids in particular (Fig. 3). In contrast, erythrocytes from Se-adequate rats which were incubated with GSH and t-BuOOH did not show any change in the fatty acid composition.

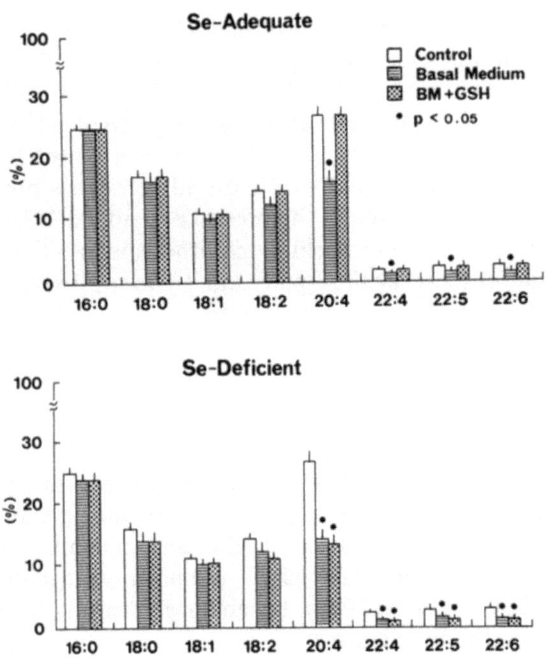

Fig. 3. Fatty acid composition of the total phospholipid fraction in erythrocyte membranes of Se-adequate and Se-deficient rats. Data are expressed, after correction for the recovery of margaric acid added as internal standard, as the percentage of total fatty acids in the total phospholipid fraction of control erythrocytes

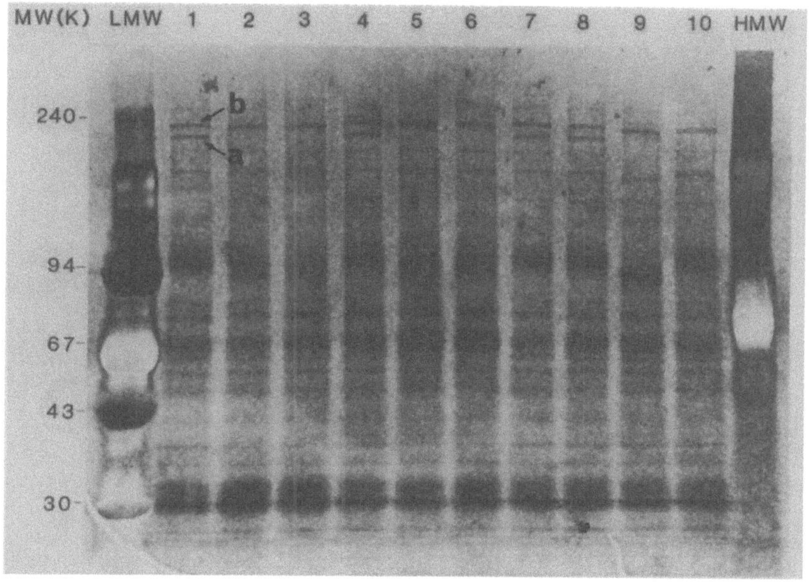

Fig. 4. SDS-polyacrylamide gel electrophoresis pattern of erythrocytes from Se-adequate and Se-deficient rats as effected by incubation with 0.6 mM t-BuOOH. *Lane 1*, Se-adequate, before incubation; *lane 2*, Se-adequate, incubated for 2 h in the basal medium (BM); *lane 3*, Se-adequate, incubated for 4 h in BM; *lane 4*, Se-deficient, before incubation; *lane 5*, Se-deficient, incubated for 2 h in BM; *lane 6*, Se-deficient, incubated for 4 h; *lane 7*, Se-deficient, incubated for 2 h with 2 mM GSH; *lane 8*, Se-adequate, incubated for 4 h with 2 mM GSH; *lane 9*, Se-deficient, incubated for 4 h with 2 mM GSH; *lane 10*, Se-deficient, incubated for 4 h with 2 mM GSH. *Arrows a* and *b* identify ankyrin (*band 2-1*) and spectrin-2 (*band 2*) respectively

Alteration of Composition of Erythrocyte Protein by Incubation with t-BuOOH. Along with the decrease of phospholipid species as described above, a specific loss of cytoskeleton protein, ankyrin, was detected on the SDS-polyacrylamide gel electrophoresis pattern of the hemolyzed erythrocytes (Fig. 4). No significant difference could be detected between intact Se-adequate (lane 1) and intact Se-deficient erythrocytes (lane 4) on the protein pattern on SDS-polyacrylamide gel. For the Se-adequate erythrocytes, incubated with t-BuOOH in the absence of added GSH, a band corresponding to ankyrin (215 kD, lane 2) disappeared selectively as early as after 2 h incubation. Similar results were obtained from Se-deficient rats incubated with the basal medium (lanes 4–6). On the other hand, Se-adequate erythrocytes incubated with t-BuOOH and externally added GSH sustained their components even after 4 h incubation (lanes 7–8). Such a protective effect of GSH, however, was not observed with Se-deficient erythrocytes (lanes 9–10) at all. These results are consistent with the observation made by Rice-Evans et al. (1985).

3.4 Proposed Function of GSH-Px in Erythrocytes

The experimental evidence presented allowed the authors to envisage the mechanism how t-BuOOH-induced hemolysis is caused in the Se-deficient eryth-

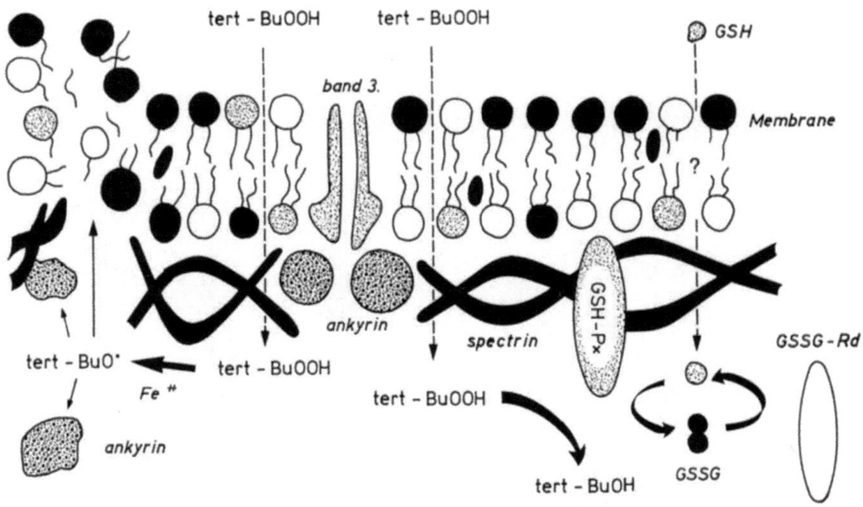

Fig. 5. Postulated mechanism of t-BuOOH-induced hemolysis and protection afforded by glutathione peroxidase for Se-deficient (*right side*) and Se-adequate erythrocytes (*left side*)

rocyte and suppressed in the Se-adequate erythrocyte (Fig. 5). The amphiphatic molecule, t-BuOOH, penetrates easily inside the erythrocyte; some may stay inside the membrane. In the Se-deficient erythrocyte, the tert-butoxyl radical, generated from t-BuOOH in the cytosol, attacks ankyrin and polyunsaturated fatty acid residues of inner membrane phospholipids, such as phosphatidylethanolamine. Membrane portions where vitamin E exists are protected from oxidative damage because the t-butoxyl radical is immediately scavenged by vitamin E; however, vitamin E does not have the ability to move rapidly enough to scavenge radicals existing far away on the inner surface of the biomembrane. Accordingly, vitamin E cannot stop chain reactions of lipid peroxidation initiated by butoxyl radicals on the inner membrane phospholipids. The ankyrin attacked by the t-butoxyl radical, on the other hand, becomes oxidized and degenerated. Erythrocytes from Se-adequate rats contain enough GSH-Px to maintain their functionality under such oxidative stress when GSH is abundant. The mechanisms discussed in our concept of erythrocyte protection by dietary selenium extend to other cells of the body as well.

Acknowledgements. This research was supported by a Grant-in-Aid 60480054 from the Ministry of Education, Science and Culture, Japan.

References

Bieri JG (1961) The nature of the action of selenium in replacing vitamin E. Am J Clin Nutr 9:89–96

Bieri JG, Poukka RKH (1970) In vitro hemolysis as related to rat erythrocyte content of α-tocopherol and polyunsaturated fatty acids. J Nutr 100:557–564

Bligh EG, Dyer WJ (1959) A rapid method of total lipid extraction and purification. Can J Biochem Physiol 37:911–917

Hamilton JW, Tappel AL (1963) Lipid antioxidant activity in tissues and proteins of selenium-fed animals. J Nutr 79:493–502

Kim CH, Yasumoto K, Suzuki T, Yoshida M (1988) Failure of glutathione in suppressing *tert*-butyl hydroperoxide-induced hemolysis in vitro of selenium-deficient rat erythrocytes. Nutr Res 8:767–775

Kim CH, Yasumoto K, Suzuki T, Yoshida M (1989) *tert*-Butyl hydroperoxide-induced hemolysis of α-tocopherol-decreased erythrocytes from selenium-deficient and selenium-adequate rats. J Nutr Sci Vitaminol 34:481–490

Levander OA (1986) Selenium. In: Mertz W (ed) Trace elements in human and animal nutrition, 2nd edn. Academic Press, London New York, pp 209–279

Merril CR, Dunau ML, Goldman D (1981) A rapid sensitive silver stain for polypeptides in polyacrylamide gels. Anal Biochem 110:201–207

Rice-Evans C, Baysal E, Hochstein P (1985) t-Butyl hydroperoxide-induced perturbations of human erythrocytes as a model for oxidant stress. Biochim Biophys Acta 815:426–432

Rouser G, Fleischer S, Yamamoto A (1970) Two-dimensional thin-layer chromatographic separation of polar lipids and determination of phospholipids by phosphorus analysis of spots. Lipids 5:494–496

Seaman C, Wyss S, Piomelli S (1980) The decline in energetic metabolism with aging of the erythrocyte and its relationship to cell death. Am J Hematol 8:31–42

Suzuki T, Kim CH, Yasumoto K (in press) Deterioration of membrane morphology, phospholipids and cytoskeletal protein in rat erythrocytes exposed to *tert*-butyl hydroperoxide: protection by exogenous glutathione fails in selenium deficiency. J Nutr Sci Vitaminol 34:491–506

Tanaka H, Mino M (1986) Membrane-to-membrane transfer of tocopherol in red blood cells. J Nutr Sci Vitaminol 32:463–474

Tappel AL (1980) Vitamin E and selenium protection from in vivo lipid peroxidation. In: Levander OA, Cheng L (eds) Micronutrient interactions: vitamin, minerals, and hazardous elements. NY Acad Sci, New York, pp 18–31

Thompson JN, Scott ML (1969) Role of selenium in the nutrition of the chick. J Nutr 97:335–342

Thompson JN, Scott ML (1970) Impaired lipid and vitamin E absorption related to atrophy of the pancreas in selenium-deficient chicks. J Nutr 100:797–809

Yamamoto Y, Niki E, Eguchi J, Kamiya Y, Shimasaki H (1985) Oxidation of biological membranes and its inhibition: free radical chain oxidation of erythrocyte ghost membranes by oxygen. Biochim Biophys Acta 819:29–36

Active Oxygen Generation by the Reaction of Selenite with Reduced Glutathione in Vitro

Y. SEKO[1], Y. SAITO[1], J. KITAHARA[1], and N. IMURA[1,2]

1 Introduction

Increased lipid peroxidation caused by selenite treatment was observed in vivo (Dougherty and Hoekstra 1982) and in vitro (Bunyan et al. 1960; Stacey and Klaassen 1981; Seko 1986). In vitro treatment of cells with selenite resulted in the decreased content of reduced glutathione and NADPH (Tsen and Collier 1960; Anundi et al. 1984), which are important in protecting cells against oxidation. However, the decrease in GSH concentration did not always relate to cell damage (Tsen and Collier 1960; Seko 1986) or lipid peroxidation (Seko 1986) in erythrocyte suspension. On the other hand, the importance of GSH (Young et al. 1981) or reactive sulfhydryls of hemoglobin (Seko 1986) have been considered to play an important role in selenite induced in vitro hemolysis. In order to estimate the relationship between GSH and the selenite-induced lipid peroxidation, we have examined the active oxygen generation in the mixture of selenite and GSH by using luminol- or lucigenin-dependent chemiluminescence (CL) as an indicator.

2 Methods

CL Caused by Selenite and GSH. GSH dissolved in distilled water was added with agitation into a solution containing selenite and lucigenin at 37 °C in a Luminescence Reader (Aloka Co. Ltd., Tokyo). Final concentration of lucigenin in 0.1 M sodium phosphate buffer (pH 7.4) was 0.08 mM and various concentrations of GSH and selenite were used in combination. Final volume of the reaction mixture was 1.0 ml. Immediately after the addition of GSH CL was monitored continuously for 4 min. The agitation was stopped 10 s after the addition of GSH. To examine the effect of oxygen depletion on the CL, N_2 gas (about 200 ml min^{-1}) was bubbled through the selenite solution for about 5 min before the addition of GSH solution. GSH solution was degassed in a reduced pressure with a disposable syringe.

[1] School of Pharmaceutical Sciences, Kitasato University, Minato-ku, Tokyo 108, Japan
[2] To whom correspondence should be addressed

CL Caused by Selenodiglutathione (GSSeSG) and GSH. Selenite was replaced in the reaction mixture with GSSeSG which was synthesized by the method of Ganther (1971), and the same procedure as above was carried out.

CL Caused by Hydrogen Selenide (H_2Se) in the Presence of O_2. H_2Se was generated by the method of Diplok et al. (1973) and trapped in phosphate-buffered saline, pH 7.4. The H_2Se solution mixed with luminol in the absence of O_2 was exposed to O_2 (air) with continuous agitation, and CL was measured.

3 Results and Discussion

CL Caused by Selenite and GSH. The time course of CL emitted from the mixture is shown in Fig. 1. The length of the lag phase, the time to reach the maximum, the intensity of CL at the maximum and total CL over 4 min varied with the concentrations of GSH and selenite added. With 0.8 mM selenite maximum CL over 4 min was observed when 4 mM GSH was added. Addition of GSH above 4 mM resulted in a decrease in the total CL.

By the depletion of any component of the mixture (O_2, selenite, GSH or lucigenin) CL was decreased to less than 10% of the CL obtained with the complete assay. These results indicate the essentiality of each component for the generation of the CL. The total CL was markedly reduced by the addition of superoxide dismutase (SOD), suggesting the generation of superoxide anion (O_2^-) in the mixture.

Red precipitates, supposed to be elemental selenium, were observed in the reaction mixture after incubation for 4 min. The precipitation of elemental selenium monitored by absorbance at 500 nm was observed simultaneously with the decrease in dissolved oxygen (DO) and the increase in CL.

CL Caused by GSSeSG and GSH. GSSeSG, a reaction product of selenite and GSH (Ganther 1968), did not cause CL in the presence of O_2. By the addition of GSH into the mixture, however, the emission of CL was observed. The presence of O_2 was confirmed to be essential for CL emission. The CL emitted from the mixture of GSSeSG and GSH in the presence of O_2 was also inhibited by SOD,

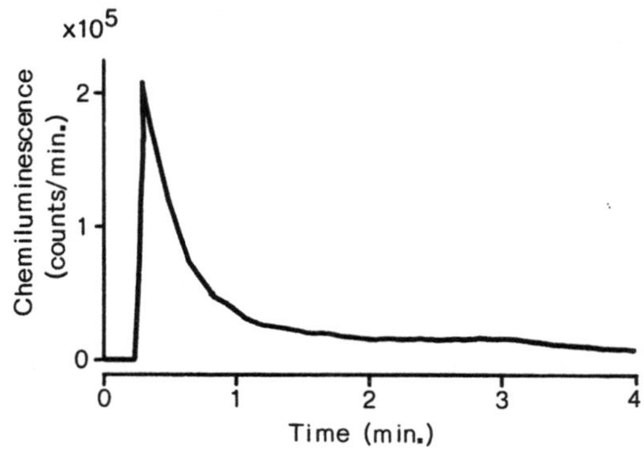

Fig. 1. The profile of the chemiluminescence emitted from the mixture of selenite (0.8 mM), reduced glutathione (4 mM) and lucigenin (0.08 mM) in the presence of oxygen

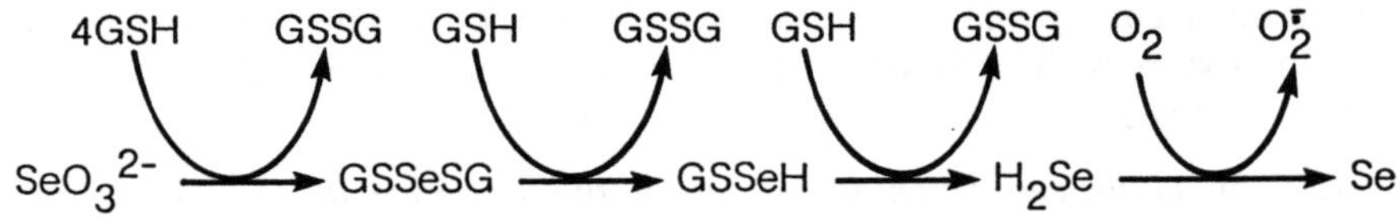

Fig. 2. Hypothetical mechanism of the generation of superoxide anion ($O_2^{\overline{\cdot}}$) in the mixture of selenite and reduced glutathione (*GSH*)

but not by heat-inactivated SOD. This fact also suggests the generation of $O_2^{\overline{\cdot}}$.

CL Caused by H_2Se with O_2. To study further the mechanism of the selenite-induced $O_2^{\overline{\cdot}}$ generation described above, the role of H_2Se, a reaction product of GSSeSG with GSH (Hsieh and Ganther 1975) was then investigated. Since lucigenin was found to be inactivated during the incubation with H_2Se in the reaction mixture, a suspension of luminol was used in place of lucigenin in the experiment with H_2Se. Though only little CL was generated by the addition of H_2Se solution to the luminol suspension in the absence of O_2, substantial CL was generated by exposing the mixture of H_2Se and luminol to air (O_2). The CL caused by H_2Se and O_2 was also inhibited by SOD, but not by heat-inactivated SOD, suggesting the generation of $O_2^{\overline{\cdot}}$.

The results described above suggest that the active oxygen, $O_2^{\overline{\cdot}}$, is generated by the reaction of O_2 with H_2Se, which is produced by the reduction of selenite with GSH, as shown in Fig. 2.

Reduction of selenite to H_2Se in cells or animal body could take place as reviewed by Ganther (1979). If the H_2Se meets the O_2 molecule, $O_2^{\overline{\cdot}}$ could be generated in cells and subsequent lipid peroxidation may occur resulting in cell damage. This hypothesis is consistent with the fact that GSH (Young et al. 1981) or reactive sulfhydryl (Seko 1986) was essential in manifestation of selenite cytotoxicity and with the increased lipid peroxidation by treatment with selenite in vitro (Bunyan et al. 1960; Stacey and Klaassen 1981; Seko 1986) and in vivo (Dougherty and Hoekstra 1982). Moreover, the decrease in DO has been observed in the mixture of selenite and GSH and was accompanied by the generation of active oxygen. The increased oxygen consumption has also been observed in isolated rat hepatocyte suspension exposed to selenite in vitro (Anundi et al. 1984). These facts may suggest the generation of active oxygen in cell suspension.

4 Conclusion

Superoxide anion, an active oxygen, is generated in the mixture of selenite and GSH in the presence of O_2. H_2Se, which is generated by the reduction of selenite with GSH, appears to be a direct reactant with O_2 to generate superoxide anion.

References

Anundi I, Ståhl A, Högberg J (1984) Effects of selenite on O_2 consumption, glutathione oxidation and NADPH levels in isolated hepatocytes and the role of redox changes in selenite toxicity. Chem-Biol Interact 50:277–288

Bunyan J, Green J, Edwin EE, Diploch AT (1960) Studies on vitamin E. 5. Lipid peroxidation in dialuric acid-induced haemolysis of vitamin E-deficient erythrocytes. Biochem J 77:47–51

Diplock AT, Caygill CPJ, Jeffery EH, Thomas C (1973) The nature of the acid-volatile selenium in the liver of the male rat. Biochem J 134:283–293

Dougherty JJ, Hoekstra WG (1982) Stimulation of lipid peroxidation in vivo by injected selenite and lack of stimulation by selenate. Proc Soc Exp Biol Med 169:209–215

Ganther HE (1968) Selenotrisulfides. Formation by the reaction of thiols with selenious acid. Biochemistry 7:2898–2905

Ganther HE (1971) Reduction of selenotrisulfide derivative of glutathione to a persulfide analog by glutathione reductase. Biochemistry 10:4089–4098

Ganther HE (1979) Chapter 5: Metabolism of hydrogen selenide and methylated selenides. Adv Nutr Res 2:107–128

Hsieh HS, Ganther HE (1975) Acid-volatile selenium formation catalyzed by glutathione reductase. Biochemistry 14:1632–1636

Seko Y (1986) Hemolysis of mouse erythrocytes by methylmercury and selenite in vitro. IV. Decrease in sulfhydryl concentration and lipid peroxidation by selenite. Teikyo Med J 9:287–297 (in Japanese with English abstract)

Stacey NH, Klaassen CD (1981) Comparison of the effects of metals on cellular injury and lipid peroxidation in isolated rat hepatocytes. J Toxicol Environ Health 7:139–147

Tsen CC, Collier HB (1960) The relationship between the glutathione content of rat erythrocytes and their hemolysis by various agents in vitro. Can J Biochem Physiol 38:981–987

Young JD, Crowley C, Tucker EM (1981) Haemolysis of normal and glutathione-deficient sheep erythrocytes by selenite and tellurite. Biochem Pharmacol 30:2527–2530

Studies on Selenite-Induced DNA Fragmentation and the Role of Poly(ADP-Ribose)Polymerase in Selenite Toxicity

J. Högberg[1,2], P. Garberg[1], and A. Ståhl[1]

1 Introduction

Selenium has been shown to act as an anticarcinogen in many experimental studies but it is also a potent toxin, and the question may be raised whether anticarcinogenic effects can be mediated by cytotoxicity. It is, for example, known that one of the toxic effects of selenium is DNA fragmentation (Shamberger 1985), and DNA fragmentation induced by other toxic substances can lead to an activation of the nuclear enzyme poly (ADP-ribose)polymerase (PARP) (Berger 1985). This enzyme utilizes NAD as a substrate, and it has been proposed that an extensive DNA fragmentation may lead to cell death via NAD depletion. It is thus possible that selenite can interact with other DNA-damaging compounds so that it selectively kills DNA-damaged cells.

Selenite metabolism and cell lysing mechanisms have been studied in some detail in suspended hepatocytes. It was shown that selenite was metabolized in isolated rat hepatocytes, but that the main end product in this system, dimethylselenide, was not formed during the first hour of incubation (Ståhl et al. 1984). Instead, there were several indications that Se metabolites entered redox cycles during the lag in metabolism: NADPH and GSH were oxidized and there was an increased oxygen consumption (Anundi et al. 1984). It was also shown that hypoxia was induced in metabolizing cells, and a need for hypoxic conditions for dimethylselenide formation was indicated (Garberg and Högberg 1987). During the second hour, when the cells were hypoxic, there was a rapid formation of dimethylselenide (Ståhl et al. 1984). Hepatocellular lysis, induced under these conditions, could be explained by reduction-oxidation cycles and their effect on redox equilibria and energy supplies (Anundi et al. 1984).

In this chapter we present data from a study on selenite-induced DNA fragmentation. It will be shown that selenite is a potent inducer of single strand breaks in DNA provided oxygen is available, and evidence for an involvement of DNA damage in selenite toxicity will be presented.

[1] Department of Toxicology, National Institute of Occupational Health, 17184 Solna, Sweden
[2] To whom correspondence should be addressed

2 Materials and Methods

Hepatocytes were isolated by collagenase perfusion (Högberg and Kristoferson 1977). Isolated cells were either incubated in suspension in round-bottomed flasks or in primary cultures on collagene-coated plastic dishes (Garberg et al. 1988). Single strand breaks in DNA (SSB) were measured as described by Ahnström and Erixon (1981). Lysis was measured as leakage of lactate dehydrogenase (LDH) (Högberg and Kristoferson 1977). Cells in culture were also examined in a phase contrast microscope, and it was found that an increased leakage of LDH was always associated with morphological changes and detachment of cells.

3 Results

When isolated hepatocytes in suspension were exposed to selenite (20 – 50 μM) there was a rapid increase in SSB. Ten μM selenite had no effect on this parameter. The SSB level reached a plateau and then declined. The changes in the SSB level were correlated to changes in the rate of oxygen uptake by the cells (Fig. 1). Incubations, where the cell suspension had been equilibrated with nitrogen instead of carbogen gas, showed that nitrogen was an effective inhibitor of SSB formation by selenite (Garberg et al. 1988). Nitrogen did not inhibit SSB formation by methylmethanesulfonate (MMS). In other experiments it was shown that inhibitors of oxygen consumption, such as mercury chloride or cyanide, also prevented SSB formation. It could also be shown that sodium selenide induced SSB and that this effect was oxygen-dependent in the same way as the effect of selenite on SSB (Garberg et al. 1988).

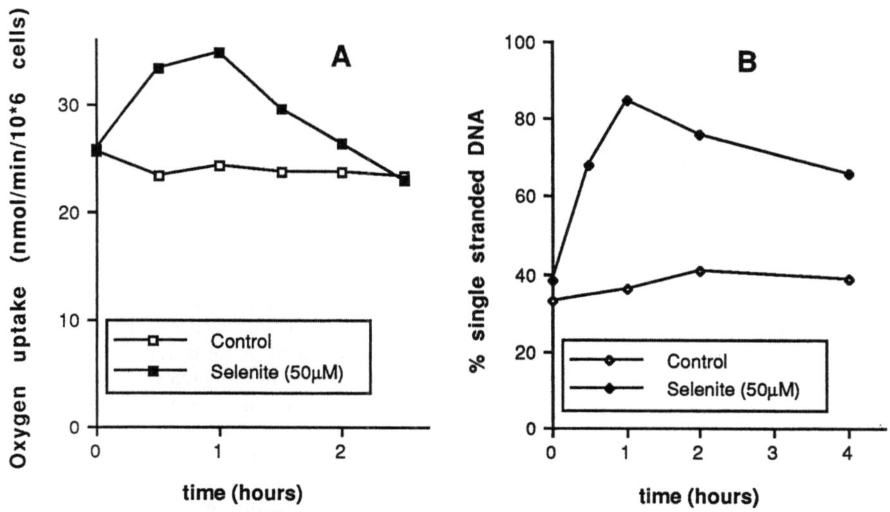

Fig. 1 a, b. Effect of selenite on oxygen consumption (**a**) and SSB levels (**b**). Isolated hepatocytes were suspended in Krebs-Henseleit buffer in the presence of selenite (50 μM). At the times indicated samples were withdrawn and analyzed for oxygen consumption (cf. Anundi et al. 1984) or SSB. Control incubates contained no selenite

In order to test the possibility that DNA fragmentation influenced the toxicological response to selenite, experiments with inhibitors of PARP, such as 3-aminobenzamide (3-ABA) and theophylline, were performed. It was found that these inhibitors did not influence the lysing effect of selenite on suspended hepatocytes, so a model system with plated hepatocytes was used for further studies. In this model the level of SSB was more slowly increased than shown in Fig. 1, and the decrease seen in cells in suspension was not prominent. The reason for this difference in SSB response is not known, but may relate to the access of oxygen, which presumably was higher in cultures of plated cells than in suspended cells; the number of cells per culture medium volume was much higher in the suspensions. The lysing capacity of selenite was then tested in this model. It was found that high concentrations (50 μM) lysed the cells within 5 h. Lower concentrations (20–30 μM), which did not lyse cells in suspension during the time period that could be studied with this model (Anundi et al. 1984), were found to lyse plated cells after about 20 h (Garberg et al. 1988).

To investigate the role of DNA fragmentation on cellular lysis two types of experiments were performed. In a first series of experiments 3-ABA was used to inhibit selenite toxicity. 3-ABA delayed or prevented the toxicity of selenite in the concentration range of 20–40 μM (Garberg et al. 1988). Theophylline also inhibited selenite toxicity, but 3-ABA was more effective. At higher concentrations

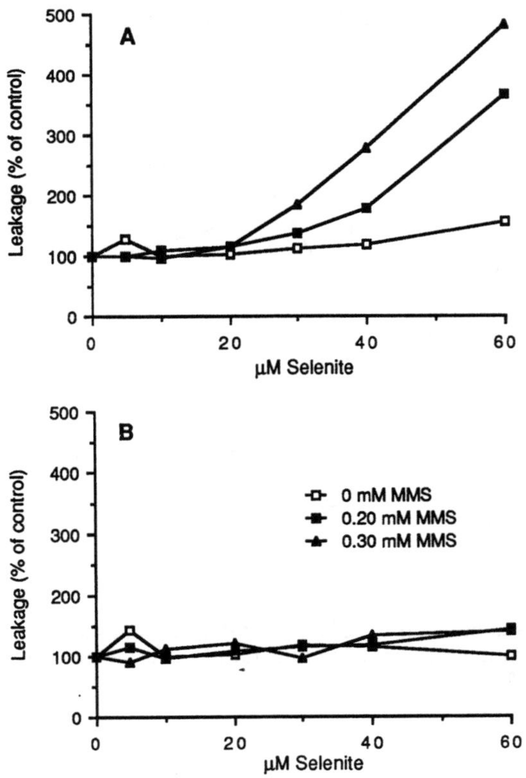

Fig. 2a, b. Selenite toxicity in hepatocytes pretreated with MMS. Isolated hepatocytes were allowed to attach to collagen-coated plastic dishes and incubated in complete medium. After 4 h the medium was changed to a complete medium with MMS (0–0.3 mM). 20 (**a**) or 44 (**b**) h after MMS additions, the medium was changed to a medium containing selenite (0–60 μM). Leakage was followed by measuring LDH activity in the medium. The figure shows the results obtained 8 h after selenite addition

3-ABA or theophylline had no effect, which is consistent with previous conclusions (Anundi et al. 1984).

In a second series of experiments cells that had been pretreated with DNA-alkylating agents were used. Cells were pretreated with either MMS or diethyl nitrosamine, and the concentrations used did not affect the number of viable cells in the absence of selenite. In the experiment described in Fig. 2 selenite was added 20 or 44 h after treatment with MMS. Figure 2 shows that MMS-pretreated cells leaked more than untreated controls 8 h after selenite addition. It also shows that the effect of the pretreatment became apparent earlier if the time interval between MMS and selenite additions was 20 instead of 44 h. To further implicate DNA fragmentation and PARP activation in the toxicity of selenite, the nicotinamide concentration in the medium was analyzed (cf. Schraufstatter et al. 1986). It was found that 30 μM selenite increased the nicotinamide concentration within 6 h (Garberg et al. 1988).

4 Discussion

This study provides evidence for an involvement of DNA damage in selenite toxicity. First, it was shown that inhibitors of PARP delayed or prevented cellular lysis in selenite-treated cells. It was thus indicated that selenite-induced DNA fragmentation activated PARP and that this activation led to NAD degradation. It has been argued that PARP activation per se cannot kill hepatocytes (Alvarez-Gonzalez et al. 1986). However, it could be shown here that DNA fragmentation was related to the redox cycling of Se metabolites. This indicates that PARP activation can be regarded as one of several potentially interacting events that eventually lead to energy depletion and redox changes. It may, for example, be suggested that the effects previously ascribed to the redox cycles, such as hypoxia, GSH oxidation and NADPH consumption, interacted with NAD depletion.

The potentiating effect of DNA-alkylating agents on selenite-induced lysis further implicates DNA alterations in selenite toxicity. Doses of the alkylating agents were used which did not kill the cells in the absence of selenite, and the interval between the addition of the DNA alkylator and selenite was sufficiently long to enable repletion of cofactors such as GSH. DNA repair, on the other hand, is a slower process and may proceed for several weeks (Brambilla et al. 1986), so the loss of interaction with time, as shown in Fig. 2, may depend on DNA repair.

The findings presented here are compatible with a selective killing of DNA-damaged hepatocytes by selenite. Selenite, given in doses that are not toxic to an animal in a conventional sense, should thus be able to kill single cells with a high proportion of unrepaired DNA damages. As it may be anticipated that such cells have a higher risk in developing malignant growth than cells with a low level of DNA damages, this effect may act to prevent tumor development.

References

Ahnström G, Erixon K (1981) Measurements of strand breaks by alkaline denaturation and hydroxyapatite chromatography. In: Friedberg EC, Hanawalt PC (eds) DNA repair. Dekker, New York

Alvarez-Gonzalez R, Eichenberger R, Althaus FR (1986) Poly(ADP-ribose) biosynthesis and suicidal NAD depletion following carcinogen exposure of mammalian cells. Biochem Biophys Res Commun 138:1051–1057

Anundi I, Ståhl A, Högberg J (1984) Effects of selenite on O_2 consumption, GSH oxidation and NADPH levels in isolated hepatocytes and the role of redox changes in selenite toxicity. Chem Biol Interact 40:277–288

Berger NA (1985) Poly (ADP-ribose) in the cellular response to DNA damage. Radiat Res 101:4–15

Brambilla G, Martelli A, Pino A, Robbiano L (1986) Sequential analysis of DNA damage and repair during the development of carcinogen-induced rat liver hyperplastic lesions. Cancer Res 46:3476–3481

Garberg P, Högberg J (1987) The role of hypoxia in selenium metabolism. Biochem Pharmacol 36:1377–1379

Garberg P, Ståhl A, Warholm M, Högberg J (1988) Studies of the role of DNA fragmentation in selenium toxicity. Biochem Pharmacol 37:3401–3406

Högberg J, Kristoferson A (1977) A correlation between GSH levels and cellular damage in isolated hepatocytes. Eur J Biochem 74:77–82

Schraufstatter IU, Hyslop PA, Hinshaw DB, Spragg RG, Sklar LA, Cochrane CG (1986) Hydrogen peroxide-induced injury of cells and its prevention by inhibitors of poly(ADP-ribose)polymerase. Proc Natl Acad Sci USA 83:4908–4912

Shamberger RJ (1985) The genotoxicity of selenium. Mutat Res 154:29–48

Ståhl A, Anundi I, Högberg J (1984) Selenite metabolism to volatile metabolites in an isolated hepatocyte model system. Biochem Pharmacol 33:1111–1117

Involvement of Cellular Sulfhydryl Compounds in the Inhibition of DNA and RNA Synthesis by Selenite

G. D. FRENKEL[1,2] and D. FALVEY[1]

1 Introduction

There have been a number of studies which have suggested that sulfhydryl (SH) compounds may play a role in the cellular effects of selenite. Our recent finding that DNA and RNA polymerases are inhibited by selenite only in the presence of an SH compound (Frenkel et al. 1987) led us to hypothesize that cellular SH compounds are involved in the inhibition by selenite of cellular DNA replication and transcription. We have investigated this hypothesis by examining the effect of selenite on DNA and RNA synthesis in cells which have been depleted of endogenous SH compounds. Our results have provided direct evidence for the involvement of cellular SH compounds in this effect of selenite.

2 Results

HeLa cells were treated with diethylmaleate (DEM), an agent which has been shown to deplete the cells of SH compounds (Vos et al. 1984). The effect of selenite on nucleic acid synthesis in control and DEM-treated cells was then compared. The results (Fig. 1) showed that treatment of the cells with DEM significantly reduced the inhibitory effect of selenite on both DNA and RNA synthesis. The inhibition of DNA synthesis in isolated nuclei by selenite was also significantly lower in nuclei isolated from DEM-treated cells (Frenkel and Falvey 1989).

A reasonable hypothesis to explain this result is that selenite first reacts with intracellular SH compounds by the reaction (Ganther 1968):

$$4\ RSH + H_2SeO_3 \rightarrow RSSeSR + RSSR + 3\ H_2O \qquad (1)$$

to form selenotrisulfides which are the actual inhibitors of nucleic acid synthesis. Because the concentration of SH compounds in DEM-treated cells is much lower than in control cells, this reaction proceeds to only a limited extent; thus, much less selenotrisulfide is formed in DEM-treated cells and there is less inhibition of nucleic acid synthesis. If this hypothesis is correct, then there should not be any difference between control and DEM-treated cells when they are exposed to

[1] Department of Biological Sciences, Rutgers University, Newark, NJ 07102, USA
[2] To whom correspondence should be addressed

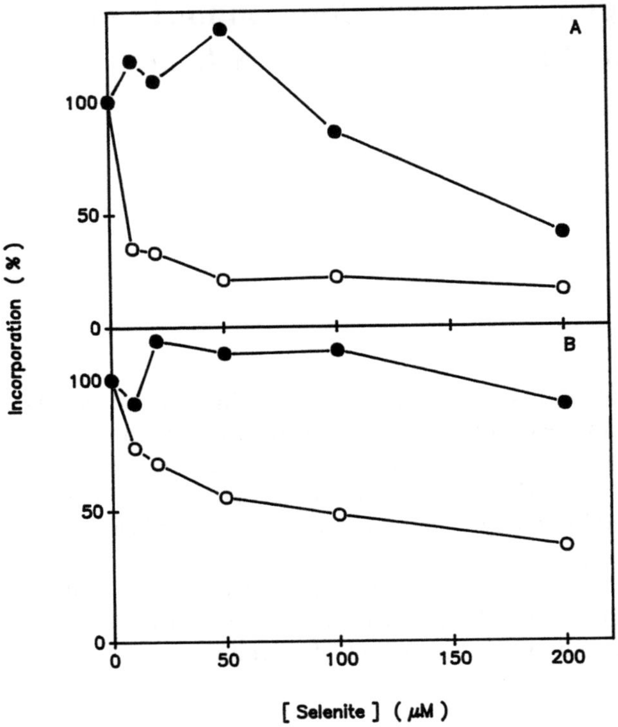

Fig. 1a, b. Inhibition of nucleic acid synthesis by selenite in DEM-treated (SH-depleted) HeLa cells. DNA and RNA synthesis was measured as the incorporation of [^{3}H]thymidine and [^{3}H]uridine, respectively, into acid-precipitable material, as described previously (Frenkel and Randles 1982). Control (○) or DEM-treated cells (●) were exposed to the indicated concentration of selenite for 1 h prior to addition of the radioactive nucleoside. DEM-treated cells were exposed to 1 mM DEM for 2 h prior to the addition of selenite. The results are presented as the percent of the incorporation in the absence of selenite. **a** DNA synthesis; **b** RNA synthesis

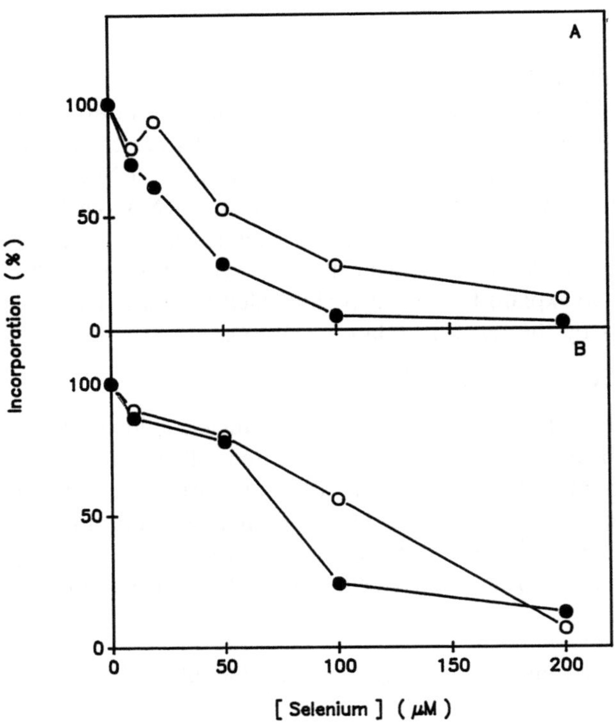

Fig. 2a, b. Inhibition of nucleic acid synthesis by the product of the reaction of selenite with glutathione. The experiment was carried out as in Fig. 1, except that instead of selenite the cells were exposed to the product of the reaction of selenite with glutathione. **a** DNA synthesis; **b** RNA synthesis

Fig. 3a, b. Inhibition of nucleic acid synthesis by selenite in BSO-treated (glutathione-depleted) HeLa cells. The experiment was carried out as in Fig. 1, with control (○) cells or cells which had been treated with BSO (●) (1 mM for 24 h). **a** DNA synthesis; **b** RNA synthesis

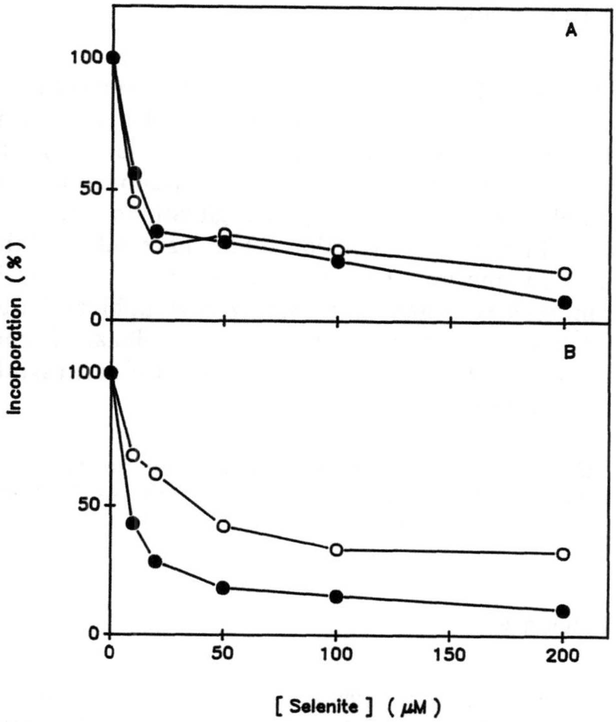

selenotrisulfide rather than to selenite. To examine this, selenite was reacted with glutathione, and the effect of the product of the reaction on DNA and RNA synthesis was determined. The results of this experiment (Fig. 2) were consistent with the prediction: nucleic acid synthesis in DEM-treated cells was inhibited to the same (or slightly greater) extent than in control cells. It should be noted that the disulfide product of reaction (1) (RSSR), which in this case is oxidized glutathione, had no comparable inhibitory effect on synthesis in either control or DEM-treated cells. Thus, the defect in DEM-treated cells appears to be the decreased production of selenotrisulfide via reaction (1).

DEM is a nonspecific reagent in that it reduces the cellular level of all nonprotein SH compounds. Since glutathione is by far the most abundant nonprotein SH compound in eukaryotic cells, it was of interest to specifically examine its possible role in the inhibition of DNA and RNA synthesis by selenite. To do this we treated the cells with buthionine sulfoximine (BSO), an agent which depletes the cells specifically of glutathione by inhibiting its biosynthesis (Griffith and Meister 1979). We found that treatment of cells with BSO did not decrease the inhibitory effect of selenite on DNA and RNA synthesis, and in the case of RNA may even have enhanced it somewhat (Fig. 3). Thus, nonspecific depletion of cells of SH compounds severely decreased the ability of selenite to inhibit cellular DNA and RNA synthesis, but specific depletion of the cells of glutathione did not.

3 Conclusions

Our results have indicated that a cellular SH compound is required for the inhibition of DNA and RNA synthesis by selenite. Most probably the selenite must react with the SH compound to form the selenotrisulfide derivative; the ultimate inhibitor of synthesis may be the selenotrisulfide itself or a product of its metabolism. Although exogenous glutathione can serve as the necessary SH compound (see Fig. 2), depletion studies have indicated that intracellular glutathione does not appear to be involved (see Fig. 3). This may indicate that, for some reason, selenite is unable to react with glutathione intracellularly. This conclusion leaves open the question of which intracellular SH compound or compounds are actually involved in the inhibition of nucleic acid synthesis by selenite; this is currently under investigation.

Acknowledgements. This work was supported in part by grant number ES-04087 from the National Institutes of Health. This is publication No. 71 from the Department of Biological Sciences, Rutgers-Newark.

References

Frenkel GD, Falvey D (1988) Evidence for the involvement of sulfhydryl compounds in the inhibition of cellular DNA synthesis by selenite. Mol Pharmacol 34:573–577

Frenkel GD, Randles K (1982) Specific stimulation of α-amanitin sensitive RNA synthesis in isolated HeLa nuclei by methyl mercury. J Biol Chem 257:6275–6279

Frenkel GD, Walcott A, Middleton C (1987) Inhibiton of RNA and DNA polymerases by the product of the reaction of selenite with sulfhydryl compounds. Mol Pharmacol 31:112–116

Ganther HE (1968) Selenotrisulfides. Formation by the reaction of thiols with selenious acid. Biochemistry 7:2898–2905

Griffith OW, Meister A (1979) Potent and specific inhibition of glutathione synthesis by buthionine sulfoximine. J Biol Chem 254:7558–7560

Vos O, Schans GP Van der, Roos-Verhey WSD (1984) Effects of BSO and DEM on thiol-level and radiosensitivity in HeLa cells. Int J Radiat Oncol Biol Phys 10:1249–1253

Part II Influences of Selenium on the Metabolism of Animals and Man

Selenium-Dependent Metabolic Modulations in Mouse Liver

R. REITER[1,2], R. OTTER[1,3], H.-M. HANEY[1], and A. WENDEL[1]

1 Introduction

With the discovery of glutathione peroxidase (Se-GSH-Px) as a selenoenzyme (Flohé et al. 1973) the key to the understanding of the effects of selenium (Se) deficiency was thought to be found. The loss of Se-GSH-Px is an early event during Se depletion and the resulting impairment in the detoxication of cellular hydroperoxides explains the increased susceptibility to peroxidative damage that is characteristic of Se-deficient animal models. However, recent studies from several laboratories have pointed out that Se deficiency is not simply Se-GSH-Px deficiency, and the importance of other Se proteins (Yang et al. 1987) or Se-dependent cellular pathways has been emphasized.

Previous work from this laboratory has focussed on the biochemical characterization of hepatic drug metabolism in Se-deficient (Se^-) mice (Reiter and Wendel 1983, 1984, 1985). In the course of these studies widespread modulations of in vitro enzyme activities were found. A causatively exclusive role of Se-GSH-Px on these enzyme modulations was challenged, based on the dose dependence and the kinetics of reversal by a single injection of Se.

The present studies continue this line of research by addressing three questions: (1) Are these enzyme modulations relevant in vivo? (2) How are these changes caused? (3) How important is Se to hepatic function?

2 Material and Methods

Dietary Se deficiency was produced by feeding weanling mice a torula yeast-based low Se diet (less than 10 ppb) for at least 4 months. The same diet but supplemented with 0.33 ppm Se was fed to a control group (Se^+). Details of the experimental procedures used have been described elsewhere (Reiter and Wendel 1984; Otter et al. 1986, 1989; Reiter et al. 1989; Vogl et al. 1987).

[1] Fakultät für Biologie, Universität Konstanz, Postfach 5560, 7750 Konstanz, FRG
[2] Present address: Boehringer Mannheim GmbH, 6800 Mannheim, FRG
[3] Present address: BASF AG, 6700 Ludwigshafen, FRG

3 Results

Prolonged dietary Se deficiency results in the loss of Se-GSH-Px activity and in widespread alterations of hepatic enzyme activities compared to control mice. Figure 1 summarizes the observed changes which can be characterized in the following way: The changes in enzyme activities were not unidirectional, e.g.,

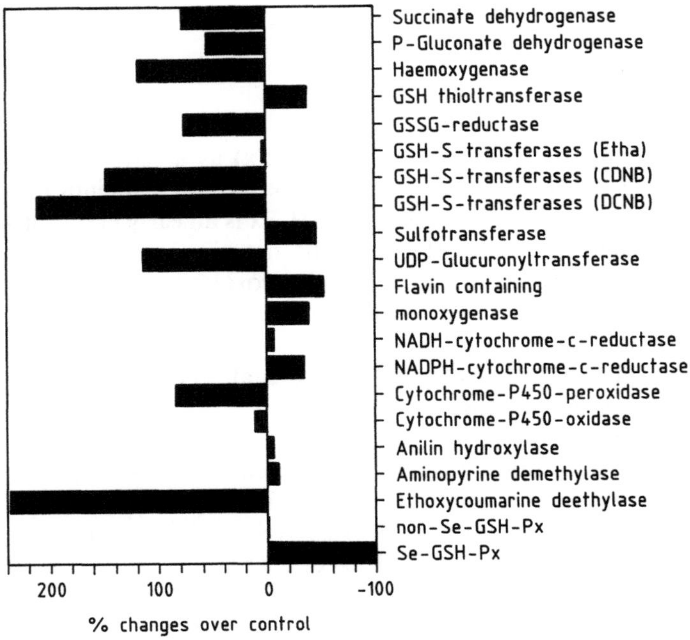

Fig. 1. Selenium-dependent modulations of enzyme activities in mouse liver

Table 1. In vivo selenium-dependent metabolic alterations in mice

	Se-deficient	Control	N
Depletion of total hepatic glutathione by phorone (300 mg kg^{-1}, i.p.) Time for half-maximal depletion (min)	3.5	15	Reiter and Wendel (1985)
Stimulation of hepatic DNA synthesis following a 40% hepatectomy: ^{3}H-methylthymidine incorporation (dpm μg^{-1} DNA)	4690 ± 1510	1810 ± 690	Reiter and Wendel (1984)
CO_2 exhalation following oral application of 100 nmol L-^{14}C-glucose: $^{14}CO_2$, excreted over 11 h (percent of dose)	78.0 ± 4.3	57.0 ± 2.3	Reiter and Wendel (1983)
CO_2 exhalation following i.p. application of L-^{14}C-leucine: $^{14}CO_2$, excreted over 48 h (percent of dose)	1.2 ± 0.1	1.6 ± 0.3	Reiter and Wendel (1983)

some enzyme activities were enhanced during deficiency, while others decreased. The affected enzymes originated in different subcellular compartments and contributed to different metabolic pathways, for example drug metabolism, the pentose phosphate shunt or the tricarboxylic acid cycle.

Alterations were also evident when studying selected metabolic performances in vivo. In the first experiment (Table 1) depletion of hepatic glutathione (GSH) by phorone was determined. Loss of GSH proceeded considerably faster in Se-deficient mice compared to control mice. In another experiment the hepatic proliferative response following a 40% hepatectomy was markedly enhanced in Se^- mice, although basal DNA synthesis rates did not differ between both dietary groups (Otter et al. 1986). Moreover, the increased carbon dioxide exhalation rates from tracer doses of labelled glucose or leucine in Se^- mice underline the relevance of Se-dependent metabolic changes in vivo. The obvious question to ask is about a common denominator of the observed changes. A possible involvement of Se-dependent site(s) in protein synthesis and/or degradation was investigated by subcellular fractionation of mouse liver. Thus, 48 h after intraperitoneal injection of ^{14}C-labelled leucine, a ^{3}H-leucine pulse was given intraportally. One hour later the animals were sacrificed and the livers were subjected to subcellular fractionation. Recoveries and purities of the individual fractions were independent of the nutritional status of the mice (Otter et al. 1989). The incorporation of the ^{3}H-label into protein was taken as an estimate of the relative protein synthesis rate, while the ratio of protein-bound $^{3}H/^{14}C$ label served as an index of relative protein degradation and/or secretion rate. As shown in Fig. 2 relative protein synthesis was significantly increased in the microsomal fraction of Se^- mouse liver, as was relative degradation and/or secretion. Furthermore, in the homogenate

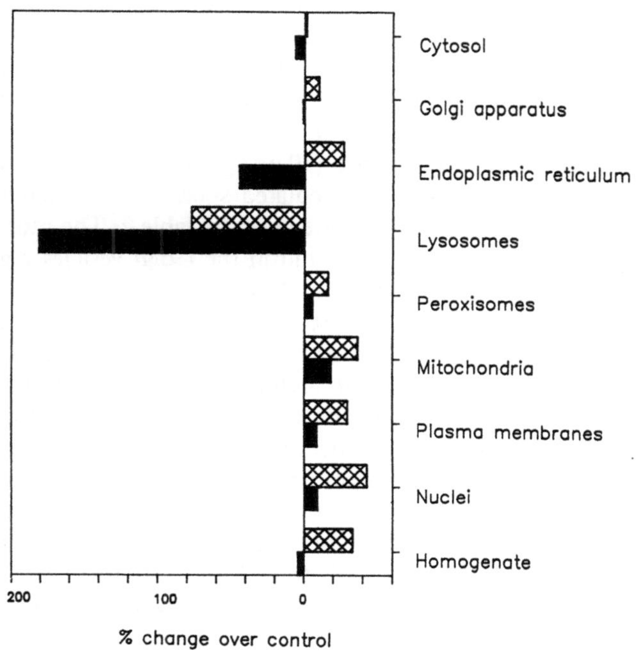

Fig. 2. Alterations in the relative rates of protein synthesis, degradation and/or secretion in subcellular fractions of mouse liver during selenium deficiency. ■ protein synthesis; □ protein degradation and/or secretion

Table 2. Estimated specific as well as total 75-Se content of subcellular compartments in deficient mouse liver after pulse-labelling with sodium selenite

Compartment	Specific and total 75-Se content of hepatic compartments					
	3.3 μg Se kg^{-1}, 1 h		10 μg Se kg^{-1}, 1 h		10 μg Se kg^{-1}, 24 h	
	(pmol mg^{-1})	(pmol g^{-1})	(pmol mg^{-1})	(pmol g^{-1})	(pmol mg^{-1})	(pmol g^{-1})
Homogenate	1.0	178	3.3	591	1.1	198
Nuclei	0.39	10.7	0.78	21.5	0.48	13.2
Plasma membranes	0.25	1.6	0.96	6.2	0.68	4.4
Mitochondria	0.49	22.8	1.79	83.2	0.68	31.6
Peroxisomes	0.38	4.9	1.11	14.3	0.54	7.0
Lysosomes	0.37	1.1	1.00	2.9	0.48	1.4
Endoplasmic reticulum	0.80	21.8	1.52	41.5	0.78	21.3
Golgi apparatus	7.50	26.3	20.74	72.6	3.50	12.3
Cytosol	0.90	49.7	4.02	221.9	1.57	86.7

and the Golgi fraction of Se^- mice a significant decrease in the relative degradation and/or secretion rate was found. These data indicate alterations in hepatic protein metabolism during Se deficiency, which were specific to certain cell compartments and involved different mechanisms, e.g., synthesis, degradation, secretion or a combination of these processes. The cause of these metabolic changes is unclear. Previously, the involvement of (a) hepatic Se pool(s), different from Se-GSH-Px, was emphasized (Reiter and Wendel 1984). Therefore, we studied the subcellular distribution of a tracer dose of Se in deficient mouse liver at dose levels below (3.3 μg Se kg^{-1}) and above (10 μg kg^{-1}), the threshold for reversal of certain enzyme modulations. The calculated specific Se contents of the hepatic compartments and their pool sizes are given in Table 2. The most prominent accumulation of protein-bound Se was found in the Golgi vesicles regardless of whether 3.3 or 10 μg Se kg^{-1} had been given. After 1 h labelling time a seven- to ninefold increase over the specific hepatic Se content was found and 24 h later still a roughly threefold accumulation over the homogenate remained. The specific cytosolic Se content was comparable or slightly higher than that of homogenate. Se accumulation in the remaining subcellular compartments decreased in the order: endoplasmic reticulum > mitochondria > nuclei, plasma membranes, peroxisomes and lysosomes. Under all conditions studied the cytosol accounted for roughly 50% of the hepatic Se. The Golgi vesicles were the second largest pool after 1 h, but declined from 19 to 8% within 24 h. Similarly, the endoplasmic reticular pool dropped from 16 to 12% during this time. Mitochondria and nuclei accounted for 17 and 7% respectively of the total hepatic Se, while the lysosomes, plasma membranes and the peroxisomes contained only traces of the trace element.

4 Discussion

In this study metabolic alterations in selenium-deficient mouse liver have been demonstrated in vitro and in vivo. They point to a rather highly controlled response to Se deficiency than to metabolic malfunctioning. Since no signs of hepatic damage were found (Vogl et al. 1987) such widespread metabolic changes may indicate an adaptation to noxious effects of Se deficiency and might be suitable for recruiting additional metabolic capacity. The fast and extensive accumulation of Se in the Golgi apparatus and the role of this cell organelle in protein modification and secretion provides a clue to a hepatic role in Se extraction and distribution during deficiency. The 60% drop in hepatic Se content over 24 h (see Table 2) could be explained by extensive hepatic processing of Se. The liver, with its high blood supply and metabolic activity, seems perfectly suited to extract Se of dietary or endogenous origin from the circulation and to convert it into a form that is readily recognizable by tissues with a more stringent need for the trace element.

Minimizing loss of Se by recycling the trace element is an important mechanism by which, in combination with the excretory/detoxication pathway yielding Se-methylation products, the liver can exert control on Se homeostasis.

Acknowledgement. Supported by the Deutsche Forschungsgemeinschaft grant We 686/8-3.

References

Flohé L, Günzler WA, Schock HH (1973) FEBS Lett 32:132
Otter R, Goldberg M, Vogl S, Wendel A (1986) Chem-Biol Interact 59:295
Otter R, Reiter R, Wendel A (1989) Biochem J 258:535
Reiter R, Wendel A (1983) Biochem Pharmacol 32:3063
Reiter R, Wendel A (1984) Biochem Pharmacol 33:923
Reiter R, Wendel A (1985) Biochem Pharmacol 34:2287
Reiter R, Otter R, Wendel A (1989) Biochem J 258:541
Vogl S, Goldberg M, Hoffmann G, Ruhenstroth-Bauer G, Otter R, Wendel A (1987) J Hepatol 4:212
Wendel A, Otter R (1987) Biochim Biophys Acta 925:94
Yang J-G, Plummer JM, Burk RF (1987) J Biol Chem 262:73

Selenium Deficiency and Thyroid Hormone Metabolism

J. R. ARTHUR [1,3] and G. J. BECKETT [2]

1 Introduction

Selenium deficiency has many effects on cell metabolism, some of these can be accounted for by loss of selenium-containing glutathione peroxidase (Se-GSH-Px) activity but for others there is, as yet, no convincing explanation. These latter effects include characteristic increases or decreases in several hepatic enzyme activities. Reiter and Wendel (1984) and Hill et al. (1987) have demonstrated that these can be reversed by small doses of dietary or parenteral selenium. Effects of selenium deficiency on neutrophil microbicidal and on plasma pyruvate kinase activity were also reversed by injection of amounts of selenium (as Na_2SeO_3) that did not affect cytosolic Se-GSH-Px activity (Arthur et al. 1987a). However, in the case of pyruvate kinase activity despite no change in muscle cytosolic Se-GSH-Px activity plasma Se-GSH-Px activity was increased by the selenium supplementation. Since plasma Se-GSH-Px is immunologically different from hepatic Se-GSH-Px, some of the diverse effects of selenium deficiency may be caused by the different responses of these forms of Se-GSH-Px to selenium depletion and repletion (Arthur et al. 1987a). Additionally, since selenium deficiency affects such a wide range of enzyme and metabolic processes some of these changes might be mediated via alterations in the metabolism of a hormone. Thyroid hormone status is known to affect the expression of many enzymes and hypothyroidism increases hepatic glutathione S-transferase (GST) activity (Arias et al. 1976) and impairs neutrophil function, similar changes which occur in selenium deficiency (Arthur et al. 1987a). This prompted our investigations of the effects of selenium deficiency on the metabolism of thyroid hormones.

2 Plasma Thyroid Hormones in Selenium Deficiency

Thyroxine (T4) is considered to be a prohormone and is synthesized only in the thyroid gland. The more active thyroid hormone 3:5,3′-triiodothyronine (T3) is produced by 5′ outer ring deiodination of T4. This occurs principally in the liver

[1] Rowett Research Institute, Bucksburn, Aberdeen AB2 9SB, Scotland
[2] University Department of Clinical Chemistry, The Royal Infirmary, Edinburgh EH3 9YW, Scotland
[3] To whom correspondence should be addressed

and the kidney but approximately 15% of T3 in the plasma is synthesized in the thyroid gland. Although selenium supplementation of chickens has been shown to increase plasma concentrations of both T4 and T3 (Jensen et al. 1986), these changes were associated with increases in food intake and thus cannot be attributed solely to a reversal of selenium deficiency. Selenium deficiency in rats increases plasma T4 concentrations and decreases plasma T3 concentrations, before any changes occur in the food intake or the weight of the animals (Beckett et al. 1987). Concentrations in plasma of reverse triiodothyronine (rT3), the inactive isomer of T3, are not affected by selenium deficiency, showing that there is not a switch from 5′ to 5 deiodination of T4 (Beckett et al. 1987). Such a change from T3 to rT3 production often occurs in cases of illness or malnutrition.

3 Extrathyroidal T3 Production in Selenium Deficiency

Three deiodinase enzymes have been characterized in the organs of the rat. In liver and kidney, type I deiodinase is thought to catalyze both 5′ outer ring and 5 inner (tyrosyl) ring deiodination to produce T3 and rT3 respectively. Type II deiodinase, which is found in the central nervous system, pituitary and brown adipose tissue, can only carry out outer ring deiodination to produce T3 and is

Table 1. 5′ Deiodinase activity in kidney and brain homogenates, plasma TSH concentrations, thyroid weights, T4, T3 contents and GSH-Px activities in selenium-deficient and selenium-supplemented rats[a]

Variable	Group		p
	Se^+	Se^-	
Kidney homogenate T3 (fmol min^{-1} mg^{-1} protein)	1.73 ± 0.16(6)	0.23 ± 0.11(6)	<0.001
Brain homogenate T3 (fmol min^{-1} mg^{-1} protein)	0.53 ± 0.10(6)	0.19 ± 0.04(6)	<0.01
Plasma TSH (ng ml^{-1})	0.68 ± 0.06(6)	1.27 ± 0.13(9)	<0.005
Thyroid T4 (nmol g^{-1} protein)	2186 ± 119(6)	1763 ± 126(6)	<0.05
Thyroid T3 (nmol g^{-1} protein)	352 ± 46(6)	232 ± 24(6)	<0.05
Thyroid GSH-Px (U mg^{-1} protein)	0.030 ± 0.003(6)	0.015 ± 0.001(6)	<0.001
Thyroid weight (mg)	18.7 ± 1.5(6)	18.4 ± 1.3(9)	ns

[a] Rats consumed selenium-deficient or selenium-supplemented diets (Arthur et al. 1987b) for 5 weeks from weaning (6 weeks for plasma TSH and thyroid weights). Results are expressed as mean ± SEM, with numbers of animals used in parentheses. The significance of differences between Se^- and Se^+ groups was determined using Student's t-test. ns denotes not significant

much less sensitive than the type I enzyme to inhibition by propyl-thiouracil. Type III deiodinase, which occurs in the brain but not the pituitary, deiodinates the inner ring of T4 and T3 to produce rT3 and T2 respectively. Hepatic type I deiodinase activity in rats is decreased up to 14-fold in prolonged selenium deficiency and this may contribute to the changes in plasma T4 and T3 described above (Beckett et al. 1987). Selenium deficiency has a similar inhibitory effect on kidney (type I) and brain (type II) deiodinase activities (Table 1). Decreased T3 production in the kidney will contribute to the fall in plasma T3 concentrations observed in selenium deficiency. However, the impairment of hepatic and renal deiodinase activities is much greater than the fall in plasma T3 concentrations. One explanation is that selenium deficiency also inhibits the catabolism of T3 to T2, which is dependent on a 5 inner ring deiodination. Decreased T3 degradation would help to maintain the concentration of T3 in plasma despite smaller amounts of the hormone being produced due to impaired 5′ deiodination.

Increased plasma T4 concentrations normally should inhibit thyroid stimulating hormone (TSH) release from the pituitary and thus thyroidal T4 synthesis. Indeed, in man primary hyperthyroidism results in undetectable levels of TSH in plasma. However, since type II deiodinase activity is inhibited in brain, it is likely that the activity is also decreased in the pituitary. Feedback inhibition of T4 on TSH release depends on intra-pituitary conversion of T4 to T3. This impairment of T3 production in the pituitary probably causes the increased plasma TSH concentrations in selenium-deficient rats (Table 1).

4 The Thyroid Gland in Selenium Deficiency

There are no differences in weights of thyroid glands of male rats consuming selenium-deficient or selenium-supplemented diets for 6 weeks from weaning. However, selenium deficiency decreases thyroidal T4 and T3 contents and Se-GSH-Px activity (Table 1). Additionally, in selenium deficiency thyroid total iodine concentration is decreased (unpublished observations). The alterations in thyroid gland T4, T3 and iodine concentration in selenium deficiency are consistent with a response to increased circulating TSH concentrations.

5 Effects of Selenium Depletion and Repletion in Rats on Thyroid Hormone Metabolism

Our previous work has shown that a severe selenium deficiency which will decrease hepatic Se-GSH-Px by over 95% and increase synthesis of hepatic GST proteins will also impair thyroid hormone metabolism (Beckett et al. 1987; Arthur et al. 1987b). Thyroid hormone metabolism is also impaired at an early stage of selenium deficiency before any changes in hepatic GST activity are detected (Beckett et al. 1988). Additionally, repletion of selenium-deficient rats with intra-

Table 2. Plasma thyroid hormone concentrations and hepatic GST activities in selenium-deficient and selenium-supplemented rats[a]

Variable	Group			
	Se^+	Se^-	Se^- (10 μg Se)	Se^- (200 μg Se)
GST (μmol CDNB min^{-1} mg^{-1} protein)	0.37 ± 0.03	0.47 ± 0.02^c	0.37 ± 0.02	0.33 ± 0.02
T4 ($nmol\,l^{-1}$)	76.7 ± 5.2	137.0 ± 6.1^b	131.7 ± 1.73^b	82.5 ± 7.0
T3 ($nmol\,l^{-1}$)	1.43 ± 0.07	1.19 ± 0.07^c	0.63 ± 0.04^b	1.45 ± 0.04

[a] Rats consumed selenium-deficient or selenium-supplemented diets for 4 weeks from weaning. Groups Se^- (10 μg Se) and Se^- (200 μg Se) also consumed a selenium-deficient diet for 4 weeks and were given respectively single intra-peritoneal injections of 10 or 200 μg Se kg^{-1} body wt (as Na_2SeO_3) 5 days before being killed. Results are mean ± SEM, 6 animals/group.
[b] $p < 0.001$, significantly different from Se^+ group.
[c] $p < 0.025$, significantly different from Se^+ group

peritoneal injections of different doses of selenium (Table 2) shows that induced GST activity can be returned to normal with a low dose of selenium (10 μg Se kg^{-1} body as Na_2SeO_3) which does not reverse the effects of deficiency on thyroid hormone metabolism, whereas 200 μg Se kg^{-1} body weight is required to normalize plasma T4 and T3 concentrations. It is therefore probable that the effects of selenium deficiency on thyroid hormone metabolism are completely independent of the effects of the deficiency on hepatic drug metabolizing enzyme activity.

6 Conclusions

An hypothesis summarizing the possible effects of selenium deficiency on thyroid hormone metabolism is shown in Fig. 1. Initially, selenium deficiency, either directly or indirectly, can decrease the activity of both type I and type II thyroid hormone deiodinases in the organs of the rat. Impaired hepatic and renal type I activity decreases T4 metabolism, thus tending to increase plasma T4 concentrations and decrease plasma T3 concentrations. Enzyme changes seen in plasma of selenium-deficient cattle are consistent with tissue hypothyroidism caused by decreased type I deiodinase activity (Arthur et al. 1988). Impaired type II deiodinase activity which occurs in the brain and presumably the anterior pituitary reduces the feedback control of T4 on TSH secretion by impairing T3 production and augments the increase in plasma T4. Lower thyroidal T4, T3 and I concentrations are in accord with the increased circulating TSH concentrations. Decreased pituitary growth hormone content in selenium-deficient rats (Ewan 1976) could also be caused by less T3 production in the pituitary since T3 is essential for growth hormone synthesis (see De Gennaro et al. 1988). We are now investigating the possibility that alterations in thyroid hormone metabolism in

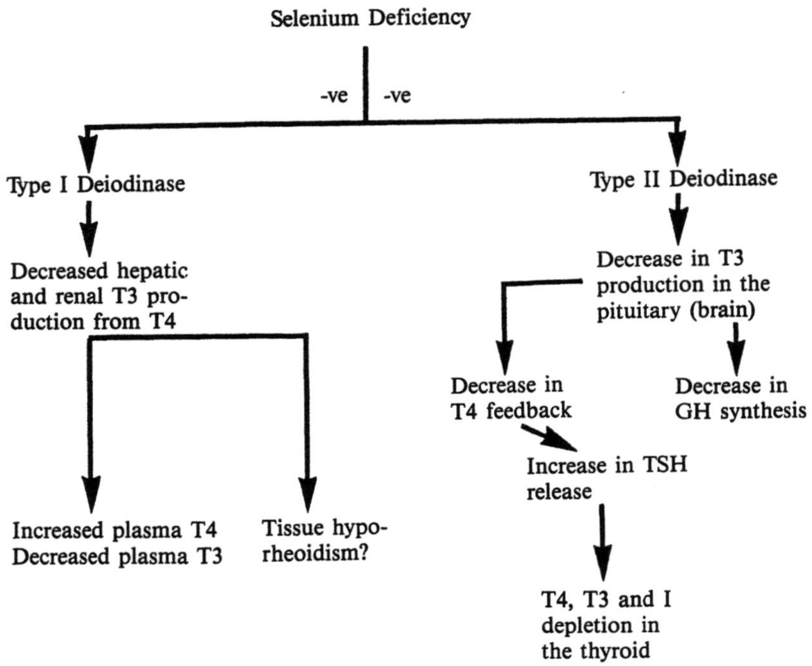

Fig. 1. Effects of selenium-deficiency on thyroid hormone metabolism

selenium deficiency are mediated through loss of hitherto unrecognized selenoproteins in the liver and thyroid glands.

Acknowledgements. We are grateful to our colleagues Fergus Nicol, Peter Rae, Alison Hutchinson and David McDougal who participated in some of the research described in this work. Reagents for radioimmunoassay of thyroid stimulating hormone were a gift from the NIH National Hormone and Pituitary Programme, University of Maryland School of Medicine and The Scottish Antibody Production Unit.

References

Arias IM, Fleischer G, Kirsch R, Mishkin S, Gatmaintan Z (1976) On the structure and function of ligandin. In: Arias IM, Jakoby WB (eds) Glutathione: metabolism and function. Kroc Found Ser, vol 6. Raven Press, New York, pp 175–188

Arthur JR, Nicol F, Boyne R, Allen KGD, Hayes JD, Beckett GJ (1987a) Old and new roles for selenium. In: Trace substances in environmental health XXI. Ed. Hemphill DD, Univ Missouri, Colombia, pp 487–498

Arthur JR, Morrice PC, Nicol F, Beddows SE, Boyd R, Hayes JD, Beckett GJ (1987b) The effects of selenium and copper deficiencies on glutathione peroxidase in rat liver. Biochem J 248:539–544

Arthur JR, Morrice PC, Beckett GJ (1988) Thyroid hormone concentrations in plasma from selenium-deficient and -sufficient cattle. Res Vet Sci 45:122–123

Beckett GJ, Beddows SE, Morrice PC, Nicol F, Arthur JR (1987) Inhibition of hepatic deiodination of thyroxine caused by selenium deficiency in rats. Biochem J 248:443–447

Beckett GJ, Hutchinson AR, Nicol F, Arthur JR (1988) A role for selenium in the 5′ deiodination of thyroxine. J Endocrinol Suppl 117:287

De Gennaro V, Cella SG, Bassetti M, Rizzi R, Cocchi D, Muller EE (1988) Impaired growth hormone secretion in neonatal hypothyroid rats: hypothalamic versus pituitary component. Proc Soc Exp Biol Med 187:99–106

Ewan RC (1976) Effect of selenium on rat growth, growth hormone and diet utilisathion. J Nutr 106:702–709

Hill KE, Burk RF, Lane JM (1987) Effect of selenium depletion and repletion on plasma glutathione and glutathione dependent enzymes in the rat. J Nutr 117:99–104

Jensen LS, Colnago GL, Takahashi K, Akiba Y (1986) Dietary selenium status and plasma thyroid hormones in chicks. Biol Trace Elem Res 10:11–18

Reiter R, Wendel A (1984) Selenium and drug metabolism-II, Independence of glutatione peroxidase and reversibility of hepatic enzyme modulations in deficient mice. Biochem Pharmacol 33:1923–1928

Glutathione Metabolism as Affected by Selenium Deficiency

K.E. Hill [1,2] and R.F. Burk [1]

Glutathione (GSH) is a tripeptide-γ-glutamylcysteinylglycine. GSH synthesized in the liver is transported across the sinusoid into the blood. The liver is the only tissue that releases GSH to the systemic circulation. Other cells can release GSSG under conditions of oxidative stress, but the liver constantly releases GSH to the blood maintaining a plasma GSH level in the micromolar range. The function of plasma GSH is currently a subject of great interest. Recently, it has been suggested that plasma GSH may be a substrate for one or more plasma enzymes (e.g., plasma GSH-Px) (Maddipati and Marnett 1987).

Selenium deficiency has several effects on GSH and glutathione-related enzymes, the most thoroughly studied effect being the loss of the selenoenzyme GSH-Px (Wendel 1980). In severe selenium deficiency liver and plasma GSH-Px activities are less than 1% of that in selenium-adequate animals. Selenium deficiency increases hepatic GSH synthesis and release and plasma GSH levels (Hill and Burk 1982). Several microsomal drug metabolizing enzyme systems are affected by selenium deficiency (Correia and Burk 1978; Reiter and Wendel 1984), among them are the glutathione S-transferases (GSH S-Tr). GSH S-Tr activity as measured by 1-chloro-2,4-dinitrobenzene is increased as much as two times control by selenium deficiency (Lawrence et al. 1978).

The mechanism by which selenium affects hepatic GSH metabolism is not known. However, the activity of γ-glutamylcysteine synthetase (the initial enzyme in the synthesis of glutathione) is doubled in selenium-deficient rat liver (Hill and Burk 1982). Rather than increasing the intracellular concentration of GSH, this doubling of the GSH synthesis rate is reflected in the rate of GSH release from the isolated perfused selenium-deficient liver (Table 1). The increased GSH release is into the caval effluent. There is no change in the biliary release of GSH or GSSG. Thus the increased caval release of GSH results in an increased plasma GSH level in selenium-deficient rat plasma compared to control (30.3 ± 4.3 nmol GSH ml^{-1} versus 12.9 ± 3.9 nmol GSH ml^{-1}, respectively).

In order to correlate changes in liver and plasma GSH-Px activity, plasma GSH, and liver GSH S-Tr activity with the development of selenium deficiency, these values were measured at weekly intervals in weanling rats fed selenium-deficient or control (0.5 mg Se kg^{-1}) diets (Hill et al. 1987). As selenium deficiency developed there was a rapid decrease in liver and plasma GSH-Px activity

[1] Division of Gastroenterology, Department of Medicine, and Center in Molecular Toxicology, Vanderbilt University, School of Medicine, Nashville, TN 37232, USA
[2] To whom correspondence should be addressed

Table 1. Effect of selenium deficiency on glutathione release from isolated perfused rat liver[a]

Diet	n	Caval release[b]	Biliary release[b]		Hepatic GSH[c]
			GSH	GSSG	
Selenium-deficient	3	12.0±0.5	1.94±0.53	1.21±0.20	6.53±1.06
Control	3	7.14±1.8	1.67±0.03	1.17±1.80	6.67±0.53

[a] Values are means ± SD, liver perfusion and glutathione assay methods are described in Cummings et al. (1988).
[b] Units are nmol GSH equivalents min^{-1} g^{-1} liver.
[c] Units are μmol GSH equivalents min^{-1} g^{-1} liver.

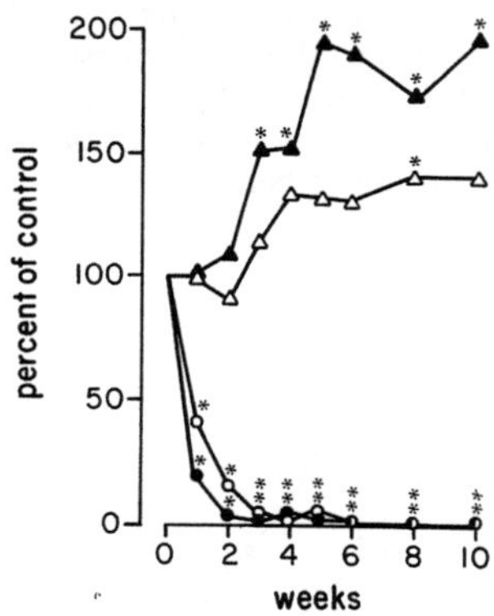

Fig. 1. Percentage changes with time in glutathione and glutathione-requiring enzymes in control and selenium-deficient rats. Plasma glutathione (▲), plasma GSH-Px activity (○), liver GSH-Px activity (●), and liver GSH S-Tr (△) are shown (n = 3). Control and selenium-deficient values are statistically different for those values marked with an *asterisk* (from Hill et al. 1987, American Institute of Nutrition)

(Fig. 1). In the selenium-deficient rat, the activity of both enzymes fell to less than 5% of control within 3 weeks of the start of the experiment. The time on the selenium-deficient diet required to observe a change in liver GSH S-Tr activity and plasma GSH was longer than that required to lower GSH-Px activity. The changes in GSH S-Tr activity and plasma GSH concentration began to occur at 3 weeks but required 5 weeks or more to reach a maximum. Thus, the changes in GSH-Px activity and GSH S-Tr activity and plasma GSH are unrelated timewise and suggest that different metabolic functions of selenium are involved.

Plasma GSH concentration is a function of hepatic release (input rate) and systemic degradation and tissue uptake (removal rate). Hepatic sinusoidal release has been reported to decrease with increasing age (Ookhtens et al. 1987). This was confirmed by measuring GSH release from isolated perfused rat liver (Fig. 2b). From 2 to 6 weeks on the diet (7 to 10 weeks of age) the hepatic release of GSH from control liver decreased with a simultaneous decrease in the plasma concentration of GSH (Fig. 2a). In the selenium-deficient liver this decrease did not occur. As a result, plasma GSH did not decrease with increasing age as seen in control. Thus, the "increase" in plasma GSH in selenium deficiency appears to be the result of a failure to decrease the hepatic synthesis of GSH with increasing age.

In order to separate further the dependence of the changes in GSH-Px activity, plasma GSH and liver GSH S-Tr activity on selenium status, selenium-deficient

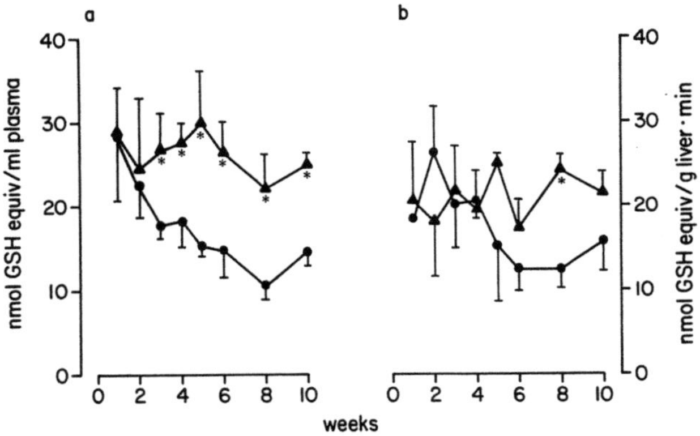

Fig. 2a, b. Change with time in plasma glutathione (**a**) and perfused liver release of glutathione (**b**) in control and selenium-deficient rats. Control rat values are represented by ● and selenium-deficient values are represented by ▲. Each point represents the mean of 3–6 animals, and 1 SD is represented by the *bracket*. Selenium-deficient values marked with an *asterisk* are statistically different from control values ($p < 0.05$) by Student's-*t*-test (from Hill et al. 1987, American Institute of Nutrition)

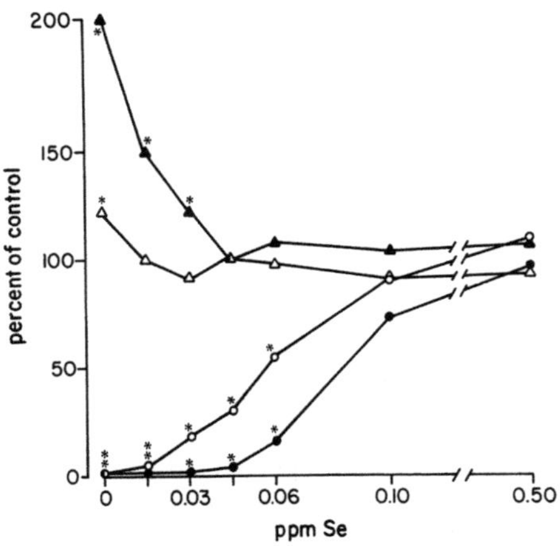

Fig. 3. Effect of feeding graded levels of selenium on plasma glutathione (▲), plasma GSH-Px activity (○), liver GSH-Px activity (●), and liver GSH S-Tr (△). Selenium was supplemented as Na_2SeO_4 in the diet of selenium-deficient rats (n = 4) for 4 weeks. The values shown are reported as a percent of age-matched control values (plasma glutathione = 12.1 ± 0.9 nmol GSH equivalents ml^{-1} plasma; plasma GSH-Px activity = 1053 ± 63 nmol NADPH oxidized (ml^{-1} plasma min^{-1}); liver cytosolic GSH-Px activity = 258 ± 46 nmol NADPH oxidized (mg^{-1} protein min^{-1}); liver GSH S-Tr activity = 816 ± 84 nmol GSH conjugate formed (mg^{-1} protein min^{-1}). Control and selenium-deficient values are statistically different for those values marked with an *asterisk* (from Hill et al. 1987, American Institute of Nutrition)

rats (6 weeks on selenium-deficient diet) were fed diet supplemented with 0 to 0.5 ppm Se as Na_2SeO_4. Figure 3 shows the changes in these values after 4 weeks of selenium supplementation. Liver GSH S-Tr activity was corrected to control levels by 0.015 ppm Se, the lowest level of supplementation. Plasma GSH was corrected to control values by 0.045 ppm Se, while 0.1 ppm Se was required to correct

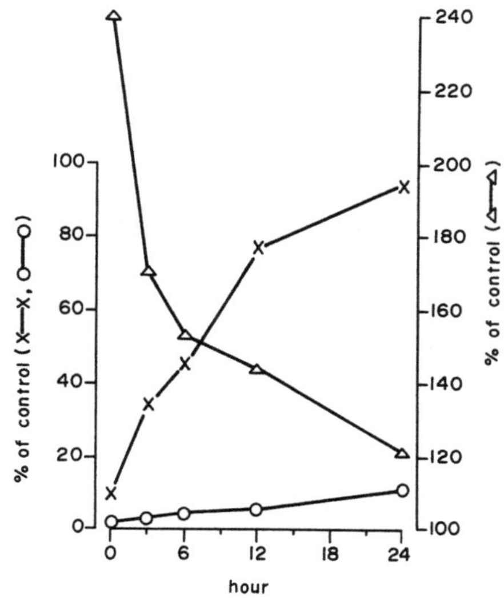

Fig. 4. Time course of changes in plasma selenoprotein P concentration (X), GSH-Px activity (○), and glutathione (Δ) following a single selenium dose (50 μg/rat). Three rats were used at each time point. Control values for: selenoprotein P = 57.6 ± 0.5 μg ml^{-1} plasma; GSH-Px activity = 757 ± 119 nmol NADPH oxidized (ml^{-1} plasma min^{-1}); GSH = 12.9 ± 1.9 nmol GSH equivalents ml^{-1} plasma. Plasma Se-P concentration was measured by RIA as described in Yang et al. (1987)

plasma and liver GSH-Px activity. Between 0.03 and 0.06 ppm Se supplementation, plasma and liver GSH-Px activity responded differently to the amount of selenium available. The order of response to selenium supplementation was liver GSH S-Tr activity > plasma GSH > plasma GSH-Px activity > liver GSH-Px activity. Thus, this experiment clearly separates the dependency of these four measurements on selenium status.

Recently, the isolation and purification of another selenium-containing protein from rat plasma has been reported (Yang et al. 1987). This protein has been designated selenoprotein P (Se-P). Monoclonal antibodies raised against Se-P were used to prepare an immunoaffinity column. Se-P was then purified from rat plasma by immunoaffinity column chromatography. The purified protein migrated as a single band (M_r = 57000) on SDS-polyacrylamide gel electrophoresis and the selenium was shown to comigrate with the protein. Using the purified monoclonal antibody, a radioimmunoassay was developed to measure Se-P. With this RIA, Se-P in selenium-deficient plasma was found to be less than 10% of the amount in control plasma. Using this same RIA, we have measured and correlated changes in plasma Se-P levels, GSH-Px activity, and GSH.

Figure 4 shows the changes in plasma Se-P concentration, GSH-Px activity, and GSH following single i.p. injections of Se (50 μg/rat) into selenium-deficient rats. Se-P appeared rapidly following the Se injection. It reached 35% of the control value within 3 h and continued to increase in a time-dependent manner until 12 h when it was 77% of control. In the next 12 h an additional increase of 17% was measured. By 24 h after the Se injection there was a small but significant increase in GSH-Px activity to 10% of control. Thus, these results demonstrate that Se-P synthesis is rapid and has priority over the synthesis of plasma GSH-Px.

This experiment also shows the rapidity with which GSH levels can be returned to control values by a single Se injection. At 24 h after the Se injection there was no statistical difference between the Se-treated selenium-deficient plasma and control plasma. These results suggest that there may be an association between Se-P and GSH in the plasma. As Se-P levels were increased there was a concomitant decrease in GSH concentration with both values having reached control levels 24 h after administration of a single selenium dose. Thus, it is possible that the function of plasma GSH may be related to selenium status by an interaction with Se-P. Further studies are required to assess the extent of association between these two plasma constituents.

In summary, selenium deficiency decreases the amount of the selenium-containing proteins GSH-Px and Se-P present in the plasma while causing an increase in plasma GSH concentration. The relationship, if any, between the change in plasma GSH levels and the changes in selenium-containing proteins is still unresolved. But, the utilization of plasma GSH by a selenoprotein (e.g., Se-P) is an attractive possibility.

Acknowledgements. The authors thank Dr. J.-G. Yang for his measurement of plasma Se-P. This work was supported by NIH Grant ES 02497.

References

Correia MA, Burk RF (1978) J Biol Chem 253:6203–6210
Cummings SW, Hill KE, Burk RF, Ziegler DM (1988) Biochem Pharmacol 967–969
Hill KE, Burk RF (1982) J Biol Chem 257:10668–10672
Hill KE, Burk RF, Lane JM (1987) J Nutr 117:99–104
Lawrence RA, Parkhill LK, Burk RF (1978) J Nutr 108:981–987
Maddipati K, Marnett LJ (1987) J Biol Chem 262:17398–17403
Ookhtens M, Lyon I, Kaplowitz N (1987) Biochem Pharmacol 36:4015–4017
Reiter R, Wendel A (1984) Biochem Pharmacol 33:1923–1928
Wendel A (1980) Enzymatic basis of detoxification. Academic Press, vol 1, pp 333–353
Yang, J-G, Morrison-Plummer J, Burk RF (1987) J Biol Chem 262:13372–13375

Selenium-Deficiency and the Metabolism of Ketone Bodies

U. Olsson[1]

1 Introduction

It was previously reported that selenium (Se) deficiency caused a significant increase of urinary acetoacetate excretion in fed rats, and 24 or 48 h of starvation enhanced this effect (Olsson 1985). Further, control rats fed the Se-deficient torula yeast-based diet but supplemented with 0.2 ppm selenium in the drinking water were quite comparable to rats fed a standard diet, thus indicating that the torula diet as such did not affect the metabolism or excretion of ketone bodies. No Se-dependent effect was found for the liver or blood content of glucose or the two ketone bodies, i.e., acetoacetate (AcAc) and 3-hydroxybutyrate (3-OHBA), while indications for a higher renal content of ketone bodies (AcAc plus 3-OHBA) was noted in the Se-deficient rat. Two days of Se supplementation to Se-deficient rats reduced the amount of urinary AcAc and 3-OHBA to 50% of the deficiency value, indicating an enzymatic role for selenium regarding ketone bodies (Olsson 1985). This role might be unrelated to the activity of GSH-Px as only a very small increase of this enzyme activity was noted after 2 days of Se supplementation (unpublished results).

The ketone bodies are synthesized in the liver from partly oxidized fatty acids in response to an extrahepatic lack of glucose, as in the case of starvation. They are then transported by blood to extrahepatic tissues, such as the brain, kidney and muscles, and used there as metabolic fuels (Robinson and Williamson 1980). The blood content of ketone bodies is thus a measure of the extrahepatic demand of alternative fuels when glucose is lacking, and the urinary excretion of ketone bodies is linearly related to the blood content (Wirthensohn and Guder 1986). The finding of enhanced urinary excretion of ketone bodies without a corresponding increase of blood or liver content in the Se-deficient rat was therefore somewhat surprising compared to commonly agreed facts on the metabolism and excretion of ketone bodies.

The present experiments were performed in order to verify and expand previous results and to determine whether the renal handling of ketone bodies rather than hepatic ketogenesis and blood delivery to extrahepatic tissues causes the enhanced urinary excretion of ketone bodies in the Se-deficient rat.

[1] Department of Genetic and Cellular Toxicology, Wallenberg Laboratory, University of Stockholm, 10691 Stockholm, Sweden

To avoid loss of ketone bodies during tissue sampling, a freeze clamp technique using liquid nitrogen was adopted in the present experiments instead of the previous technique (Olsson 1985) with dry ice as freezing agent. Another difference from previous experiments was that 0.1 ppm instead of 0.2 ppm selenium was added to the drinking water of control rats. All rats were starved for 48 h before sampling or used in experiments. There was no significant Se-dependent difference in body weight (determined as 499±32 and 476±11 g for Se-supplemented and Se-deficient rats respectively; means±SE of five and six rats respectively), and the activity of Se-dependent glutathione peroxidase in the postmicrosomal liver fraction from Se-deficient rats was less than 3% of the supplemented controls.

2 Results and Discussion

Rats in previous experiments (Olsson 1985) were used when 20–25 weeks old, that is after 15–20 weeks of Se-deficient or Se-supplemented diet regimen. To determine the probable dependency of ketone body excretion with time of diet regimen, rats were repeatedly starved, and collected urine was enzymatically analyzed for the content of AcAc and 3-OHBA.

The total urinary excretion of ketone bodies did not show a straightforward increase with time of feeding the Se-deficient diet regimen (Fig. 1), as expected, but was dramatically raised and highly variable after 2 and 5 weeks. The extremely variable excretion of ketone bodies within the Se-deficient group at 2 and 5 weeks was associated with the combined effect of ketone body concentration and urinary volume, i.e., a high concentration was connected with high urinary volume, and vice versa. This initial variability within groups was a probable consequence of adaptation to the diet regimen. However, after about 10 weeks the urinary volume as well as the ketone body excretion became less variable and the total urinary excretion of ketone bodies in the Se-deficient group was about ten times higher than in the controls. At the end of the experiment the urine volume was 15±5 ml for the Se-supplemented and 22±6 ml for the Se-deficient group during the overnight period (16 h) of urine collection.

The sum of ketone bodies in Fig. 1 conceals the actual difference in excretion of individual ketone bodies. In fact, at the end of the experiment, Se-deficient rats excreted 7 times more 3-OHBA and 18 times more AcAc than the Se-supplemented controls. This difference was built up with time of diet regimen as shown in Fig. 2. The ratio of urinary 3-OHBA to AcAc was significantly correlated with time of diet regimen, as indicated by the linear regression of Fig. 2. While the ratio was significantly increased in the Se-supplemented controls, it was decreased in the Se-deficient group, both with a probability of 99%. In quantitative terms, the most prominent effect was the increasing excretion of AcAc with time of Se-deficient diet regimen. However, this does not explain the main question of why the Se-deficient rats excreted much more AcAc (18 times) as well as 3-OHBA (7 times) than their corresponding Se-supplemented controls.

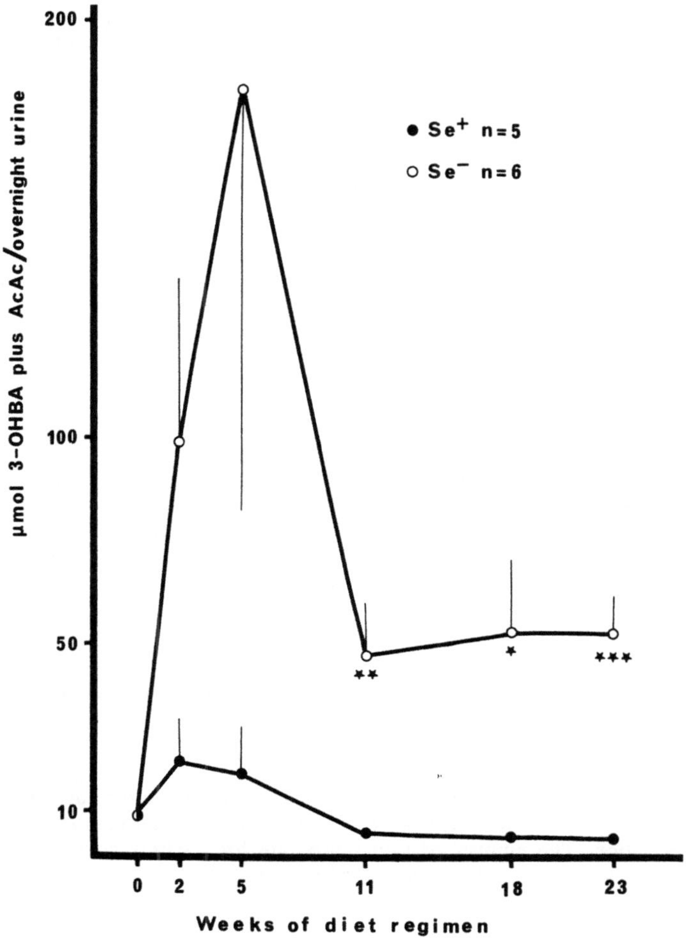

Fig. 1. Total urinary excretion of the ketone bodies, 3-hydroxybutyrate (3-OHBA) and acetoacetate (AcAc), in relation to time of feeding a Se-deficient (○) or Se-supplemented (●) diet regimen to rats. Values for the last 16 h (overnight) of an 48-h period of starvation. Significant differences between groups indicated as $*p<0.05$, $**p<0.01$, and $***p<0.001$

In order to trace the site of disturbance, the content of ketone bodies in the ketone synthesizing liver, the transporting blood and the ketone-utilizing kidney was determined (Fig. 3). However, no Se-deficient difference was found in these tissues (or in hind limb muscles, not shown). The absence of Se-dependent liver and blood content of ketone bodies clearly indicates that ketogenesis was not enhanced in Se deficiency. The sum of ketone bodies per gram tissue decreased in the order from the liver via the blood to the kidney. This is in contrast to the urine content where the per-ml value for the Se-deficient group increases several times above the Se-supplemented controls. Previous results (Olsson 1985) of enhanced urinary excretion of ketone bodies without increased liver or blood content in the Se-deficient rat, as compared to Se-supplemented controls, was thus

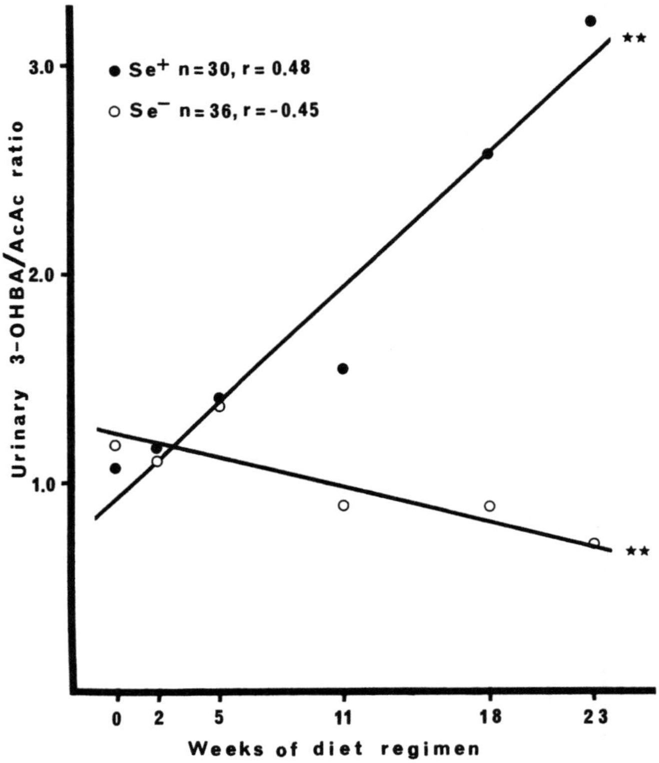

Fig. 2. Linearity of urinary ketone body ratio (3-hydroxybutyrate/acetoacetate) to time of feeding a Se-deficient (○) or Se-supplemented (●) diet regimen as determined after 48 h of starvation. The number of values used are for six time points of six (○) or five (●) rats. Significance of correlation indicated as $^{**}p<0.01$

confirmed while the previous indication of increased renal content of ketone bodies was not. The conclusion from this is that the renal handling or utilization of ketone bodies seems to be of much more interest than the tissue contents for the understanding of urinary ketone body excretion in Se deficiency.

The renal utilization of ketone bodies accounts for a larger part of their clearance from blood than does the renal excretion of them (Angielsky and Lukovicz 1978). Nevertheless, the urinary excretion of ketone bodies increases with increasing blood content (Wirthensohn and Guder 1986). As Se deficiency caused increased excretion of ketone bodies without a corresponding increase of blood or kidney content, a possible failure in the renal capacity to utilize ketone bodies came into focus. Thus, the activity of the two key enzymes in the utilization of ketone bodies, i.e., of 3-oxoacid CoA-transferase and acetoacetyl-CoA thiolase (Robinson and Williamson 1980), was determined. However, neither the transferase nor the thiolase activity of kidney, heart, muscle or liver homogenates was affected by Se deficiency (unpublished results). Furthermore, the activity of these two enzymes in kidney preparations from torula-fed rats did not deviate significantly from the values reported for rats fed a standard diet (Williamson et al.

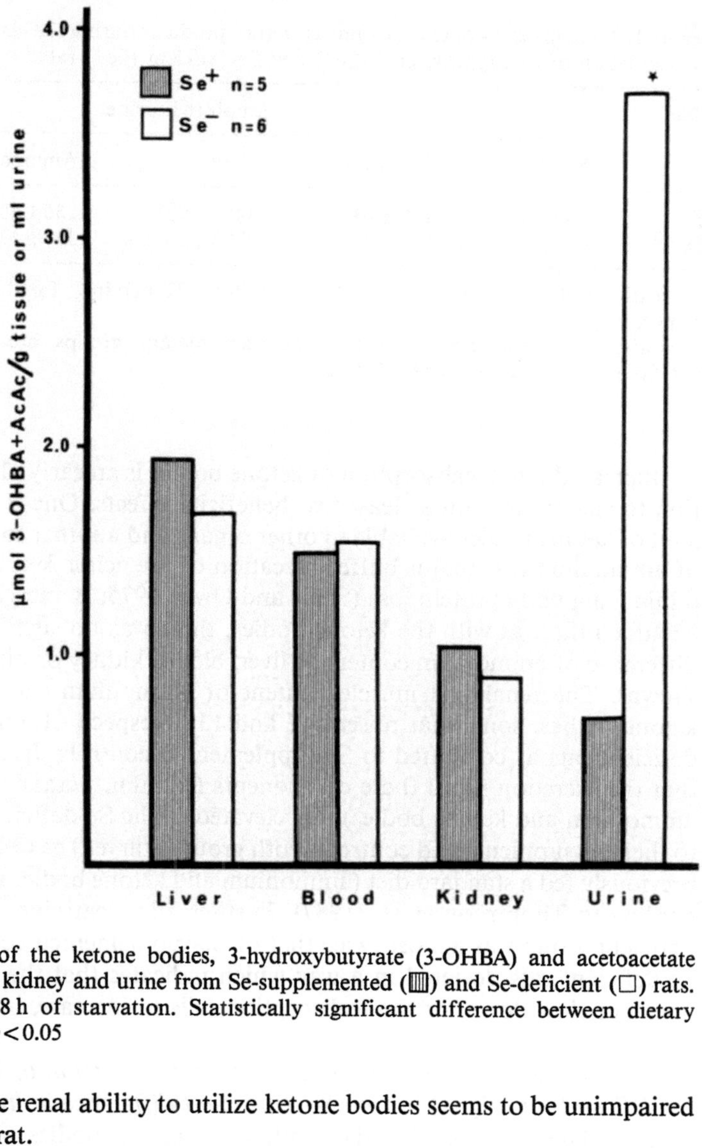

Fig. 3. Total content of the ketone bodies, 3-hydroxybutyrate (3-OHBA) and acetoacetate (AcAc), in liver, blood, kidney and urine from Se-supplemented (▥) and Se-deficient (□) rats. Samples taken after 48 h of starvation. Statistically significant difference between dietary groups indicated as $*p<0.05$

1971). Therefore, the renal ability to utilize ketone bodies seems to be unimpaired in the Se-deficient rat.

In addition to ketone bodies, some urinary samples were also analyzed for the content of sodium, potassium, urea, ammonium and creatinine. In the present study the excretion of these compounds was between two and four times higher in the Se-deficient group than in the supplemented controls (Table 1), indicating a Se-dependent difference of renal status. For comparison, it is worth mentioning a recent study (Thompson et al. 1987) on rats fed a "standard" diet, which shows that starvation causes a significant decrease in the total urinary excretion of sodium, potassium, urea and creatinine when measured during a 24-h period. Creatinine clearance, which is often used as a measure of the renal glomerular filtration rate, was also significantly decreased in response to starvation (Thompson et al. 1987).

Table 1. Total urinary content of some excretory products during the last 16 h of a 48-h period of starvation of Se-supplemented (Se^+) and Se-deficient (Se^-) rats

Diet	µmol/16 h urine				
	Sodium	Potassium	Urea	Ammonium	Creatinine[a]
Se^+	88 ± 37	166 ± 70	1484 ± 607	166 ± 66	43 ± 17
Se^- [b]	365 ± 112	516 ± 11**	4292 ± 186*	339 ± 47	109 ± 4*

[a] Creatinine clearance was 0.86 ± 0.27 and 1.58 ± 0.22 ml min^{-1} for the Se^+ and Se^- group respectively.
[b] Statistically significant differences between dietary groups are *$0.01 < p < 0.02$ and **$0.001 < p < 0.01$ means ± SE of 3 rats

Increased renal reabsorption of ketone bodies is an early physiological adaptation to starvation with at least two beneficial effects. One is to maintain a high level of ketone bodies available to other organs and another to prevent large losses of ammonium (the major buffering cation of the acidic ketone bodies) and thus minimizing body protein loss (Sapir and Owen 1975; Barac-Nieto 1986). It could be added that, as with the ketone bodies, there was no significant Se-dependent difference of ammonium content in liver, blood, kidney or hind limb muscle (not shown). The renal and muscle content of ammonium was, in contrast to the ketone bodies, somewhat raised (12 and 13% respectively) in samples from Se-deficient rats as compared to Se-supplemented controls. It is of interest to note that the excretion of all these components (sodium, potassium, urea, creatinine, ammonium and ketone bodies) was elevated in the Se-deficient rats as compared to their Se-supplemented controls (both groups starved) as well as the starved rats, previously fed a standard diet (ammonium and ketone bodies not determined), according to Thompson et al. (1987). Further, the creatinine clearance of the Se-deficient group was almost twice that of Se-supplemented controls. In short, the Se-deficient rats deviated in a way which indicates that the renal filtration rate does not slow down, or the reabsorption fails to increase, in response to starvation.

In summary, there was an *increased urinary excretion of ketone bodies in the Se-deficient rat* (Fig. 1) but *no Se-dependent difference in liver, blood or kidney* (Fig. 3). This indicates that the synthesis of ketone bodies was not affected, nor was the blood delivery of ketone bodies to the kidney. There must thus be *a renal impairment in the handling of ketone bodies.* But this was not due to a decreased ability to utilize these compounds as *the activity of the key enzymes in the utilization of ketone bodies was not affected.* In addition to the increased excretion of ketone bodies, there was also an *increase of excreted sodium, potassium, urea, ammonium and creatinine* in the Se-deficient rats and a probable *increase of the creatinine clearance* (Table 1). Altogether these Se-dependent differences of compound excretion give the impression of a more general, rather than ketone-specific, waste of compounds in the Se-deficient rat. That is, the increased excretion of ketone bodies might be a side effect to a more basic metabolic impairment.

One such possible impairment would be if the renal prostanoid-thromboxane system of vasodilators and vasoconstrictors is affected in Se-deficiency. This suggestion is based on two supportive circumstances. Firstly, it is well known that *renal prostanoids regulate the filtration rate and excretion or reabsorption of sodium, potassium, urea and creatinine in the kidney* (Ballerman et al. 1986; Schlondorff and Ardaillou 1986). Thus, if the ratio of renal vasodilators to vasoconstrictors increases in the Se-deficient rat this could account for the increased excretion of these compounds as well as of ketone bodies. And secondly, this assumption is partly supported by several in vitro results, which indicate that *there is a connection between Se deficiency and changed synthesis of prostacyclin and thromboxane in platelets and aortas* (Schoene 1985; Funk et al. 1987). One may conclude from this, that if Se deficiency affects the synthesis of prostanoids in platelets and aorta, by decreasing the content of prostacyclin and increasing that of thromboxane, it is likely that the renal prostanoid-thromboxane system is also affected.

However, the effects of Se deficiency on the content of prostacyclin and thromboxane in platelets and aorta must not be directly applied to the kidney, as the metabolic profile of renal prostanoids differ in several respects from that of platelets and aorta. One main difference is that prostacyclin and thromboxane are not the dominating species in the kidney as in platelets and aorta and, further, the proportion of prostanoids changes in different parts of the kidney (Ballerman et al. 1986). Tissue differences of the prostanoid profile might be of great importance because prostanoids can influence the synthesis or metabolism of one another (Granström and Kumlin 1987). Changes in the renal prostanoid-thromboxane system is an attractive hypothesis to explain part of the findings. However, several other metabolic changes in Se deficiency may be equally important.

In this context it is of interest to note that ketogenesis might be present also in the kidney (Weideman and Krebs 1969) and not exclusively localized to the hydroxymethylglutaryl CoA pathway of the liver (Robinson and Williamson 1980), as generally assumed. An alternative pathway for renal synthesis of AcAc was suggested by Brady et al. (1982) and characterized as a direct deacylation of acetoacetyl-CoA. Although not generally accepted, the possibility of renal synthesis of AcAc is a highly interesting concept as it could well play a role in the time-dependent decrease of urinary 3-OHBA to AcAc ratio of the Se-deficient rat. It might also contribute to the increased total ketone body excretion of the Se-deficient rat. Future work will hopefully reveal the mechanisms behind the increased urinary excretion of ketone bodies in the Se-deficient rat.

Acknowledgement. Thanks are due to Dr. Brita Beije for help during the preparation of this paper. This work was supported by the Swedish Natural Science Research Council and the Ekhaga Foundation.

References

Angielsky S, Lukovicz J (1978) The role of the kidney in the removal of ketone bodies under different acid-base status of the rat. Am J Clin Nutr 31:1635–1641

Ballerman BJ, Levenson DJ, Brenner BM (1986) Renin, angiotensin, kinins, prostaglandins, and leukotrienes. In: Brenner BM, Rector FC (eds) The kidney, vol I. Saunders, Philadelphia, pp 281–340

Barac-Nieto M (1986) Renal reabsorption and utilization of hydroxybutyrate and acetoacetate in starved rats. Am J Physiol 251:F257–F265

Brady PS, Scofield RF, Ohgaku S, Schumann WS, Bartsch GE, Margolis JM, Kumaran K, Horvat A, Mann S, Landau BR (1982) Pathways of acetoacetate's formation in liver and kidney. J Biol Chem 257:9290–9293

Funk CD, Boubez W, Powell WS (1987) Effects of selenium-deficient diets on the production of prostaglandins and other oxygenated metabolites of arachidonic acid and linoleic acid by rat and rabbit aortae. Biochim Biophys Acta 921:213–220

Granström E, Kumlin M (1987) Metabolism of prostaglandins and lipoxygenase products: relevance for eicosanoid assay. In: Benedetto C, McDonald-Gibson RG, Nigram S, Slater TF (eds) Prostaglandins and related substances: a practical approach. IRL Press, Oxford, pp 5–27

Olsson U (1985) Impaired ketone body metabolism in the selenium deficient rat. Possible implications. Metabolism 34:993–998

Robinson AM, Williamson DH (1980) Physiological roles of ketone bodies as substrates and signals in mammalian tissues. Physiol Rev 60:143–187

Sapir DG, Owen OE (1975) Renal conservation of ketone bodies during starvation. Metabolism 24:23–33

Schlondorff D, Ardaillou R (1986) Prostaglandins and other arachidonic acid metabolites in the kidney. Kidney Int 29:108–119

Schoene NW (1985) Selenium-dependent glutathione peroxidase and eicosanoid production. In: Lands WEM (ed) Biochemistry of arachidonic acid metabolism. Nijhoff, Boston, pp151–160

Thompson CS, Mikhailidis DP, Jeremy JY, Bell JL, Dandova P (1987) Effect of starvation on biochemical indices of renal function in the rat. Br J Exp Pathol 68:769–775

Weideman MJ, Krebs HA (1969) The fuel of respiration of rat kidney cortex. Biochem J 112:149–166

Williamson DH, Bates MW, Page MA, Krebs HA (1971) Activities of enzymes involved in acetoacetate utilization in adult mammalian tissues. Biochem J 121:41–47

Wirthensohn G, Guder WG (1986) Renal substrate metabolism. Physiol Rev 66:469–497

Transport of Selenomethionine and Methionine Across the Intestinal Brush Border Membrane

S. WOLFFRAM[1,2], B. BERGER[1], and E. SCHARRER[1]

1 Introduction

Selenoamino acids are generally considered to be the main selenium source in naturally occurring food and feedstuffs. Early studies by Spencer and Blau (1962) and McConnell and Cho (1965) have shown active transport of selenomethionine (Se-Met) but not of selenocystine across the intestinal wall of hamsters. Furthermore, some evidence for a common transport site for Se-Met and its sulfur analog methionine (Met) was presented (McConnell and Cho 1965). The goal of our investigations was to characterize transport of Se-Met and Met across the intestinal brush border membrane. Particularly, experiments on Na^+-dependence, substrate specificity, and transport kinetics of these two amino acids were performed. For the experiments isolated brush border membrane vesicles (BBMV) from pig jejunum were used.

2 Materials and Methods

Pig jejunal BBMV were isolated by a Ca^{2+} precipitation technique combined with differential centrifugation (Kessler et al. 1978). The intestines were obtained from freshly killed animals at the local slaughterhouse. The utility of the BBMV was routinely checked by determination of the activity of marker enzymes as well as by means of Na^+-dependent D-glucose uptake. The specific activity of the BBM marker enzyme alkaline phosphatase was enriched in our BBM preparations 11- to 20-fold with respect to the original homogenate. Cross-contamination with basolateral membranes was negligible as shown by the 1.5-fold enrichment of the activity of Na,K-ATPase, a marker enzyme of basolateral membranes. D-glucose uptake by the BBMV was strongly stimulated by an inwardly directed Na^+ gradient, whereas uptake in the absence of Na^+ occurred by simple diffusion.

Uptake of ^{75}Se-labeled L-Se-Met and ^{3}H-labeled L-Met under various incubation conditions was measured by a rapid filtration technique (Wolffram et al. 1988).

[1] Institute of Veterinary Physiology, Winterthurerstraße 260, 8057 Zürich, Switzerland
[2] To whom correspondence should be addressed

3 Results and Discussion

Transport of Se-Met and Met by the BBMV at incubation periods of 15 s to 3 min was clearly stimulated by an inwardly directed Na^+ gradient ($Na^+_{out} > Na^+_{in}$) compared with K^+-gradient conditions, whereas the equilibrium uptake (60-min incubation period) was similar under both conditions. Furthermore, uptake of Se-Met under both incubation conditions exceeded uptake of Met at all incubation periods. However, Na^+-dependent uptake (uptake under Na^+-gradient conditions minus uptake under K^+-gradient conditions) was similar for both substrates (Fig. 1). To determine whether uptake of Se-Met and Met was due to substrate uptake into the vesicular lumen or to binding to the membranes, equilibrium uptake of Se-Met and Met was determined at various osmolarities of the incubation medium. For comparison, D-glucose uptake was also investigated. In the presence of intact vesicles, increasing the osmolarity of the incubation medium results in shrinkage of the vesicles and a reduction of the intravesicular lumen, thus reducing equilibrium uptake. For all substrates investigated, an inverse linear relationship between medium osmolarity and uptake was found. This

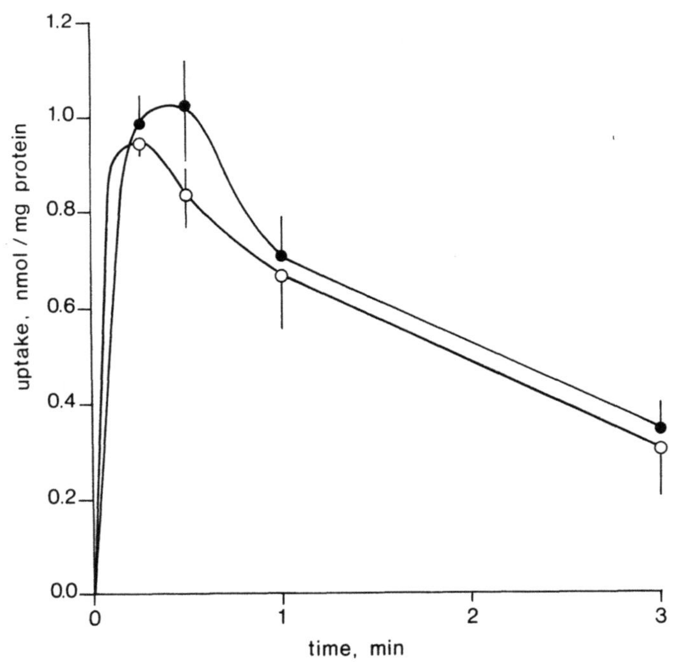

Fig. 1. Time course of Na^+-dependent uptake of Se-Met and Met. Values are means±SEM from 3 different membrane preparations each determined in duplicate; vesicles were prepared in (mmol l^{-1}): 300 D-mannitol, 20 HEPES adjusted with Tris to pH 7.4, and incubated in media of the following composition (final concentrations, mmol l^{-1}): 100 NaCl or KCl, 100 D-mannitol, 20 HEPES adjusted with Tris to pH 7.4, and 0.1 Se-Met (●) or Met (○); for further details see text

demonstrates substrate uptake into an osmotically reactive space, namely, the vesicular lumen. Furthermore, extrapolation to an infinite osmolarity, at which measured substrate uptake theoretically represents only binding, revealed an extensive binding of Se-Met, a moderate binding of Met, and a negligible binding of D-glucose (results not shown). Thus, the higher uptake rates for Se-Met compared with Met observed in the experiment first mentioned is mainly due to greater binding of Se-Met.

In the next experiments initial (1.5-s incubation) Na^+-dependent uptake of Se-Met and Met as a function of the extravesicular substrate concentration was examined. Na^+-dependent uptake was obtained by subtracting uptake under K^+-gradient conditions from total uptake under Na^+-gradient conditions. Figure 2 shows a Hofstee plot of the Na^+-dependent component of Se-Met and Met uptake at a concentration range of 0.025 to 2.5 mmol l^{-1}. For both amino acids a single carrier-mediated transport component was calculated under these conditions, indicated by the straight lines. The transport kinetics of both substrates are similar with respect to substrate affinity (K_m) as well as with respect to maximal transport velocity (V_{max}). The respective values (means ± SEM from six individually analyzed membrane preparations) are shown in the inset of

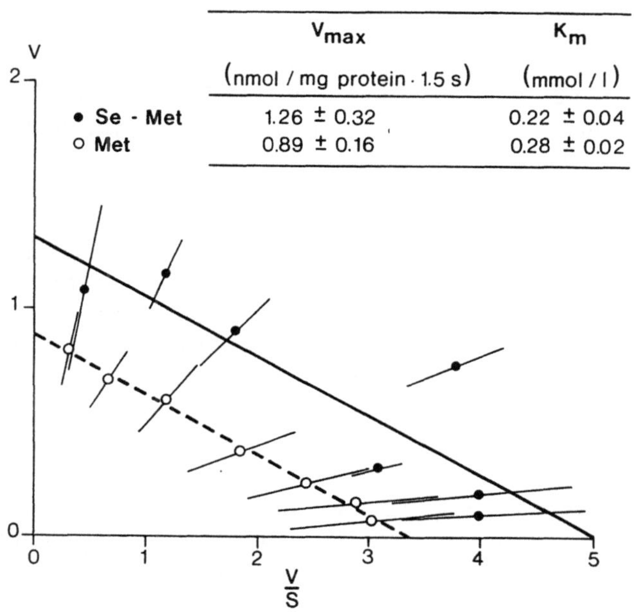

Fig. 2. Hofstee plot of Na^+-dependent Se-Met and Met uptake (V) as a function of extravesicular substrate concentration (S). Values are means ± SEM from 6 different membrane preparations each determined in triplicate; the *straight lines* were calculated by linear regression analysis; vesicles were prepared in (mmol l^{-1}): 330 D-mannitol, 20 HEPES adjusted with Tris to pH 7.4, and incubated in media of the following composition (final concentrations, mmol l^{-1}): 100 NaCl or KCl, 110 D-mannitol, 0.025–2.5 L-Se-Met (●) or L-Met (○), 20 HEPES adjusted with Tris to pH 7.4; osmolarity was kept constant by addition of D-mannitol; for further details, see text. *Inset*: Kinetic parameters of Se-Met and Met uptake

Table 1. Influence of various amino acids on uptake of Se-Met (0.1 mmol l^{-1}) and on Met (0.1 mmol l^{-1}) by jejunal BBMV[a]

Inhibitor (conc.: 10 mmol l^{-1})	Inhibition, percent of control uptake of	
	Se-Met	Met
Selenomethionine	–	95
Methionine	87	–
Cysteine	86	90
Alanine	65	95
Leucine	84	93
Phenylalanine	77	82
Serine	42	55
Lysine	7	13
Glutamic acid	7	8

[a] Values are means from 4 determinations of a single membrane preparation; the variation between determinations was within 5%

Fig. 2. It remains to be investigated in future experiments, how the electrical potential difference across the BBM affects these values.

Additional experiments were performed to examine the substrate specificity of Se-Met and Met uptake. At a substrate concentration of 0.1 mmol l^{-1} and an inhibitor concentration of 10.0 mmol l^{-1} uptake of both amino acids in the presence of an inwardly directed Na^+ gradient (100 mmol l^{-1}) was strongly inhibited to a similar extent by the neutral amino acids cysteine, alanine, leucine, phenylalanine, and serine, whereas the basic amino acid lysine and the acidic amino acid glutamic acid only slightly decreased Se-Met and Met uptake, respectively (Table 1). Additionally, Met caused about 90% inhibition of Se-Met uptake and vice versa when present simultaneously in the incubation medium (Table 1). This mutual cis-inhibition between Se-Met and Met suggests transport of both amino acids by a common transport site. Further evidence for a common transport mechanism for Se-Met and Met is derived from trans-stimulation experiments. Thereby, Met uptake at a concentration of 0.2 mmol l^{-1} by BBMV preloaded with 10.0 mmol l^{-1} Se-Met was determined under Na^+-equilibrium conditions ($Na^+_{out} = Na^+_{in} = 100$ mmol l^{-1}). In a typical experiment preloading the vesicles resulted in a 2.7-fold stimulation of Met uptake after an incubation of 15 s (992 and 374 pmol mg^{-1} protein with preloaded and control vesicles, respectively), whereas uptake after an incubation of 60 min was similar under both conditions (526 and 435 pmol mg^{-1} protein with preloaded and control vesicles, respectively). These results clearly show that in the absence of energy in the form of a transmembrane Na^+ gradient, preloading resulted in a transient accumulation of Met energized by the outwardly directed concentration gradient of Se-Met. This trans-stimulation together with the mutual cis-inhibition between Se-Met and Met unequivocally proves transport of both substrates by a common transport site.

4 Conclusions

From the results presented it is concluded that transport of Se-Met and Met across the intestinal BBM occurs by a common Na^+-dependent mechanism, specific for neutral amino acids. Replacement of sulfur in Met by selenium apparently does not significantly affect the transport characteristics of this amino acid.

References

Kessler M, Acuto O, Storelli C, Murer H, Müller M, Semenza G (1978) A modified procedure for the rapid preparation of efficiently transporting vesicles from small intestinal brush border membranes. Biochim Biophys Acta 506:136–154

McConnell KP, Cho GJ (1965) Transmucosal movement of selenium. Am J Physiol 208:1191–1195

Spencer RP, Blau M (1962) Intestinal transport of selenium-75 selenomethionine. Science 136:155–156

Wolffram S, Grenacher B, Scharrer E (1988) Transport of selenate and sulphate across the intestinal brush-border membrane of pig jejunum by two common mechanisms. Q J Exp Physiol 73:103–111

Plasma Malondialdehyde and Exhaled Hydrocarbons: Are They Reliable Indicators of Lipid Peroxidation and Glutathione Peroxidase Activity?

C. R. WADE[1,2] and A. M. VAN RIJ[1]

1 Introduction

Free radical-mediated oxidation of polyunsaturated fatty acids (PUFA) results in the formation of cytotoxic and bioactive lipid peroxides which are reduced by glutathione peroxidase to harmless lipid alcohols. There are two principle methods that are commonly used for the measurement of lipid peroxides, the malondialdehyde thiobarbituric (MDA-TBA) method and the breath hydrocarbon method. However, interpretation of results obtained using both of these methods is complicated by several factors outlined in this chapter that are frequently not taken into account by investigators in the field. This has led to many conflicting reports in the literature which has fuelled the controversy surrounding the reliability and specificity of these two methods and their scientific value.

2 Lipid Peroxides and Their Secondary Decomposition Products

Lipid peroxides formed as a result of the abstraction of a hydrogen atom form a methylene group in an unsaturated fatty acid undergo complex secondary oxidation and cyclization reactions to form mono- and bicyclic peroxides. The primary and secondary peroxides produced are thermally unstable and undergo metal catalyzed decomposition reactions to form numerous secondary monomeric and volatile low and high molecular weight compounds. Most efforts to quantify lipid peroxides have centered around two of the products formed from their decomposition, malondialdehyde (MDA) and the volatile hydrocarbon gases ethane and pentane, which are derived from linolenic acid and linoleic acid hydroperoxides respectively.

3 The Malondialdehyde-Thiobarbituric Acid (MDA-TBA) Method

The MDA-TBA test is the most commonly used method for the measurement of lipid peroxides. Under the acid and heating conditions of the test lipid peroxides

[1] Surgery Department, University of Otago Medical School, P.O. Box 56, Dunedin, New Zealand
[2] To whom correspondence should be addressed

decompose to MDA which subsequently reacts with TBA to from a red chromogen (MDA-TBA) that can be measured spectrophotometrically at 532 nm, or by fluorimetry.

The specificity of the TBA test has been strongly questioned in this laboratory (Wade and van Rij 1985, 1988) and by other workers (Wong et al. 1987) as several non-lipid compounds that are frequently present in biological samples are also TBA-reactive. Furthermore, the TBA reactivity of lipid peroxides is strongly influenced by the acidity of the reaction mixture, the presence of Fe ions and the method of sample preparation (Wade and van Rij 1988). Lack of specificity, standardization and optimization of the TBA-reaction conditions are reflected in the wide variance of reported values for plasma lipid peroxide levels in the healthy (range 0.6–47.2 nmol ml^{-1}; Wong et al. 1987).

To overcome these problems we have published an HPLC procedure specific for the MDA-TBA chromogen (Wade et al. 1985). Our next step has been to optimize the reaction conditions, to exclude compounds that give rise to artifactual MDA and to investigate the source of plasma MDA. From these studies we have proposed a standardized TBA test that is highly specific for plasma lipid peroxides (Wade and van Rij 1988). Formation of MDA-TBA is optimized by the addition of Fe to the reaction mixture and in vitro autoxidation of PUFA during the test is prevented by the addition of butylated hydroxytoluene (BHT).

We have now applied this assay in our clinical studies. Contrary to published reports from workers using less specific methods, we have been unable to demonstrate increased plasma lipid peroxides in diabetes, peripheral vascular disease or rheumatoid arthritis (Table 1). However, significantly decreased levels of peroxides were observed in malnourished patients suggesting that dietary lipids or lipid peroxides may contribute to the circulating plasma levels. Further evidence for this has become apparent when post-prandial plasma samples were analyzed from subjects who had consumed a high fat meal (150.0 ± 29.0 ngm ml^{-1}, mean ± SD, $p < 0.001$). The significance of this post-prandial effect is not known and requires further investigation.

Table 1. Plasma lipid peroxides

Groups	No.	Mean ± SD[a]	p value[b]
Healthy	78	104 ± 15	
Diabetes	11	108 ± 22	ns
PVD[c]	16	111 ± 24	ns
Rheumatoid arthritis	12	98 ± 18	ns
Malnourished	7	73 ± 21	$p < 0.001$

[a] ngm ml^{-1}.
[b] Mann-Whitney test between healthy and diseased groups.
[c] Peripheral vascular disease.

4 Breath Hydrocarbon Measurements

Although there are very few reports of the application of this technique in man we have recently published a method for the collection of the small amounts of ethane and pentane exhaled in human breath (Wade and van Rij 1985). The method involves rebreathing in a closed circuit spirometer for a 2-h period to allow accumulation of the exhaled hydrocarbons. We have found that several physiological and environmental factors complicate the measurement of hydrocarbons in man. Firstly, previous exposure to hydrocarbon-contaminated environments results in high tissue levels which must be washed out by scrubbing with hydrocarbon-free air before rebreathing. This effect is most obvious in smokers and precludes them from further study due to the very high tissue levels of hydrocarbons in body fat which can take up to 5 days to clear. Secondly, the presence of ethane and pentane in diets high in PUFA content may also effect exhalation measurements as will the decomposition of dietary lipid peroxides absorbed postprandially after the ingestion of high fat meals. The third and potentially most serious factor confounding measurements of exhaled levels of hydrocarbons is that they are metabolized in humans. We have determined that pentane metabolism is rapid and can be inhibited by diethyldithiocarbamate (disulfiram; van Rij and Wade 1987). By contrast, ethane metabolism is slow allowing the calculation of production using the initial linear phase of ethane exhalation. In a recent study, using the rebreathe technique, selenium supplementation ($n = 6$, 200 μg daily for 28 days) did not statistically affect hydrocarbon production.

5 Conclusions

Measurement of lipid peroxides by the MDA-TBA or exhaled hydrocarbon methods is complicated by several factors, including lack of specificity and standardization, diet, hydrocarbon contamination of environments and metabolism. However, studies performed using the specific techniques outlined in this work along with careful consideration of experimental protocols may allow a more meaningful assessment of lipid peroxidation in health and disease and the protective role of glutathione peroxidase.

References

Rij AM van, Wade CR (1987) The metabolism of low molecular weight hydrocarbon gases in man. J Free Radiat Res 4:99–103

Wade CR, Rij AM van (1985) In vivo lipid peroxidation in man as measured by the respiratory excretion of ethane, pentane and other low molecular weight hydrocarbons. Anal Biochem 150:1–7

Wade CR, Rij AM van (1988) Plasma thiobarbituric acid reactivity: reaction conditions and the role of iron, antioxidants and lipid peroxy radicals on the quantitation of plasma lipid peroxides. Life Sci 43:1085–1093

Wade CR, Jackson PG, Rij AM van (1985) Quantitation of malondialdehyde (MDA) in plasma, by ion-pairing reverse phase HPLC. Biochem Med 33:291–296

Wong SHY, Knight JA, Hopfer SM, Zaharia O, Leach CN Jr, Sundermann FW (1987) Lipoperoxides in plasma as measured by liquid chromatographic separation of the malondialdehyde-thiobarbituric adduct. Clin Chem 33:214–220

Biochemical Studies on Selenium and Keshan Disease – The Oxidant Stress and Defence Capacity in Blood of Selenium-Deficient Children

LIAN-ZHEN ZHU[1,5], JIAN-HUA PIAO[1], YIMING XIA[1], YUN-YU CHENG[2], JI-YOUNG CHEN[3], JIA-YU ZHENG[3] and CUN-LONG NIU[4]

1 Introduction

In China, the daily selenium intake of humans varied widely, from 7 μg to 38 mg (Yang 1987), thus the individual selenium status is quite different. The population in Keshan disease areas where an endemic cardiomyopathy, so-called Keshan disease (KD), is prevalent has a selenium-deficient status (Chen 1981). Selenium deficiency alters the response of the organism to a number of stresses, hence the biochemical effects of selenium deficiency on Keshan disease was studied by the Investigation Team of Keshan Disease in Chuxiong Yi Autonomous Prefecture of Yunnan Province in 1984–1986. One part of this study was to compare the status of oxidant stress and the defence system of children with differing selenium status and of KD patients. This part was assessed according to three aspects:

1. The antioxidant defence system in blood;
2. The antioxidant ability of erythrocytes in vitro;
3. Free radicals in erythrocytes.

2 Experimental Procedures

2.1 Selected Population

Six groups of normal children from 6 to 12 years of age with differing selenium status were selected from six villages: Bixi, Chuxiong, Daizhai, Samachang, Tianfeng and Donggua in Yunnan Province of China. Keshan disease patients and patients with infectious disease were in-patients in Chuxiong Institute of Prevention and Cure of Keshan Disease and Chuxiong State Hospital. Their selenium status was indicated by their hair selenium level.

[1] Institute of Nutrition and Food Hygiene, Chinese Academy of Preventive Medicine, 29 Nanwei Road, Beijing, China
[2] Sichuan Provincial Antiepidemic and Health Station, Chengdu, Sichuan Province, China
[3] Analytical and Testing Center, Sichuan University, Chengdu, Sichuan Province, China
[4] Chuxiong Institute of Prevention and Cure of Keshan Disease, Chuxiong, Yunnan Province, China
[5] To whom correspondence should be addressed

2.2 The Assay of Enzyme Activities

The activities of antioxidant enzymes, glutathione peroxidase (GPx) and glutathione S-transferase (GST) were determined according to the methods of Paglia and Valentine (1967), Habig et al. (1974) and Marcus et al. (1978) respectively. The activity of glutathione reductase (GR), which is involved in GSH metabolism, was determined according to the method of Worthington and Rosemeyer (1976).

2.3 The Antioxidant Ability of Erythrocytes

The antioxidant ability of erythrocytes was tested by incubating erythrocytes with 0.45 mM dihydroxyfumaric acid (DHF) which generates O_2^- by autoxidation, or with 4.11 mM vitamin C which can provoke the formation of free radicals in the presence of transition metals at 37 °C for 0.5 or 1 h respectively. For examining the damage of the erythrocytes, the percentage of hemoglobin oxidized, the rate of hemolysis and the deformation of erythrocytes observed by scanning electron microscope were determined.

2.4 The Detection of Free Radicals

Free radicals in erythrocytes were detected with electron spin resonance (ESR) according to the method of Lohmann et al. (1979). The results were expressed as spin concentration.

3 Results and Discussion

3.1 Antioxidant Defence System in Blood

As can be seen from Table 1, only the activity of GPx varied with selenium status. No significant difference was found in GR activities of six groups of normal children. However, it is worthy to note that the GR activity of KD patients increased about 60%, while the GPx activity decreased. These changes may influence the GSH metabolism of KD patients.

Lawrence et al. (1978) found that the GST activity is higher in selenium-deficient male rat liver cytosol than in controls, but in our study the difference of GST activities of erythrocytes among all groups was not significant.

3.2 The Antioxidant Ability of Erythrocytes in Vitro

The results show that the lower the selenium status, the more the hemoglobin oxidized by O_2^- (Table 1). The percentage of hemoglobin oxidized is also inversely related to GPx activity. The correlation coefficient is -0.7.

Table 1. Hair Se level, erythrocyte GPx activity and oxidation of hemoglobin by O_2^- of children with differing selenium status and KD patients

Sampling sites	Hair Se level ($\mu g\ g^{-1}$)	GPx activity[a]	Hb oxidized[b] (%)
Bixi	0.273 ± 0.010(31)[c]	33.27 ± 1.07(24)	2.39 ± 0.60(21)
Chuxiong	0.236 ± 0.011(56)	27.98 ± 0.96(53)	1.67 ± 0.30(30)
Daizhai	0.228 ± 0.008(25)	25.84 ± 1.53(24)	1.07 ± 0.27(24)
Samachang	0.131 ± 0.006(20)	13.57 ± 0.80(38)	4.43 ± 0.12(10)
Tianfeng	0.081 ± 0.006(35)	8.32 ± 0.51(60)	7.63 ± 0.66(33)
Donggua	0.073 ± 0.004(37)	6.35 ± 0.34(59)	10.28 ± 0.97(21)
KD patients		6.67 ± 0.98(25)	11.67 ± 0.97(25)

[a] Specific activity = μmol NADPH oxidized $min^{-1}\ g^{-1}$ Hb
[b] 10 μl erythrocytes were incubated with 0.45 mM DHF at 37 °C for 30 min
[c] Mean ± SE (number of samples)

There is no difference in hemolysis of erythrocytes. The rate of hemolysis in all groups was about 5%, regardless of whether the glucose was present or not. This result indicated that the hemolysis induced by ascorbic acid in this study was independent of selenium status. The results also showed that there was no apparent vitamin E deficiency among the children investigated. Our previous animal experiment showed that vitamin E deficiency caused more than 90% of hemolysis induced by ascorbic acid, whereas the hemolysis decreased to 5% when 25 mg vitamin E kg^{-1} diet was supplemented (He and Zhu 1987). Furthermore, the plasma vitamin E level showed that the vitamin E status of all children in our study is normal, although that of Se-adequate children is higher than of Se-deficient children. The plasma vitamin E level of Se-adequate children is 11 $\mu g\ ml^{-1}$, and that of Se-deficient children is more than 6 $\mu g\ ml^{-1}$ (Li and Ren 1989). This means that the major responsibility for the decrease of antioxidant ability of children in the KD areas of Yunnan Province is selenium deficiency.

After oxidation by O_2^-, the erythrocytes changed in shape and the Se-deficient erythrocytes appeared more crystael cells than the Se-adequate erythrocytes.

3.3 *Free Radicals in Erythrocytes*

The results in Tables 2 and 3 indicate that the spin concentration in lyophilized erythrocytes or blood did not relate to GPx activities significantly. However, it is interesting that the spin concentrations in lyophilized erythrocytes or blood of subacute KD patients and of patients with infectious disease are significantly higher than that of normal children. Ouyang and Yang (1989) reported that the free-radical level in myocardium of KD patients was also higher than that of normal persons killed by accident, the mean values were 1.61 and 1.08 respectively.

Most KD patients had symptoms of intestinal disorder or respiratory infection, suggesting that infection might at least partly cause the overproduction of free radicals in the body of Se-deficient children in KD areas, which may result

Table 2. Free radicals and GSH-Px activities in lyophilized erythrocytes of children with differing selenium status and patients in Yunnan Province

	GSH-Px activities (μmol NADPH oxidized $min^{-1} g^{-1}$ Hb)	Free radicals (spin concentration)
Daizhai	25.84 ± 1.53(24)[a]	2.81 ± 0.47(13)
Chuxiong Yi Autonomous Prefecture		
Chuxiong	28.72 ± 1.70(14)	1.88 ± 0.33(10)
Tianfeng	6.18 ± 0.59(19)	2.23 ± 0.39(14)
Intestinal infectious disease patients	9.59(6)	3.08(6)
Subacute Keshan disease patients	4.67(4)	3.94(2)

[a] Mean ± SE (number of samples)

Table 3. Free radicals and GSH-Px activities in lyophilized whole blood of children with differing selenium status and patients in Sichuan Province

	GSH-Px activities (μmol NADPH oxidized $min^{-1} g^{-1}$ Hb)	Free radicals (spin concentration)
Nanbu	13.39 ± 0.49(33)[a]	10.90 ± 2.04(15)
Liangshan Autonomous Prefecture		
Mianning	18.63 ± 1.44(36)	8.23 ± 1.84(33)
Dechang	8.61 ± 1.02(41)	8.87 ± 1.34(29)
Infectious disease patients	11.82(3)	17.01(5)
Subacute KD patients	6.25(3)	17.30 ± 1.20(4)

[a] Mean ± SE (number of samples)

in a loss of balance between oxidant stress and antioxidant defence. It is suggested that this may contribute to the outbreak of Keshan disease.

4 Summary

1. In the antioxidant defence system, only the activity of GPx varies with selenium status. The GR activity in erythrocytes of subacute KD patients is 60% higher than that of normal children.
2. The antioxidant ability in erythrocytes of Se-deficient children is reduced due to the decrease of GPx activity.
3. The free-radical levels in erythrocytes or blood of KD patients and of patients with infectious disease are higher than that of normal children.

References

Chen XS, Chen XC, Yang GQ, Wen ZM, Chen JS, Ge KY (1981) Relation of selenium deficiency to the occurrence of Keshan disease. In: Spallholz JE, Martin JL, Ganther HE (eds) Selenium in biology and medicine. AVI Press, New York, p171

Habig WH, Pabst MJ, Jakoby WB (1974) Glutathione S-transferase. J B C 249:7130–7139

He YP, Zhu LZ (1987) Effect of dietary selenium and vitamin E on the antioxidative ability in rat. Chin Biochem J 3:301–306 (in Chinese)

Lawrence RA, Parkhill LK, Burk RF (1978) Hepatic cytosolic nonselenium-dependent glutathione peroxidase activity: its nature and the effect of selenium deficiency. J Nutr 108:981–987

Li GY, Ren YY (1989) Disturbance of lipid metabolism in children in Keshan disease epidemic areas of Yunnan Province. In: Yu WH, Wang F (eds) Collected works of comprehensive scientific survey on Keshan disease in Chuxiong Prefecture. PMPH Press, Beijing, p 247 (in Chinese)

Lohmann W, Schreiber J, Strobelt W, Muller-Eckhardt Ch (1979) On the possible involvement of ascorbic acid and copper proteins in leukemia: I. Electron spin resonance (ESR) investigation on native blood, erythrocytes and leukocytes. Blut 39:317–326

Marcus CJ, Habig WH, Jakoby WB (1978) Glutathione transferase from human erythrocyte. Arch Biochem Biophys 188:287–293

Ouyang YP, Yang DS (1989) The metabolism of free radicals in the myocardium of patients with Keshan disease. In: Yu WH, Wang F (ed) Collected works of comprehensive scientific survey on Keshan disease in Chuxiong Prefecture. PMPH Press, Beijing, p 217 (in Chinese)

Paglia DE, Valentine WN (1967) Studies on the quantitative and qualitative characterization of erythrocyte glutathione peroxidase. J Lab Clin Med 70:158–169

Worthington DJ, Rosemeyer MA (1976) Glutathione reductase from human erythrocyte. Eur J Biochem 67:231–238

Yang GQ (1987) Research on selenium-related problems in human health. In: Combs GF Jr, Spallholz JE, Levander OA, Oldfield JE (eds) Selenium in biology and medicine, part A. AVI Press, New York, p 9

Effect of Growth Phase on Deposition of Selenium (Se) in Tissues of Rats Fed Elevated Dietary Levels of Se as Either L-Selenomethionine or Sodium Selenate

A. D. SALBE[1] and O. A. LEVANDER[1,2]

1 Introduction

Epidemiologic and intervention studies investigating possible cancer chemopreventive effects of selenium (Se) have pointed out the need for accurate indicators of Se status which would monitor the large body pools of Se (Willett et al. 1983; Salonen et al. 1984; Clark and Combs 1986), namely those in the liver and skeletal muscle (Behne and Wolters 1983). It is beneficial in such studies to utilize tissues obtained noninvasively, such as hair and nails, to assess Se status (Van Noord et al. 1987; Willett 1986; Chen et al. 1980). The influence of various dietary and physiological factors on hair and nail Se levels has scarcely been investigated, however. We have previously reported that hair and nail Se content is affected by the level and chemical form of dietary Se (Salbe and Levander 1987). Here, we report on the effect of growth on Se deposition in hair, nail, and other tissues of adult and weanling rats.

2 Materials and Methods

Animals and Diets. Twenty-four male weanling Sprague-Dawley rats were fed a Se-deficient basal diet containing 30% torula yeast, 60.2% sucrose, 5% corn oil, 3.5% AIN mineral mix minus Se, 1.0% AIN vitamin mix (including 50 IU all-rac-α-tocopheryl acetate/kg diet), 0.063% choline, and 0.3% D,L-methionine. Rats were fed this diet for 15 weeks at which time the mean weight ± SD of the animals was 312 ± 25 g. At that point, these 24 adult rats, as well as 24 additional male weanling rats (mean initial weight ± SD, 45 ± 6 g), were randomly divided into groups of four. The torula yeast-based diet described above was supplemented with Se at 0.5, 1.5, or 2.5 $\mu g\ g^{-1}$ diet as L-selenomethionine (SeMet) or sodium selenate (Na_2SeO_4) and the diets were fed to the 12 groups of rats for 4 weeks.

Tissue Preparation and Analysis. Rats were anesthetized with ether and bled from the abdominal aorta. Blood was collected in syringes containing 0.1 volume of

[1] US Department of Agriculture, Agricultural Research Service, Beltsville Human Nutrition Research Center, Vitamin and Mineral Nutrition Laboratory, Beltsville, MD 20705, USA
[2] To whom correspondence should be addressed

3.2% trisodium citrate and centrifuged at 4 °C, 2500 × g, for 15 min. Livers were excised, rinsed in ice-cold 1.15% KCl, and homogenized in 4 volumes of KCl with a Teflon homogenizer. Skeletal muscle tissue was excised from both hind legs of the animals, put through a tissue press, and homogenized in 9 volumes of KCl with a Polytron mixer. Hair samples were obtained by shaving the dorsal side of the carcass with a pair of animal clippers. The hair was washed with 50 ml deionized H_2O while shaking on a wrist action shaker for 20 min, suction filtered, and air dried. Nails were obtained from the carcass by pulling from all four paws with forceps. Nails were debrided, washed with deionized H_2O while shaking, suction filtered, and air dried.

Plasma Se was analyzed by Zeeman atomic absorption spectroscopy (Lewis et al. 1986). Se levels in all other tissues were analyzed by perchloric acid digestion and fluorimetry (Hoffman et al. 1968).

Statistical Analyses. Log transformed data were analyzed by analysis of variance and Duncan's multiple-range test (Steel and Torrie 1960).

3 Results

Weanling rats gained five times as much weight as adult rats during the 4-week feeding period (Table 1). There were no significant differences in weight gain as a result of chemical form or dietary intake of Se. Hair Se levels in adults were only 10–20% of the amount found in weanling rats, whereas nail Se retention in the adults was about half that of weanlings. Adult rats deposited approximately half as much Se in skeletal muscle as weanling rats, whereas liver Se levels often appear similar between the two growth phases. Red cell Se levels were approximately 35% lower in adult as compared to weanling rats. In contrast to all the other tissues analyzed, plasma Se levels were higher in adult, as compared to weanling, rats. At each dietary level, and in both growth phases, feeding rats SeMet caused greater tissue Se deposition than feeding Na_2SeO_4. The percent increase caused by the organic vs. the inorganic form was similar in both growth phases.

4 Discussion

Although the use of hair and nails to assess Se status is gaining acceptance among nutrition researchers (Hebert and Miller 1988; Willett 1986; Chen et al. 1980), the results of this and a previous study from our laboratory (Salbe and Levander 1987) suggest that hair and nails may not always accurately reflect dietary Se intake. In addition to dietary intake level, both the chemical form of the element and growth rate affect the deposition of Se in hair and nails.

Whereas muscle Se content of adult rats was half that of weanlings, and liver Se retention was similar for both groups of rats, hair Se levels in adult rats were only 10–20% of hair Se levels in weanling rats. Turnover rates in hair seem to

Table 1. The effect of growth on tissue selenium (Se) concentrations of weanling and adult rats fed diets containing sodium selenate (Na_2SeO_4) or L-selenomethionine (SeMet) for 4 weeks[a]

Form of Se and animal age	μg Se g^{-1}diet	Weight gain (g)	Tissue selenium concentrations (μg^{-1}g)[b]					
			Hair	Nail	Muscle	Liver	Red cell	Plasma
Na_2SeO_4								
Weanling	0.5	136 ± 11^a	0.49 ± 0.12^a	0.46 ± 0.04^a	0.10 ± 0.01^a	0.76 ± 0.02^a	0.43 ± 0.02^a	0.24 ± 0.04^a
	1.5	126 ± 20^a	1.45 ± 0.42^b	1.06 ± 0.35^b	0.16 ± 0.01^b	1.20 ± 0.13^b	0.62 ± 0.05^b	$0.28 \pm 0.01^{a,b}$
	2.5	139 ± 7^a	3.51 ± 0.56^d	1.36 ± 0.19^c	0.18 ± 0.02^b	1.87 ± 0.46^c	1.00 ± 0.17^c	$0.35 \pm 0.06^{b,c}$
Adult	0.5	27 ± 5^m	0.10 ± 0.02^m	0.27 ± 0.10^m	0.08 ± 0.01^m	0.77 ± 0.02^m	0.30 ± 0.03^m	0.43 ± 0.04^m
	1.5	23 ± 3^m	$0.15 \pm 0.07^{m,n}$	$0.39 \pm 0.08^{m,n}$	$0.09 \pm 0.01^{m,n}$	$0.87 \pm 0.04^{m,n}$	0.37 ± 0.04^m	0.59 ± 0.21^m
	2.5	26 ± 8^m	0.28 ± 0.17^o	$0.63 \pm 0.24^{n,o}$	0.11 ± 0.01^n	1.25 ± 0.22^n	$0.78 \pm 0.47^{n,o}$	0.54 ± 0.08^m
SeMet								
Weanling	0.5	141 ± 9^a	1.42 ± 0.21^b	$1.15 \pm 0.02^{b,c}$	0.40 ± 0.06^c	0.91 ± 0.03^a	0.70 ± 0.14^b	$0.35 \pm 0.10^{b,c}$
	1.5	135 ± 10^a	2.63 ± 0.34^c	1.89 ± 0.26^d	0.84 ± 0.03^d	1.57 ± 0.07^c	1.14 ± 0.10^c	$0.38 \pm 0.01^{b,c}$
	2.5	146 ± 12^a	5.58 ± 1.50^e	3.24 ± 0.12^e	1.42 ± 0.07^e	2.93 ± 0.10^d	2.14 ± 0.14^d	0.45 ± 0.03^c
Adult	0.5	30 ± 6^m	$0.22 \pm 0.06^{n,o}$	0.63 ± 0.11^o	0.17 ± 0.01^o	1.04 ± 0.08^n	$0.47 \pm 0.11^{m,n}$	0.45 ± 0.04^m
	1.5	32 ± 5^m	0.46 ± 0.09^p	1.21 ± 0.48^p	0.43 ± 0.07^p	2.01 ± 0.81^o	$0.83 \pm 0.28^{o,p}$	0.71 ± 0.16^n
	2.5	26 ± 10^m	0.53 ± 0.21^p	1.33 ± 0.21^p	0.52 ± 0.11^p	2.08 ± 0.14^o	0.90 ± 0.04^p	0.53 ± 0.08^m

[a] Means within a column not sharing a common superscript are significantly different as a result of chemical form and dietary level within a growth phase for weanling (superscript $^{a-e}$) or adult (superscript $^{m-p}$) rats ($p<0.001$, ANOVA and Duncan's multiple-range test).
[b] Means ± SD of 4 animals per group. Weight gain and hair, nail, muscle, and red cell Se levels were significantly lower, whereas plasma Se levels were significantly higher in adult, as compared to weanling, rats ($p<0.001$, ANOVA). There were no significant differences in liver Se levels as a result of growth phase.

be much slower than turnover rates in the tissues containing the major body pools of Se and hair Se content is therefore not a uniform marker of these pools. Plasma Se levels, which were greater in adult as compared to weanling rats, also appear to be of questionable value as a marker of the large body pools of Se.

Se is known to be more toxic to younger, as compared to older, animals (Rosenfeld and Beath 1964), and increased tissue deposition of Se is thought to contribute to this higher susceptibility to toxicity. The results of the present study, which found markedly greater tissue stores in weanling as compared to adult rats, support this hypothesis.

Our results indicate the need to consider phase of growth if hair and nail Se levels are to be used as indices of Se status, either for nutrition monitoring purposes or for epidemiologic studies attempting to relate dietary Se intake with certain human diseases.

Acknowledgements. Supported in part by NCI-Y01-CN-40620.

References

Behne D, Wolters W (1983) Distribution of selenium and glutathione peroxidase in the rat. J Nutr 113:456–461

Chen X, Yang G, Chen J, Chen X, Wen Z, Ge K (1980) Studies on the relations of selenium and Keshan disease. Biol Trace Element Res 2:91–107

Clark LC, Combs GF Jr (1986) Selenium compounds and the prevention of cancer: research needs and public health implications. J Nutr 116:170–173

Hebert JR, Miller DR (1988) Methodologic considerations for investigating the diet-cancer link. Am J Clin Nutr 47:1068–1077

Hoffman I, Westerby RJ, Hidiroglou M (1968) Precise fluorometric determination of selenium in agricultural materials. J Assoc Off Agric Chem 51:1039–1042

Lewis SA, Hardison NW, Veillon C (1986) Comparison of isotope dilution mass spectrometry and graphite furnace atomic absorption spectrometry with Zeeman background correction for determination of plasma selenium. Anal Chem 58:1272–1273

Noord PAH van, Collette HJA, Maa MJ, Waard F de (1987) Selenium levels in nails of premenopausal breast cancer patients. Intl J Epidemiol (Suppl) 16:318–322

Rosenfeld I, Beath OA (1964) Selenium: geobotany, biochemistry, toxicity and nutrition. Academic Press, London New York

Salbe AD, Levander OA (1987) Hair and nails as indicators of Se status in rats fed elevated dietary levels of Se as L-selenomethionine or sodium selenate. Fed Proc 46:113 Abstr

Salonen JT, Alfthan G, Huttunen JK, Puska P (1984) Association between serum selenium and the risk of cancer. Am J Epidemiol 120:342–349

Steel RGD, Torrie JH (1960) Principles and procedures of statistics. McGraw-Hill, New York

Willett WC (1986) Epidemiologic studies in nutrition: utility and limitations. J Nutr 116:2557–2558

Willett WC, Polk BF, Morris JS, Stampfer MJ, Pressell S, Rosner B, Taylor JO, Schneider K, Hames CG (1983) Prediagnostic serum selenium and risk of cancer. Lancet 2:130–134

Selenium Intake Induces Growth Retardation Through Reversible Growth Hormone and Irreversible Somatomedin C Suppression

O. THORLACIUS-USSING[1,5], A. FLYVBJERG[2], U. TARP[3], K. OVERVAD[4], and H. ØRSKOV[2]

1 Introduction

Growth retardation is an outstanding symptom in animals exposed to toxic doses of selenium and is one of the main symptoms in stocks with alkali disease. Even Moxon (1937) demonstrated a convincing correlation between the dose of selenium consumed by experimental animals and the level of growth retardation. These results have later been confirmed by many others and the National Research Council concluded in 1976 that growth retardation may be the best indicator of selenium toxicity. The mechanism behind this symptom has, however, not been clearly established although reduced food intake seems to be a part of this mechanism.

The aim of this chapter is to review a series of experiments conducted in our department to elucidate some of the aspects of selenium-induced growth retardation (Thorlacius-Ussing et al. 1987, 1988a, b). These studies have mainly focused on the growth hormone (GH) and somatomedin C (SMC) secretion since earlier histochemical studies have demonstrated the presence of catalytic selenium bonds in the secretory granules of particularly the growth hormone producing cells in the anterior pituitary (Thorlacius-Ussing and Danscher 1985).

2 Growth, Growth Hormone and Somatomedin C Level in Selenite-Treated Rats

In the first study nursing rats were treated with drinking water containing 5 ppm selenium as selenite from the day of delivery. Fifteen male rats, 21-days-old, were removed from their mothers and continued with the same drinking water as their mothers, whereas controls received normal water during the entire experimental period. This treatment induced a highly significant growth retardation. At the end of the experiment (day 56) the weight was 119±9 g in selenium-treated and

[1] Institute of Neurobiology, University of Aarhus, 8000 Aarhus, Denmark
[2] Second University Clinic of Internal Medicine and Institute of Experimental Clinical Research, Aarhus Municipal Hospital, 8000 Aarhus, Denmark
[3] Department of Rheumatology, Orthopedic Hospital, 8000 Aarhus, Denmark
[4] Danish Cancer Society, Department of Nutritional Research, 8000 Aarhus, Denmark
[5] To whom correspondence should be addressed

223±9 g in controls. In order to estimate the GH production the rats were anaesthetized with sodium barbital and 1 µg of growth hormone releasing factor 44 (GRF 44) was injected in the jugular vein. The GRF 44-stimulated GH secretion, 5 min after injection of GRF 44, was in the selenium-treated rats reduced by 77%. The SMC in serum was reduced even more, i.e., 83%. From this experiment it was concluded that 5 ppm selenium in drinking water as selenite induced a highly significant reduction in the stimulated GH release and serum SMC.

3 The Effect of GH Treatment on Growth, GH and SMC

Treatment with exogenous GH in post-weanling rats exposed to 3.3 ppm selenium as selenite was able to restore normal growth in the rats, since the growth rate in controls and those treated with both selenite and GH was practically identical. Whereas the GH treatment was able to restore growth, it was unable to restore normal SMC production. Indeed, the SMC level remained at an identically low level in the control group and the group treated with both selenite and GH. From this experiment it was concluded that GH treatment was able to restore normal growth in the rats but not normal SMC production.

4 The Effect of Discontinuation of Selenium Treatment on Growth, GH and SMC

In this third experiment three groups of rats were exposed to selenite in the same manner as in the previous experiment. One group was continuously exposed for 6 weeks, the second group only for the first 3 weeks and the last untreated group served as controls. Discontinuation of selenite treatment resulted in an instant growth spurt followed by a growth rate significantly lower than the controls and significantly higher than the rats continuously treated. At the time of discontinuation of treatment the GH release was, as expected, lower in the selenite-treated animals compared to the controls, but this difference had disappeared 3 weeks after cessation of the selenite treatment. This was, however, not the case for SMC which stayed at a low level independent of the cessation of selenite treatment.

5 SMC Level in Humans Exposed to Organic Selenium

The question is whether similar effects appear in humans exposed to selenium. We had the opportunity to study blood samples obtained from a clinical trial in which patients with rheumatic disease were randomized to receive either 256 µg organic selenium from yeast or a placebo over 6 months. Compared to paired controls, the selenium-treated patients had a significantly lower SMC level during the entire period (Fig. 1).

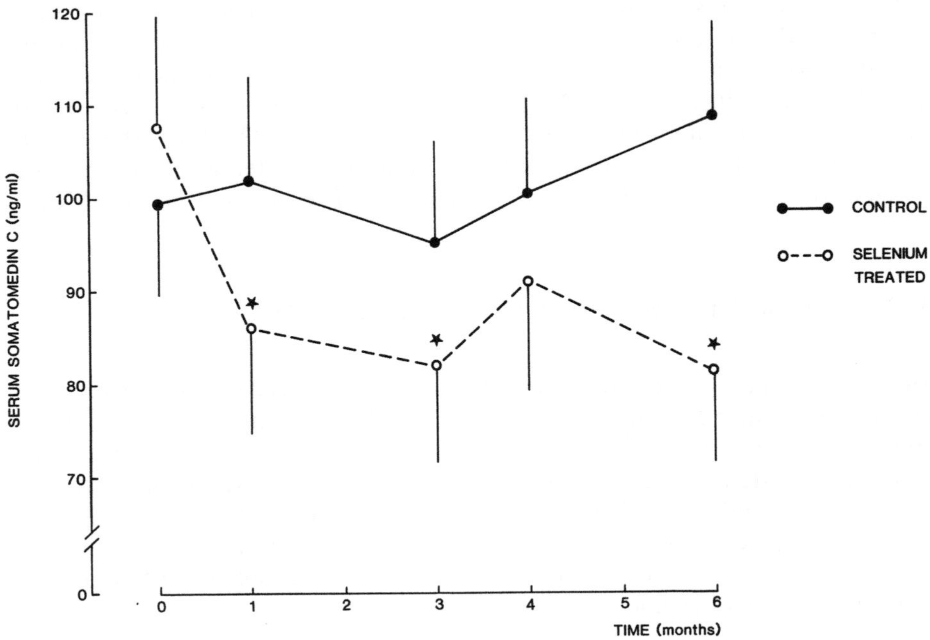

Fig. 1. The serum SMC$\pm$SC in rheumatic patients treated with 256 µg organic selenium for 6 months. (○) Selenium treated; (●) controls; (*) $p<0.05$, between selenium-treated and untreated patients

6 Concluding Remarks

The present series of studies indicate that in young experimental animals selenium in the form of selenite is able to induce a substantial decrease in GH secretion and serum SMC, resulting in pronounced growth retardation. Whereas GH secretion returns to normal after cessation of selenite treatment and normal growth can be restored in selenium-exposed rats by simultaneous treatment with GH, serum SMC stayed at a low level independent of these factors. These observations may suggest that the SMC production is very sensitive to the toxic effects of selenite and the suppression of SMC may indicate the degree of selenite toxicity.

The results from the clinical trial may suggest that some of the effects observed in experimental animals could occur also in humans. Further studies are, however, needed before drawing any final conclusion on this subject. These studies seem especially important when considering the selenium supplementations in children or pregnant women since these two groups are very sensitive to changes in the SMC production.

Acknowledgements. The studies in this review were supported by the Aarhus University Research Foundation, the Danish Medical Research Council, the Nordic Insulin Foundation and NOVO's Foundation.

References

Moxon AL (1937) Alkali disease or selenium poisoning. S D Agric Exp Stn Bull 311:1–91
National Research Council (1976) Selenium. Natl Acad Sci, Washington DC, pp 115–116
Thorlacius-Ussing O, Danscher G (1985) Selenium in the anterior pituitary of rats exposed to sodium selenite: light and electron microscopic localization. Toxicol Appl Pharmacol 81:67–74
Thorlacius-Ussing O, Flyvbjerg A, Esmann J (1987) Evidence that selenium induces growth retardation through reduced growth hormone and somatomedin C production. Endocrinology 120:659–663
Thorlacius-Ussing O, Flyvbjerg A, Damm Jørgensen K, Ørskov H (1988a) Growth hormone restores normal growth in selenium-treated rats without increase in circulating somatomedin C. Acta Endocrinol (Copenh) 117:65–72
Thorlacius-Ussing O, Flyvbjerg A, Ørskov H (1988b) Growth in young rats after termination of sodium selenite exposure: studies of growth hormone and somatomedin C. Toxicology 48:167–176

Selenium in the Central Nervous System of the Rat After Exposure to 75-Se-L-Selenomethionine

H. GRØNBÆK[1,3] and O. THORLACIUS-USSING[2]

1 Introduction

The toxicity of selenium was well known long before selenium was established as an essential trace element to man and animals. Especially the central nervous system seems to be sensitive to this toxicity, since neurological symptoms like ataxia, motor incoordination, hyperreflexia, paralysis, depression, anorexia and lack of vitality are outstanding symptoms in animals exposed to toxic amounts of selenium (Moxon 1937).

In 1973 Brown and Burk observed a remarkably high retention of 75-Se sodium selenite in the central nervous system of selenium-deficient rats. These results have later been confirmed by others (Trapp and Millam 1974). In 1981 Ammar and Couri demonstrated inorganic sodium selenite to be much more toxic to the central nervous system compared to selenomethionine.

Only little information is, however, available concerning accumulation of organic selenium in the central nervous system of the rat. The aim of this very preliminary study is to investigate the accumulation of selenium in the central nervous system of rats exposed to toxic amounts of L-selenomethionine.

2 Materials and Methods

For this preliminary study we used 13 male Wistar rats divided into two groups.

Nine rats (group A) had at $t = 0$ an intraperitoneal injection of 75-Se-L-selenomethionine (1.7 mg Se kg^{-1} and 862 kBq kg^{-1} body weight). Less than 10 min after the injection the whole body content of selenium was estimated. This value was referred to as 100%. Two, 24 and 48 h later three rats in each group were recounted to determine the whole body content of selenium and subsequently exsanguinated by v. porta puncture. Cerebellum, the cerebral hemisphere and the spinal cord were removed and the wet weight recorded before the selenium content was determined using a well-counter.

[1] Institute of Neurobiology, University of Aarhus, 8000 Aarhus C, Denmark
[2] Department of Surgery I, Aarhus County Hospital, 8000 Aarhus C, Denmark
[3] To whom correspondence should be addressed

Four rats (group B) were exposed to L-selenomethionine in the drinking water for 6 days. The drinking water contained 3.0 mg Se l^{-1} and 3.7 Mbq l^{-1}. At day 6 the rats were whole body counted to determine the total content of labelled selenium and afterwards the rats were exsanguinated. Cerebellum, the cerebral hemisphere and the spinal cord were removed, and the wet weight recorded before the content of labelled selenium was determined using a well-counter. All estimates are corrected for background radiation and physical decay of the isotope by simultaneous counting of a standard with a known selenium content.

3 Results

Selenium accumulated in the central nervous system of rats exposed to L-selenomethionine independent of the route of administration. In study A the highest selenium concentration was observed 2 h after the injection. Most selenium was observed in cerebellum [0.79±0.09 (SD) μg Se g^{-1} wet weight] followed by the spinal cord (0.7±0.1) and the cerebral hemisphere (0.62±0.04) (Table 1).

Generally, the selenium content in the individual organs decreased slightly during the first 48 h. Most selenium disappeared from medulla since only 63% of the initial amount was observed 48 h after the injection. In contrast, cerebellum contained 76% and the cerebral hemisphere 84% of the initial amount at the end of the study. The whole body content, however, decreased to 56% during the experimental period.

In study B cerebellum had the highest selenium concentration [0.49±0.05 (SD) μg Se g^{-1} wet weight] after 6 days of exposure followed by the cerebral

Table 1. Selenium concentration in the central nervous system[a]

Time (h)	Cerebellum	Cerebral hemisphere	Spinal cord	Whole body
2	0.79±0.09	0.62±0.04	0.70±0.1	1.61±0.03
24	0.68±0.01	0.58±0.02	0.45±0.01	1.06±0.06
48	0.60±0.06	0.52±0.05	0.44±0.04	0.90±0.09

[a] Selenium concentration in the organs investigated after an intraperitoneal injection of 1.7 mg Se kg^{-1} (μg Se g^{-1} wet weight ± SD)

Table 2. Selenium concentration in the central nervous system[a]

Cerebellum	0.49±0.08 (6 days)
Cerebral hemisphere	0.41±0.05 (6 days)
Spinal cord	0.36±0.02 (6 days)
Whole body	0.79±0.05 (6 days)

[a] Selenium concentration (μg Se g^{-1} wet weight ± SD) in the central nervous system after 6 days of exposure to L-selenomethionine (3.0 mg Se l^{-1} drinking water)

hemisphere [0.41 ± 0.08 (SD) μg Se g^{-1}] and the spinal cord [0.36 ± 0.02 (SD) μg Se g^{-1}]. The whole body content was estimated to be 0.79 ± 0.05 (SD) μg Se g^{-1} body weight at the end of the study (Table 2).

4 Discussion

This very preliminary study demonstrates that selenium accumulates in the central nervous system of rats after intraperitoneal and oral exposure to toxic doses of L-selenomethionine.

The highest concentration was observed in cerebellum followed by the nearly identical levels in the spinal cord and the cerebral hemisphere independent of the route of administration. The cerebellum seems to accumulate selenium to a higher extent than other parts of the brain. Trapp and Millam (1974) also observed the highest concentration of selenium in cerebellum in selenium-deficient rats given an intraperitoneal tracer dose of sodium selenite.

The mechanism behind the toxicity of selenium is unknown, indeed, the concentration of selenium seems to be at least partly regulated even during exposure to toxic doses, since the central nervous system contained less selenium than other organs including the blood (Grønbæk and Thorlacius-Ussing unpublished results).

The turnover of selenium in the central nervous system, however, seems to be slower than in other organs since the level of selenium in the central nervous system decreases much slower compared to whole body. This may point to an incorporation into important proteins in the central nervous system, since selenomethionine is known to be incorporated in other proteins than glutathione peroxidase (Beilstein and Whanger 1986). Indeed, Prohaska and Ganther (1976) have demonstrated the existence of selenium-containing proteins besides glutathione peroxidase in the central nervous system.

The slow disappearance of selenomethionine from the central nervous system in this preliminary study suggests the metabolism of at least organic selenium in the central nervous system to be different from most of the organs earlier focused on in the study of selenium metabolism. Long-term studies are in progress for further elucidation of this point.

References

Ammar EM, Couri D (1981) Acute toxicity of sodium selenite and selenomethionine in mice after ICV or IV administration. Neurotoxicology 2:383–386

Beilstein MA, Whanger PD (1986) Chemical forms of selenium in rat tissues after administration of selenite and selenomethionine. J Nutr 116:1711–1719

Brown DG, Burk RF (1973) Selenium in tissues and sperm of rats fed a torula yeast diet. J Nutr 102:102–108

Moxon AL (1937) Alkali disease or selenium poisoning. S D Agric Exp Stn Bull 311:1–91

Prohaska JR, Ganther HE (1976) Selenium and glutathione peroxidase in developing rat brain. J Neurochem 27:1379–1387

Trapp GA, Millam J (1974) The distribution of 75-Se in brains of selenium deficient rats. J Neurochem 24:593–595

Effect of Selenium on Adriamycin-Induced Cardiotoxicity in Mice

XING CHENG[1] and GUANG-LU XU[1,2]

1 Introduction

One of the most effective antitumor drugs is adriamycin (ADR) which produces a dose-related cardiomyopathy that can limit its clinical use. The ADR-induced, subacute cardiomyopathy seen in the mouse is morphologically similar to that in man (Rosenoff et al. 1975). Several studies have shown that ADR administration in mice leads to an increase in the lipid peroxide (LPO) content of the heart (Myers et al. 1977; Lentzhofer et al. 1983) and produces a decrease in cardiac glutathione peroxidase (GSH-Px) activity (Doroshow et al. 1980). ADR administration in selenium (Se)-deficient mice results in an enhanced toxicity (Doroshow et al. 1980).

The present studies were undertaken (1) to investigate ADR-induced subacute cardiotoxicity in mice in synchrony with its relationship to the endogenous defense of the heart against reactive oxygen metabolites, and (2) to evaluate the ability of Se pretreatment to provide protection against cardiomyopathy.

2 Materials and Methods

In our study 162 Swiss male mice weighing 30 g were divided into ADR, ADR with Se pretreatment (ADR+Se) and control groups of 54 mice each, and supplied with a stock diet ad libitum. The ADR group was given an i.p. injection of 17.5 mg ADR kg^{-1} body wt. The ADR+Se group received 80 μg Se kg^{-1} body wt. i.p. daily for 7 days before ADR, and then every day until sacrifice. The control group was injected with an equal volume of saline. Each group was subdivided into three subgroups of 18 mice each, and the mice were killed at 24, 48 and 96 h after the first dose of ADR, respectively.

Samples of blood and heart were collected, and each sample consisted of the pooled blood or hearts of three mice for conducting histopathologic studies stained with hematoxylin and eosin, and diastase-periodic acid-Schiff (D-PAS) methods, for histochemical examinations of succinic dehydrogenase (SDH) by the

[1] Research Laboratory of Keshan Disease, Xi'an Medical University, Xi'an 710061, Shaanxi, The People's Republic of China
[2] To whom correspondence should be addressed

method of nitroneotetrazolium chloride and lipid of bromine sudan black, and for measurements of the following substances: Se by 2,3-DAN fluorometric method, GSH-Px by NADPH-coupled assay, superoxide dismutase (SOD) by fericytochrome C assay, LPO by thiobarbituric acid reaction and plasma vitamin E fluorometrically.

3 Results

The ADR-treated mice, as compared to the control group, showed no significant alterations in cardiac SOD activity and plasma vitamin E level at 24, 48 and 96 h (Tables 1 and 2). No significant changes were seen in cardiac Se content at 24 and 96 h, but it was reduced apparently at 48 h (Table 3). ADR caused remarkable depressions in cardiac GSH-Px activity (Table 4) and elevations in cardiac LPO level at 24 and 48 h (Table 5). The decreased GSH-Px activity persisted for 72 h, and was restored to the control level at 96 h (Table 4), while the cardiac LPO reached a peak level (Table 5), indicating that the observed increase in lipid peroxide in the heart of ADR-treated mice might not be the result of a decrease in GSH-Px activity. The hearts in part of the mice (4/7) had histochemical evidence of early myocardial injury associated with increased permeability of cell membranes

Table 1. Cardiac SOD (U mg^{-1} protein)

Group	Hours after ADR administration*		
	24	48	96
Control	23.1 ± 0.7^a	23.8 ± 2.2^a	21.3 ± 0.7^a
ADR	22.5 ± 2.4^a	23.0 ± 2.3^a	22.0 ± 1.0^a
ADR + Se	22.5 ± 1.0^a	23.8 ± 1.3^a	21.6 ± 1.3^a

* $\bar{X} \pm SD$; N = 6. Means with different superscripts in the same vertical column are significantly different ($p < 0.05$ or 0.01)

Table 2. Plasma vitamin E (μg ml^{-1})

Group	Hours after ADR administration*		
	24	48	96
Control	3.1 ± 0.7^a	3.0 ± 0.9^a	3.1 ± 0.7^a
ADR	2.8 ± 0.7^a	3.2 ± 0.4^a	2.9 ± 0.9^a
ADR + Se	3.2 ± 0.6^a	3.0 ± 0.7^a	3.0 ± 0.8^a

* $\bar{X} \pm SD$; N = 6. Means with different superscripts in the same vertical column are significantly different ($p < 0.05$ or 0.01)

Table 3. Cardiac Se (ng g^{-1} wet weight)

Group	Hours after ADR administration*		
	24	48	96
Control	116 ± 5^a	132 ± 6^a	142 ± 14^a
ADR	111 ± 5^a	118 ± 12^b	146 ± 14^a
ADR + Se	154 ± 13^b	170 ± 15^c	189 ± 16^b

* $\bar{X} \pm SD$; N = 6. Means with different superscripts in the same vertical column are significantly different ($p < 0.05$ or 0.01)

Table 4. Cardiac GSH-Px (mU mg^{-1} protein)

Group	Hours after ADR administration*		
	24	48	96
Control	54 ± 2^a	56 ± 2^a	52 ± 3^a
ADR	43 ± 2^b	40 ± 5^b	52 ± 4^a
ADR + Se	68 ± 4^c	65 ± 6^c	79 ± 4^b

* $\bar{X} \pm SD$; N = 6. Means with different superscripts in the same vertical column are significantly different ($p < 0.05$ or 0.01)

Table 5. Cardiac LPO (pmol MDA mg^{-1} protein)

Group	Hours after ADR administration*		
	24	48	96
Control	599 ± 73^a	527 ± 70^a	656 ± 77^a
ADR	755 ± 80^b	1117 ± 240^b	1626 ± 198^b
ADR + Se	$696 \pm 100^{a,b}$	771 ± 133^c	1481 ± 103^b

* $\bar{X} \pm SD$; N = 6. Means with different superscripts in the same vertical column are significantly different ($p < 0.05$ or 0.01)

as evidenced by D-PAS positivity. Light microscopic cardiomyopathic changes were not noted. No appreciable changes in SDH reaction and lipid staining intensity were seen.

In the ADR + Se group, significant increases of cardiac Se content and GSH-Px activity were noted (Tables 3 and 4), and cardiac SOD activities and plasma vitamin E levels were unaffected in the 24, 48 and 96 h mice (Tables 1 and 2). At 24 h, there was an insignificant decrease in LPO concentration (Table 5). The cardiac LPO was significantly reduced at 48 h, but returned to a higher level as that in ADR mice at 96 h (Table 5). Se pretreatment produced protection against

ADR-induced early myocardial injury as indicated by the lack of apparent D-PAS staining material noted in sarcoplasma. The intensities of staining of SDH and lipid were quite comparable to those of the ADR group.

4 Conclusions

The results indicate that (1) an ADR-induced early cardiac lesion in mice might be associated with increased cardiac lipid peroxidation, (2) the fact that there exists no unequivocally time-dependent correlation between the inhibition of GSH-Px and elevation of lipid peroxidation caused by ADR suggests that both changes appear to be the result of an ADR-generated increase of free-radical metabolites, (3) pretreatment with Se prevents mice from ADR-induced, early myocardiac damage, and (4) the preventive effect of Se would not be simply explained as the detoxication of LPO via the action of enhanced Se-GSH-Px activity.

Acknowledgement. We are indebted to N.-Z. Chen, W.-L. Xue, Y.-Q. Xue, C.-H. Tang and J.-P. Tian for their valuable advice and technical assistance.

References

Doroshow JH, Locker GY, Myers CE (1980) Enzymatic defenses of the mouse heart against reactive oxygen metabolite alternations produced by doxorubicin. J Clin Invest 65:128–135
Lentzhofer R, Magometschnigg D, Dudezak R, Cerni C, Bolebruch C, Moser K (1983) Indication of reduced doxorubicin-induced cardiac toxicity by additional treatment with antioxidative substances. Experientia 39:62–64
Myers CE, McGuire WP, Liss R, Ifrim I, Grotzinger K, Young RC (1977) Adriamycin: the role of lipid peroxidation in cardiac toxicity and tumor response. Science 197:165–167
Rosenoff SH, Olson HM, Young DM, Bostick F, Young RC (1975) Adriamycin-induced cardiac damage in the mice: a small animal model of cardiotoxicity (Brief communication). J Natl Cancer Inst 55:191–194

Effects of Selenium Deficiency on Mouse Articular Cartilage

K. PLOETZ[1] and E. WALLACE[1,2]

1 Introduction

An edemic form of osteoarthritis, Kashin-Beck disease, has been reported to be associated with low levels of selenium (Se) in the soil, foods, and hair of individuals from areas where this condition is common (Li et al. 1987; Mo 1987). However, other agents have been implicated in its etiology (Sokoloff 1985). Ren (1982) and Mo (1987) fed rats diets low in Se and found no necrotic areas in the articular cartilage. Nevertheless, as we had observed deformities in Se-deficient mice in our previous work, we determined to examine this tissue. Light and transmission electron microscopy showed it was affected by Se deficiency.

2 Materials and Methods

Swiss-Webster mice were fed a commercial Se-deficient diet, <0.006 ppm Se (Dyets, Bethlehem, PA) for three generations following the protocol of Wallace et al. (1983). Control animals were given 0.1 ppm Se as sodium selenite in the drinking water. The animals were sacrificed between 6 and 7 months of age. Fresh epididymal sperm was examined by phase microscopy. The testes were prepared for light microscopy (LM) by standard procedures and stained with hematoxylin and eosin (H&E). The knee joints were fixed in buffered formalin, decalcified in formic acid (Sokoloff pers. commun.), embedded, and stained with H&E. Knee articular cartilage was removed, fixed in Karnovsky's solution, decalcified in NaEDTA for 3 weeks, embedded in Epon 812, sectioned, and stained with lead citrate and uranyl acetate by standard techniques for transmission electron microscopy (TEM).

3 Results and Discussion

The appearance of fresh sperm and seminiferous tubules from the second and third generation Se-deficient mice was typical of that from Se-deficient mice ob-

[1] Biology Department, William Paterson College of New Jersey, Wayne, NJ 07470, USA
[2] To whom correspondence should be addressed

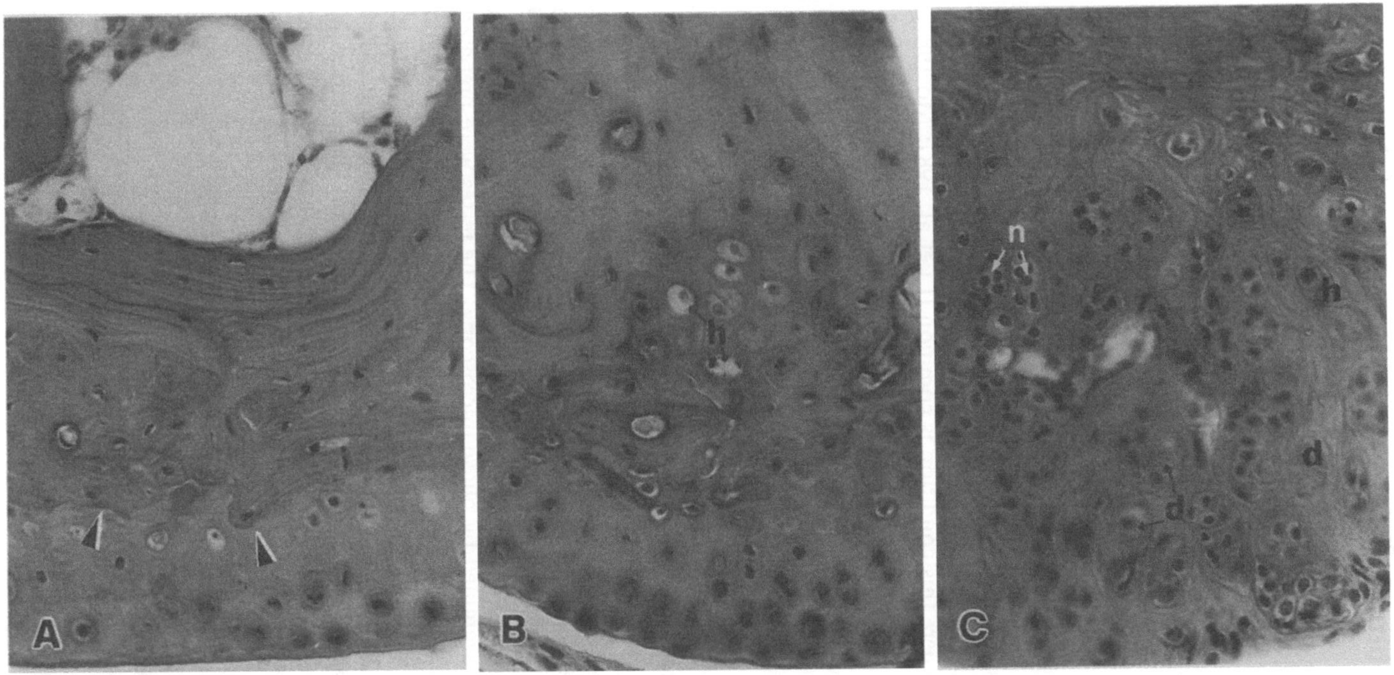

Fig. 1A–C. Photomicrographs of mouse articular cartilage, H&E (600×). Cartilage from the deficient animals (**B** and **C**) has more cell layers, and chondrocyte nests (*n*) frequently contain three or more cells. The demarkation (*arrowheads*) between cartilage and bone is clear in the control (**A**) but cannot be seen in the cartilage from Se-deficient mice as clusters of chondrocytes are found beneath bone. Patchy areas of necrosing (*d*) and hypertrophic (*h*) chondrocytes are also seen in the deficient state

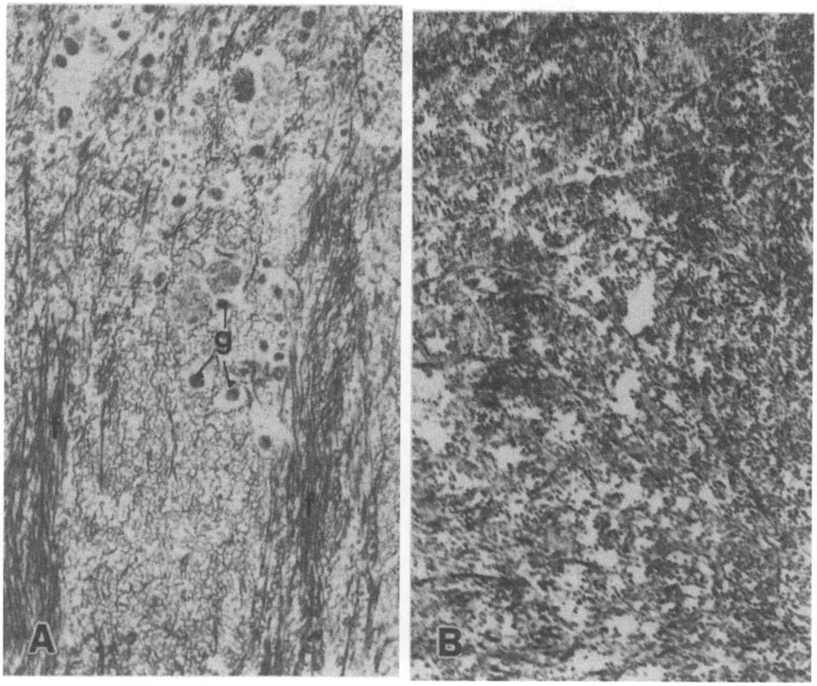

Fig. 2A, B. Transmission electron micrograph of interterritorial matrix (5400×). In the control (**A**) the collagen fibrils (*f*) are arranged in bundles interspersed with matrix granules (*g*). In contrast, in the deficient state (**B**) the fibrils are shorter, thinner, dispersed randomly, and there are fewer granules

served in earlier studies (Wallace et al. 1983, 1987) confirming the Se-deficient status of the animals used in this study.

Observation by LM showed articular cartilage of control mice four to five cell layers thick with well-defined areas of transformation (calcifying cartilage) before the bony trabeculae (Fig. 1 A). In contrast, the joint cartilage of third generation Se-deficient mice (Fig. 1 B, C) contained at least ten layers of cells and no definite transformation zone was seen before the trabecular bone shaft. Instead of the usual clear demarkation between cartilage and bone, the transition zone was disorganized; clusters of hypertrophic chondrocytes were found below ossified tissues.

TEM showed that cartilage from third generation Se-deficient mice (Fig. 2 B) had fewer fibrils than their control animals (Fig. 2 A) and these fibrils were shorter, thinner, and in disarray. There were fewer matrix granules giving the matrix a fenestrated appearance. Necrosing chondrocytes (Fig. 3 B) were frequently observed in cartilage from Se-deficient mice, as were hypertrophic chondrocytes. Other observations in the Se-deficient articular chondrocytes were an increased number of multivesiculated bodies, heterochromatic, irregularly shaped nuclei, increased lipid deposits in nuclei, swollen mitochondria, large deposits of glycogen and fibrils in maturing cells, and cytolysosomes. All of these findings

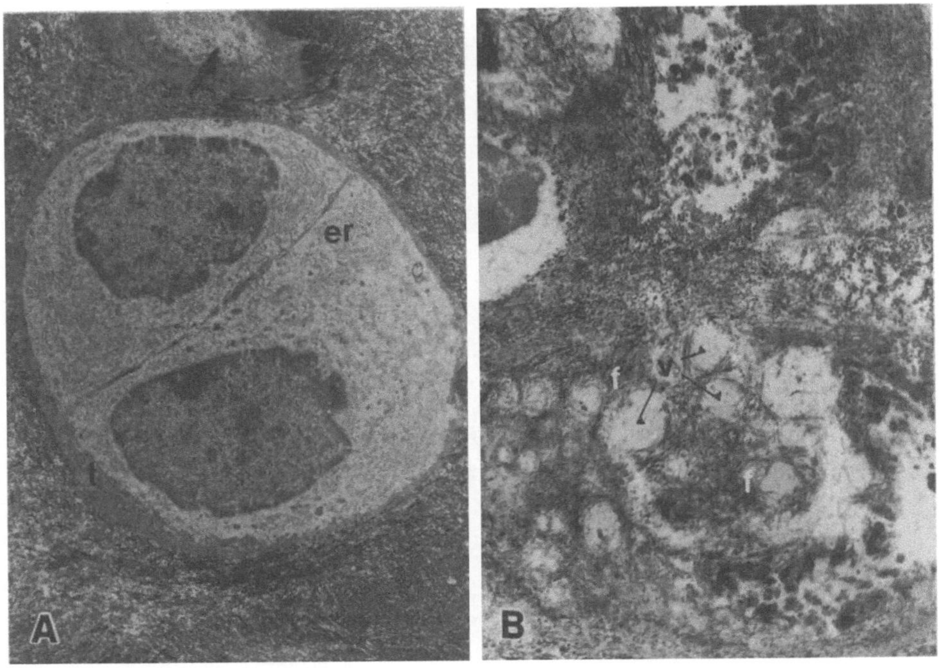

Fig. 3 A, B. Transmission electron micrographs of mouse articular chondrocytes (5400×). The nest of two maturing chondrocytes from the control (**A**) has intact nuclei, well-preserved cellular organelles (*er*), and clearly defined territorial matrix (*t*). A nest of three chondrocytes from a Se-deficient mouse (**B**) is necrosing as shown by disintegration of the nucleus and cytoplasmic organelles and an increased number of multivesiculated bodies (*v*) and fibrils (*f*)

are characteristic of some cells in normal mature cartilage. However, a greater number of necrosing cells were seen in the cartilage from Se-deficient mice.

We attempted to compare the cartilage from our Se-deficient mice with the description given by Mo (1987) of cartilage from individuals with Kashin-Beck disease. Our LM findings are similar: diminished metachromasia, patchy areas of hypertrophy and necrosis, smaller chondrocytes. The major difference is that our mouse tissue had more rather than fewer chondrocytes. Similarly, by TEM, we saw many signs of dying cells and loss of matrix integrity. Most importantly, in the articular cartilage, we found multiple focal chondronecrosis which Mo states is the most important pathological feature of Kashin-Beck disease.

Se-deficient rats raised by Mo (1987) and Ren (1982) did not show the necrotic alterations we found in the articular cartilage of our Se-deficient mice. Ren fed a Se-deficient diet to young rats for 30 days and found atrophy of the epiphyseal growth plate, which we did not examine. We did not see marked alterations in cartilage until the third generation. The different results could be due to species differences or the degree of Se deficiency. Wei et al. (1986) found no requirement for Se when growing rabbit articular or growth plate chondrocytes in culture. The effect of Se on cartilage may be more complex and indirect than can be demonstrated in tissue culture.

4 Summary and Conclusions

The mice used in this study were moderately Se-deficient as shown by their testicular histology. Light microscopy of knee articular cartilage showed evidence of both proliferation and degeneration. Ultrastructural changes as described in Section 3 are indicative of chondronecrosis. The pattern of multiple focal hypertrophic and necrosing chondrocytes is similar to that described by Mo (1987) and Sokoloff (1985) for Kashin-Beck disease. We believe the alterations in the knee articular cartilage of these mice are due to their Se-deficient status. Other joints and epiphyseal cartilage should be studied as well as cartilage from other species.

Acknowledgements. This work was supported by the A.R.T. program of William Paterson College and funds from the William Paterson College Center for Applied Science.

References

Li JY, Ren S-X, Cheng D-Z, Wan H-J, Liang S-T, Zhang F-J, Gao F-M (1987) Distribution of selenium in the microenvironment related to Kashin-Beck disease. In: Coombs GF, Spallholz JE, Levander OA, Oldfield JE (eds) Selenium in biology and medicine. AVI Van Nostrand Reinhold Company, New York, pp 911–923

Mo DX (1987) Pathology of Kaschin-Beck disease. In: Coombs GF, Spallholz JE, Levander OA, Oldfield JE (eds) Selenium in biology and medicine. AVI Van Nostrand Reinhold Company, New York, pp 924–933

Ren HZ (1982) A histological study concerning the effects of trace element selenium on the development of metaphyseal cartilage of rat. Chin J Pathol 11:220–221, 44

Sokoloff L (1985) Endemic forms of osteoarthritis. Clin Rheum Dis 11:187–202

Wallace E, Calvin HI, Cooper GW (1983) Progressive defects observed in mouse sperm during the course of three generations of selenium deficiency. Gamete Res 4:377–387

Wallace E, Calvin HI, Ploetz K, Cooper GW (1987) Functional and developmental studies on the role of selenium in spermatogenesis. In: Coombs GF, Spallholz JE, Levander OA, Oldfield JE (eds) Selenium in biology and medicine. AVI Van Nostrand Reinhold Company, New York, pp 181–196

Wei X, Wright GC, Sokoloff L (1986) The effect of sodium selenite on chondrocytes in monolayer culture. Arthritis and Rheum 29:660–664

Myocardial and Pancreatic Damage in Selenium-Vitamin E Deficient Mice

J. F. Van Vleet[1], V. J. Ferrans[2]

1 Introduction

Selenium-vitamin E (Se-E) deficiency in animals produces a wide spectrum of target organ lesions including involvement of the heart, skeletal muscles, smooth muscle, liver, pancreas, stomach, adipose tissue, skin, brain, spinal cord, testicle, embryo, bone marrow and blood, eye, lung, kidney, and teeth (Van Vleet and Ferrans 1986). Cardiac lesions are a major finding in calves, lambs, pigs, turkey poults, and ducklings with Se-E deficiency (Hadlow 1973; Robinson and Maxie 1985; Scott et al. 1967; Van Vleet et al. 1970). In many other species, myocardial necrosis is inconsistently observed with Se-E deficiency. In most cases the lesions are detected microscopically but are not apparent grossly. Affected species include mice, rats, dogs, foals, mink, goats, guinea pigs, rabbits, rottnest quokka, and monkeys (DeWitt and Schwartz 1958; Lin and Chen 1982; Mason 1973; Mason and Horwitt 1972; Shakibi and Stone 1987).

Recent reports have established that selenium deficiency is associated with the development of congestive cardiomyopathy in Chinese patients with the naturally occurring form of the deficiency (Keshan disease) and in USA patients maintained on long-term parenteral hyperalimentation (Chen 1982; Chen et al. 1980; Ge et al. 1983; Gu 1983; Li et al. 1985; Yu 1982). Keshan disease is an endemic cardiomyopathy that occurs in a belt running from the northeast to the southwest of China and results from consumption of products with low selenium concentration from the soil-plant-animal-man food chain in affected areas. Patients have low blood and hair selenium content. Cases are generally found in peasants, mostly in children and women of child-bearing age. In fatal cases, the hearts show biventricular dilatation; mural thrombi may be present. Histologically, myocardial necrosis with contraction bands and mitochondrial calcification is seen in early, acute lesions and postnecrotic fibrosis is present in chronic cases. Administration of selenium supplements, such as sodium selenite tablets or soybean supplements, has provided protection in endemic areas of China.

Congestive cardiomyopathy has also been reported in a few human patients with low selenium status following long-term parenteral hyperalimentation (Col-

[1] Department of Pathobiology, School of Veterinary Medicine, Purdue University, West Lafayette, IN 47907, USA
[2] Pathology Branch, National Heart, Lung and Blood Institute, National Institutes of Health, Bethesda, MD 20892, USA

Table 1. Experimental design – Experiment 1

Mouse strain	Weeks on BD + Ag diet[a]		
	15	20	25
CD-1	15[b]	–	–
$C_{57}BL$	15	15	15
DBA	15	15	15

[a] BD + Ag diet: torula yeast-based semipurified basal diet with 0.1% Ag as silver acetate.
[b] Female weanling mice.

Table 2. Experimental design – Experiment 2

Mouse strain	Diet			
	BD + Ag[a]	BD + Ag + Se[b]	BD + Ag + E[c]	BD + Ag + Se + E[d]
CD-1	15[e]	15	15	15
$C_{57}BL$	15	15	15	15

[a] BD + Ag diet: torula yeast-based semipurified basal diet with 0.1% Ag as silver acetate.
[b] Se (1 mg kg^{-1}).
[c] E (250 mg kg^{-1}).
[d] Se (1 mg kg^{-1}), E (250 mg kg^{-1}).
[e] Female weanling mice.

Table 3. Mortality – Experiment 1

Mouse strain	Mortality (%)		
	15 weeks	20 weeks	25 weeks
CD-1	90	–	–
$C_{57}BL$	13.5	27	93
DBA	12	47	100

Table 4. Cardiac lesions – Experiment 1

Mouse strain	Myocardial necrosis/fibrosis					
	15 weeks		20 weeks		25 weeks	
	Frequency (%)	Severity[a]	Frequency (%)	Severity	Frequency (%)	Severity
CD-1	100	2.27	–	–	–	–
$C_{57}BL$	100	2.56	100	2.38	100	2.64
DBA	93	2.20	100	3.07	100	3.00

[a] Severity scores: 0 = no damage; 1 = focal damage; 2 = mild, diffuse damage; 3 = moderate, diffuse damage; 4 = marked, diffuse damage.

lip and Chen 1981; Fleming et al. 1982; Johnson et al. 1981). Also, cardiomyopathy may develop in human patients in whom vitamin E deficiency is presumed to be induced by chronic intestinal lipid malabsorption syndromes, such as in cystic fibrosis, Byler's disease, and Basson-Kornzweig syndrome (Dische and Porro 1970; Hide and Martlow 1977; Nezelof and LeSec 1979; Saito et al. 1982).

In the present study, selenium-vitamin E (Se-E) deficiency was studied in three mouse strains to seek a suitable model for the cardiac lesions of Keshan disease in human patients.

2 Materials and Methods

The deficiency disease was induced by feeding ad libitum a torula yeast-based semipurified basal diet with 0.1% silver added as silver acetate (BD+Ag) for 15 weeks. In the first experiment (Table 1), female weanling mice of the CD-1, C_{57}BL, and DBA strains were allotted to six groups of 15 each and fed either BD+Ag or a complete commercial control diet for 15 weeks. In the second experiment (Table 2), eight groups of 15 each of either CD-1 or C_{57}BL female weanling mice were fed ad libitum either BD+Ag, BD+Ag+Se (1 mg kg^{-1}, BD+Ag+E (250 mg kg^{-1}), or BD+Ag+Se (1 mg kg^{-1})+E (250 mg kg^{-1}) for 15 weeks. The clinical and pathological features of Se-E deficiency were characterized by macroscopic and microscopic evaluation. The cardiac alterations were evaluated ultrastructurally.

3 Results

The results of these studies show that considerable variation exists among the three mouse strains in their susceptibility to induction of Se-E deficiency and in the types of lesions produced in deficient animals. Susceptibility was highest in

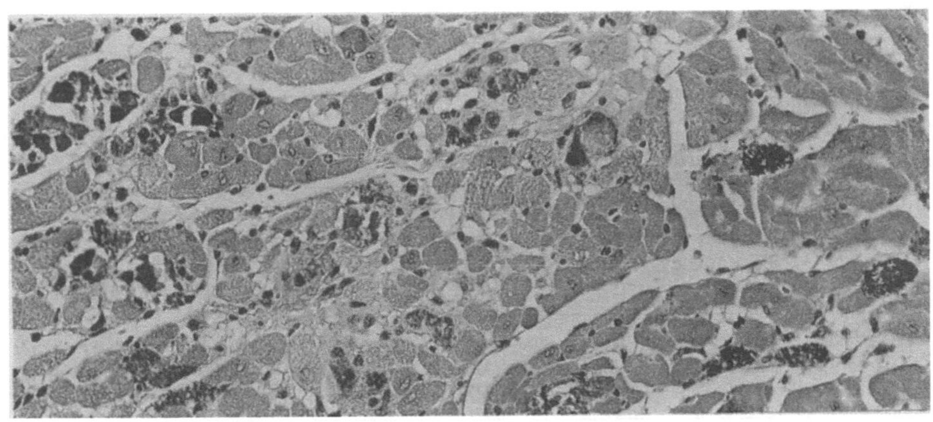

Fig. 1. Disseminated myocardial fibers show necrosis with dark, granular, mineralized deposits. Left ventricle wall of Se-E deficient mouse fed BD+Ag. Hematoxylin and eosin stain; ×250

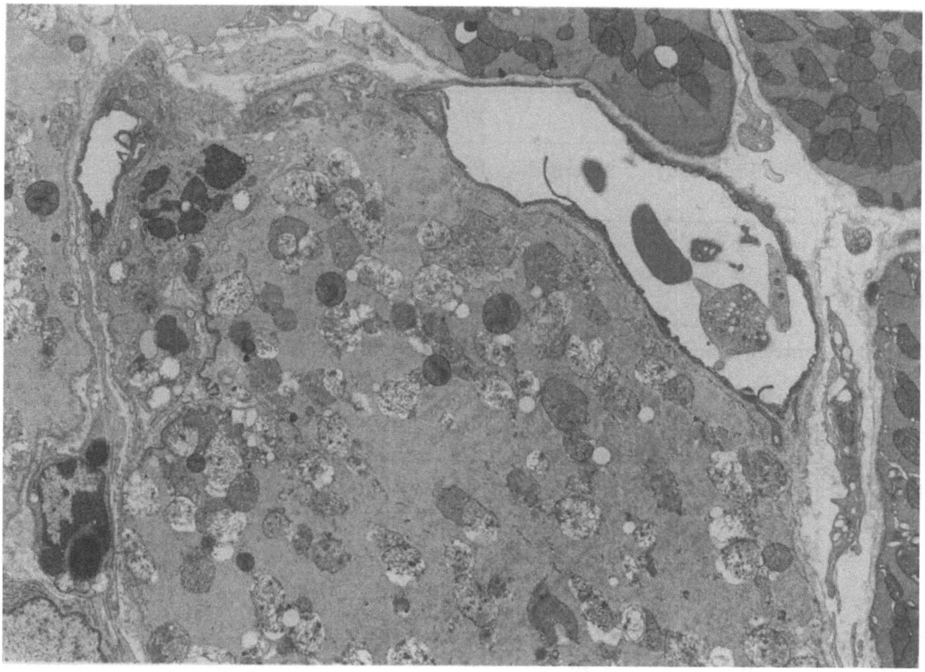

Fig. 2. Electron micrograph shows necrotic cardiac muscle cells and several intact myocytes (*right*). The necrotic fibers show swollen, disrupted mitochondria and dense, fibrillar masses of disrupted, contractile material. Left ventricular wall of Se-E deficient mouse fed BD+Ag; ×7500

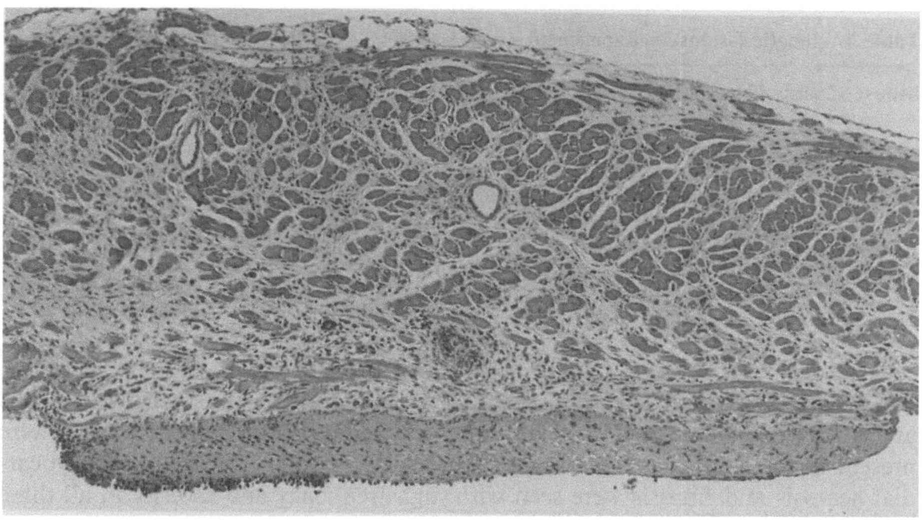

Fig. 3. Right ventricular wall has marked, diffuse, myocardial fibrosis with a flat, fibrin thrombus overlying the endocardium. Se-E deficient mouse fed BD+Ag. Hematoxylin and eosin stain; ×40

Table 5. Cardiac lesions – Experiment 1

Mouse strain	Endocardial thrombosis					
	15 weeks		20 weeks		25 weeks	
	Frequency (%)	Severity[a]	Frequency (%)	Severity	Frequency (%)	Severity
CD-1	3.5	0.04	–	–	–	–
$C_{57}BL$	60	0.90	100	1.92	93	1.64
DBA	60	1.00	27	0.27	40	0.60

[a] Severity score: 0 = no thrombi; 1 = few, small thrombi; 2 = multiple, large thrombi.

Table 6. Pancreatic lesions – Experiment 1

Mouse strain	Pancreatic necrosis					
	15 weeks		20 weeks		25 weeks	
	Frequency (%)	Severity[a]	Frequency (%)	Severity	Frequency (%)	Severity
CD-1	65	1.90	–	–	–	–
$C_{57}BL$	10	0.17	8	0.08	14	0.14
DBA	47	1.07	73	2.20	100	3.73

[a] Severity scores: 0 = no damage; 1 = focal damage; 2 = mild, diffuse damage; 3 = moderate, diffuse damage; 4 = marked, diffuse damage.

Table 7. Hepatic lesions – Experiment 1

Mouse strain	Hepatic necrosis		
	15 weeks	20 weeks	25 weeks
	Frequency (%)	Frequency (%)	Frequency (%)
CD-1	45	–	–
$C_{57}BL$	20	31	87
DBA	27	20	60

CD-1 mice (90% mortality) and relatively low (15–20% mortality) for $C_{57}BL$ and DBA mice (Table 3). Gross and microscopic lesions of Se-E deficiency were present in the heart, pancreas, liver, and skeletal muscles. Disseminated myocardial necrosis and fibrosis were seen with high frequency and severity in all three strains (Table 4; Figs. 1–3). Endocardial thrombosis occurred with high frequency and severity in DBA and $C_{57}BL$ strains but was only rarely seen in CD-1 mice (Table 5; Fig. 3). Pancreatic necrosis and fibrosis occurred with high frequency and severity in DBA and CD-1 strains but only with very mild frequency and

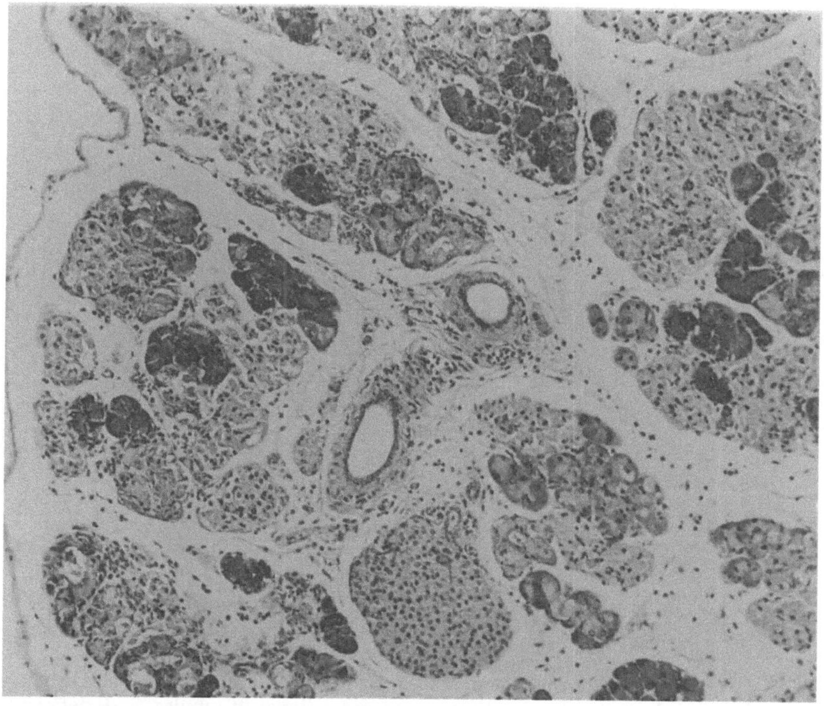

Fig. 4. Pancreas shows marked necrosis of the glandular acini and preservation of the duct system and islet tissue. A few scattered, dark acini remain intact. Se-E deficient mouse fed BD + Ag. Hematoxylin and eosin stain; ×100

Table 8. Skeletal muscle lesions – Experiment 1

Mouse strain	Skeletal muscle necrosis					
	15 weeks		20 weeks		25 weeks	
	Frequency (%)	Severity[a]	Frequency (%)	Severity	Frequency (%)	Severity
CD-1	38	1.04	–	–	–	–
$C_{57}BL$	27	0.47	77	1.54	93	1.73
DBA	7	0.07	7	0.13	40	0.73

[a] Severity scores: 0 = no damage; 1 = focal damage; 2 = mild, diffuse damage; 3 = moderate, diffuse damage; 4 = marked, diffuse damage.

Table 9. Mortality – Experiment 2

Mouse strain	Mortality (%)			
	BD + Ag	BD + Ag + Se	BD + Ag + E	BD + Ag + Se + E
CD-1	80	0	7	7
$C_{57}BL$	27	0	0	0

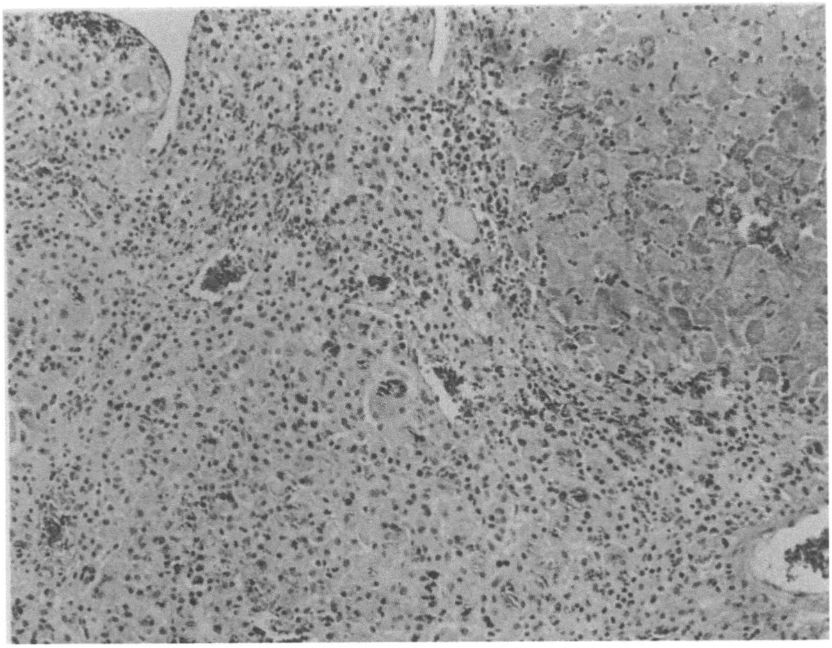

Fig. 5. Severe hepatic necrosis is evident as areas of acute damage (*right*) and extensive adjacent areas of chronic injury with parenchymal loss and collapse of lobules. Se-E deficient mouse fed BD+Ag. Hematoxylin and eosin stain; ×250

Table 10. Cardiac lesions – Experiment 2

Mouse strain	BD + Ag		BD + Ag + Se		BD + Ag + E		BD + Ag + Se + E	
	Frequency (%)	Severity[a]	Frequency (%)	Severity	Frequency (%)	Severity	Frequency (%)	Severity
CD-1	100	2.40	40	0.53	0	0.00	0	0.00
$C_{57}BL$	100	2.33	27	0.53	7	0.07	0	0.00

[a] Severity scores: 0 = no damage; 1 = focal damage; 2 = mild, diffuse damage; 3 = moderate, diffuse damage; 4 = marked, diffuse damage.

Table 11. Summary of differences among mouse strains to develop lesions of Se-E deficiency[a]

Strain	Mortality	Myocardial necrosis/fibrosis	Endocardial thrombosis	Hepatic necrosis	Pancreatic necrosis	Skeletal muscle necrosis
$C_{57}BL$	+	+ + +	+ +	+ +	±	+ / + +
CD-1	+ + +	+ + +	±	+ +	+ + +	+ / + +
DBA	+	+ + +	+ +	+ +	+ + +	±

[a] Severity scores: –, absent; +, mild; + +, moderate; + + +, marked.

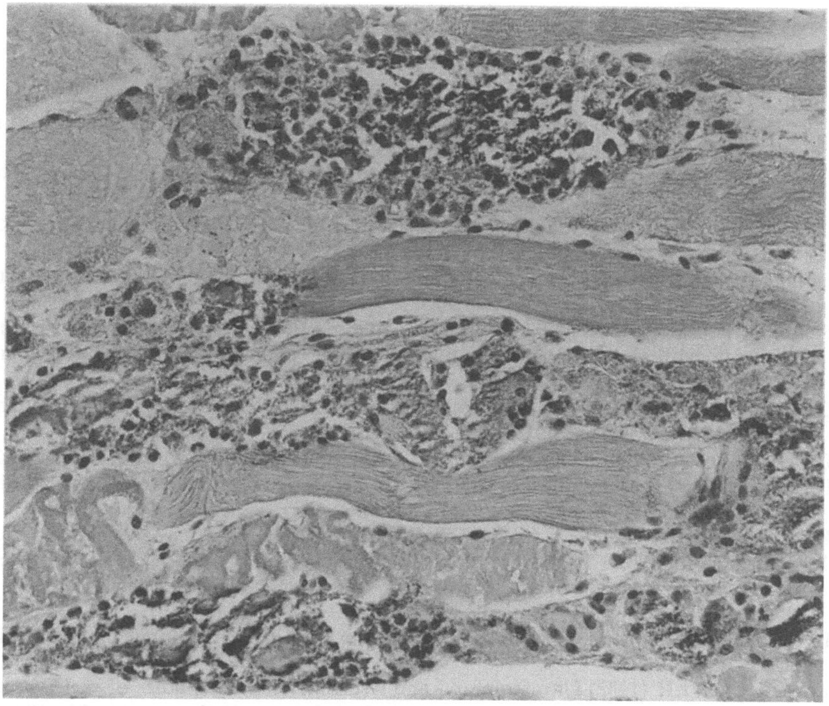

Fig. 6. Scattered necrotic skeletal muscle fibers show hypereosinophilic masses of disrupted sarcoplasm with dark, granular, mineralized deposits and invading macrophages. Se-E deficient mouse fed BD+Ag. Hematoxylin and eosin stain; ×400

severity in $C_{57}BL$ mice (Table 6; Fig. 4). Hepatic necrosis was observed with moderate frequency in all three strains (Table 7; Fig. 5). Skeletal muscle necrosis occurred at low to moderate frequency and severity in $C_{57}BL$ and CD-1 mice but had very low frequency and severity in DBA mice (Table 8; Fig. 6). The protection study revealed that supplements of either E or Se+E gave complete protection against development of the deficiency disease and Se supplementation provided nearly complete protection (Tables 9 and 10).

4 Conclusions

These studies show that it is possible to induce Se-E deficiency in weanling mice fed semipurified deficient diets with added silver. Considerable variation occurs among mouse strains in the susceptibility to Se-E deficiency and in the frequency and severity of target organ lesions (Table 11). Some strains developed severe pancreatic damage as described in Se-E deficient chickens. All strains had extensive myocardial damage but the $C_{57}BL$ strain appears to offer the best model of the cardiac lesions of Keshan disease as survival was prolonged and pancreatic damage was minimal in this strain.

References

Chen XS (1982) Selenium and Keshan disease. Ann NY Acad Sci 393:224–225

Chen XX, Yang G, Chen J, Chen X, Wen Z, Ge K (1980) Studies on the relations of selenium and Keshan disease. Biol Trace Element Res 2:91–107

Collip PJ, Chen SY (1981) Cardiomyopathy and selenium deficiency in a two-year-old girl. N Engl J Med 304:1304–1305

DeWitt WB, Schwartz K (1958) Multiple dietary necrotic degeneration in the mouse. Experientia 14:28–30

Dische MR, Porro RS (1970) The cardiac lesions in Bassen-Kornzweig syndrome. Am J Med 49:568–571

Fleming CR, Lie JT, McCall JT, O'Brien JR, Baillie EE, Thistle JL (1982) Selenium deficiency and fatal cardiomyopathy in a patient on home parenteral nutrition. Gastroenterology 83:689–693

Ge K, Xue A, Bai J, Wang S (1983) Keshan disease: an endemic cardiomyopathy in China. Virchows Arch Pathol Anat 401:1–15

Gu B (1983) Pathology of Keshan disease. A comprehensive review. Chin Med J 96:251–261

Hadlow WJ (1973) Myopathies of animals. In: The striated muscle. Int Acad Pathol Monogr No 12. Williams & Wilkins, Baltimore, pp 364–409

Hide DW, Martlow R (1977) Cystic fibrosis and myocardial fibrosis. Arch Dis Child 52:163

Johnson RA, Baker SS, Fallon JT, Maynard EP, Ruskin JN, Wen Z, Ge K, Cohen HJ (1981) An occidental case of cardiomyopathy and selenium deficiency. N Engl J Med 304:1210–1212

Li G, Wang F, Kang D, Li C (1985) Keshan Disease: an endemic cardiomyopathy in China. Hum Pathol 16:602–609

Lin CT, Chen LH (1982) Ultrastructural and lysosomal enzyme studies of skeletal muscle and myocardium in rats with long-term vitamin E deficiency. Pathology 14:375–382

Mason K (1973) Effects of nutritional deficiency on muscle. In: Bourne GH (ed) The structure and function of muscle, 2nd ed, vol IV. Academic Press, London New York, pp 155–206

Mason KE, Horwitt MK (1972) Effects of deficiency in animals. In: Sebrell WH Jr, Harris RS (eds) The vitamins. Chemistry, physiology, pathology, methods. Academic Press, London New York, pp 272–292

Nezelof C, LeSec G (1979) Multifocal myocardial necrosis and fibrosis in pancreatic diseases of children. Pediatrics 63:361–368

Robinson WF, Maxie MG (1985) The cardiovascular system. In: Jubb KVF, Kennedy PC, Palmer N (eds) Pathology of domestic animals, 3rd edn, vol 3. Academic Press, London New York, pp1–81

Saito K, Matsumoto S, Yokoyama T, Okaniwa M, Kamoshita S (1982) Pathology of chronic vitamin E deficiency in fatal familial intrahepatic cholestasis (Byler's disease). Virchows Arch Pathol Anat 396:319–330

Scott ML, Olson G, Krook L, Brown WR (1967) Selenium-responsive myopathies of myocardium and of smooth muscle in the young poult. J Nutr 91:573–583

Shakibi JG, Stone WL (1987) A quantitative analysis of the myocardial interstitial tissue space and myocardial fiber diameter in rats with vitamin E and selenium deficiency. Jpn Ht J 28:97–105

Van Vleet JF, Ferrans VJ (1986) Comparative pathology of selenium and vitamin E deficiency and excess. In: Scarpelli DG, Migaki G (eds) Nutritional dieseases: research directions in comparative pathology. Liss, New York, pp 359–396

Van Vleet JF, Carlton W, Olander HJ (1970) Hepatosis dietetica and mulberry heart disease associated with selenium deficiency in Indiana swine. J Am Vet Assoc 157:1208–1219

Yu W-H (1982) A study of nutritional and bio-geochemical factors in the occurrence and development of Keshan disease. Jpn Circ J 46:1201–1207

Part III Biochemical Pharmacology of Selenium Compounds

Metabolism and Disposition of Ebselen

H. SIES[1]

1 Introduction

Ebselen [2-phenyl-1,2-benzisoselenazol-3(2H)-one; PZ-51] represents a new class of anti-inflammatory agents as an example of seleno-organic compounds in the therapy of hydroperoxide-linked pathological conditions (see Parnham and Graf 1987 for review). Ebselen exhibits GSH peroxidase-like activity (Müller et al. 1984; Wendel et al. 1984; Maiorino et al. 1988) thus imitating the natural, selenium-dependent enzyme, glutathione peroxidase (glutathione: hydrogen-peroxide oxidoreductase, EC 1.11.1.9), but accepting not only glutathione (GSH) but also dithioerythritol (Müller et al. 1985) or N-acetylcysteine (Cotgreave et al. 1987) as reductant. In addition, a glutathione-independent antioxidant capacity was found during microsomal lipid peroxidation (Müller et al. 1984; Hayashi and Slater 1986). Ebselen was also able to inhibit ADP-iron-induced lipid peroxidation in isolated hepatocytes dependent on the presence of intracellular glutathione (Müller et al. 1985). In the isolated perfused rat liver, ebselen was capable of diminishing ethanol-induced ethane release, whereas a sulfur analog was not (Cadenas et al. 1982).

Incubation of resident mouse macrophages with ebselen revealed a partial inhibition of prostaglandin E2 production, implying a hydroperoxide scavenging GSH peroxidase-like activity of this compound (Parnham and Kindt 1984). Recently, it was found that ebselen decreased significantly the cytotoxicity of quinones in Ehrlich ascites cells, an effect attributable to the antioxidant properties of ebselen (Doroshow 1986). The anti-inflammatory effects of ebselen (Parnham et al. 1986; Wendel and Tiegs 1986) were suggested to be attributable to an inhibition of lipoxygenase and/or an isomerization of lipoxygenase products by ebselen (Kuhl et al. 1985; Safayhi et al. 1985).

Our recent observations regarding the metabolism of ebselen by the rat liver will be presented here. This work (Müller et al. 1988) revealed a novel type of glucuronide, a Se-glucuronide. Metabolic studies with intact animals and the detection of metabolites in plasma and urine are reported elsewhere (Fischer et al. 1988). Since a participation of cytochrome P-450 was suggested by the studies of Müller et al. (1988), we have also studied the interaction of ebselen with cytochrome P-450 (Kühn-Velten and Sies 1989).

[1] Institut für Physiologische Chemie I, Universität Düsseldorf, Moorenstr. 5, 4000 Düsseldorf, FRG

2 Materials and Methods

Chemicals. Ebselen and its analogs and derivatives were a kind gift from E. Graf (Rhône Poulenc/Nattermann, Cologne Research Center, FRG). All chemicals and biochemicals used were of analytical grade and purchased from Boehringer (Mannheim, FRG) and Merck (Darmstadt, FRG).

Liver Perfusion. Livers from male Wistar rats (180–220 g body weight) fed on stock diet (Altromin, Lage, FRG) were perfused at 37 °C without recirculation of the perfusate, using a bicarbonate-buffered salt solution equilibrated with O_2/CO_2 (19:1, v/v) as described (Sies 1978). The perfusate was supplemented with taurocholic acid (5 µM). Perfusate flow (4–5 ml min^{-1} g^{-1} wet wt) was maintained constant throughout the individual experiments. The concentration of O_2 and the pH in the effluent perfusate were monitored using appropriate electrodes, and care was taken to maintain an effluent O_2 concentration above 0.2 mM to avoid pericentral hypoxia. Further details, including biliary sampling and preparation for HPLC as well as mass spectrometry are described elsewhere (Müller et al. 1988).

2.1 Spectral Measurements of Microsomal Suspensions

Microsomal preparations were obtained as described previously (Kühn-Velten and Sies 1989), and difference spectra were recorded with a Shimadsu UV300 spectrometer in the single wavelength/double beam mode. Further details including the evaluation of cytochrome P-450 contents and titration of sulfhydryl groups are also given in Kühn-Velten and Sies (1989).

3 Results

3.1 Antioxidant Effect and Protection Against Lipid Peroxidation in Intact Cells

In an assay of lipid peroxidation using rat liver microsomes, the lag phase preceding the onset of ascorbate-iron-ADP-induced lipid peroxidation is increased by the addition of ebselen, whereas the sulfur analog is inactive (Fig. 1; Table 1). This pertains not only to the low-level chemiluminescence, but also to other parameters of lipid peroxidation like the evolution of ethane and n-pentane or the production of thiobarbiturate-reactive material (Müller et al. 1984). This effect is comparable to that of other antioxidants such as SH-containing compounds or OH-containing compounds (Cadenas and Sies 1985; Cadenas et al. 1989).

The GSH peroxidase-like activity (Müller et al. 1984; Wendel et al. 1984) is thought to be responsible for the protection of isolated hepatocytes against oxidative challenge. As shown in Fig. 2 significant protection was afforded against

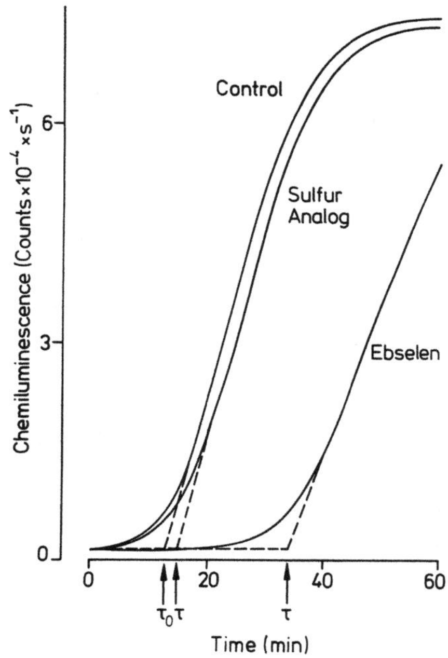

Fig. 1. Antioxidant activity of ebselen. Rat liver microsomal lipid peroxidation was detected by low-level chemiluminescence. Ebselen prolonged the lag phase before the onset of lipid peroxidation, whereas the sulfur analog did not (modified from Müller et al. 1984)

Table 1. Protective effects of ebselen and other antioxidants on microsomal lipid peroxidation evaluated by low-level chemiluminescence[a]

Compound	τ/τ_0
Ebselen (PZ-51)	2.8
Sulfur analog (PZ-25)	1.1
SH-containing compounds	
Glutathione	3.6
Dithioerythritol	2.8
Diethyldithiocarbamate	8.6
OH-containing compounds	
(+)-Cyanidanol-3	3.1
5-Hydroxyindole	8.3
5-Hydroxytryptophan	2.9
3,4-Dihydroxyphenylalanine (DOPA)	3.7

[a] The lag phase was estimated as illustrated in Fig. 1. Optimal concentrations of the antioxidants for maximal efficiency and experimental conditions as given in Müller et al. (1984) and Cadenas et al. (1989). See Cadenas and Sies (1985) for discussion.

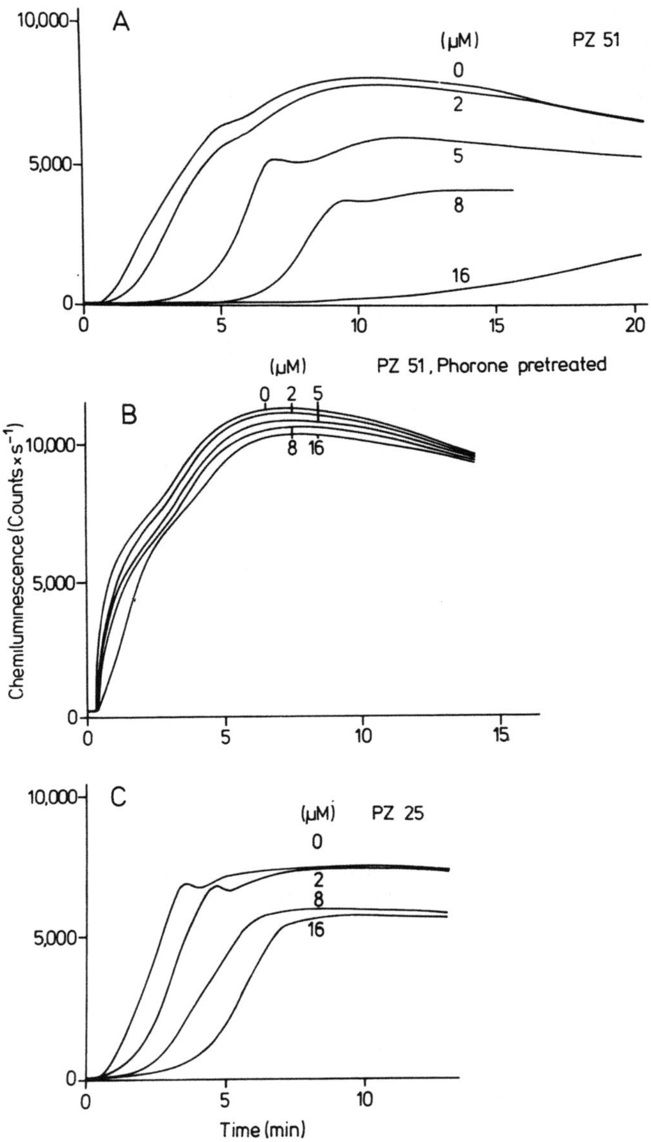

Fig. 2. Time course of ADP-Fe-induced generation of low-level chemiluminescence and the effects of different concentrations of ebselen (*PZ-51*) and its sulfur analog (*PZ-25*) in isolated rat hepatocytes. **A** and **C** Hepatocytes from untreated rats; **B** hepatocytes from phorone-pretreated rats (Müller et al. 1985)

ADP-iron-induced cell damage in control cells, whereas cells previously rendered deficient in GSH were not protected by ebselen (Müller et al. 1985). Whether this effect is causally related to the anti-inflammatory properties of the compound is a possibility, but other effects as described in the introduction may also come into play. At present, most of the observed effects can be explained by the GSH peroxidase-like activity. For example, the removal of enzyme-activating hydroperoxides will diminish flux through the cyclooxygenase reactions (Hemler et al. 1979).

3.2 Metabolism of Ebselen

Ebselen is metabolized by rat liver cells (Müller et al. 1988). The detected metabolites all share the common characteristic that the selenazole ring has been opened by cleavage of the Se-N bond. Apparently, the putative intermediate product, a selenodisulfide with glutathione [S-(2-phenylcarbamoyl-phenylselenyl)-glutathione], is labile. As shown in Fig. 3, the metabolism after ring opening involves methylation of the selenium moiety to form metabolite (I), 2-methylselenobenzanilide, or glucuronidation to form metabolite (IV), 2-glucuronylselenobenzanilide. While the latter is released into the bile, the former (I) undergoes further metabolism or is excreted across the sinusoidal membrane into the effluent.

The metabolism of (I) includes hydroxylation at the phenyl ring in the para-position to form metabolite (II) and the glucuronidation of metabolite (II) to form metabolite (III). The latter may again be excreted across the canalicular membrane into bile. This conclusion is supported by other experiments in which metabolite (I) as synthesized was infused and hydroxylation and subsequent glucuronidation were observed (II and III), but no metabolite (IV) was detected.

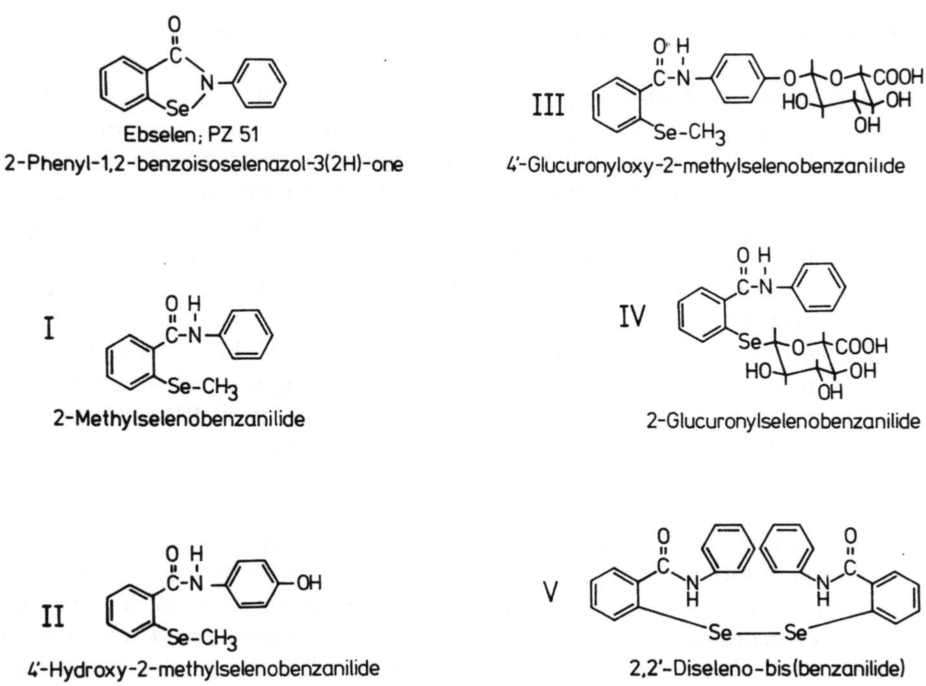

Fig. 3. Chemical structures of ebselen and metabolites of ebselen I–V (Müller et al. 1988)

Ebselen may share the methylation pathway described previously for inorganic selenium compounds as well as a number of organoselenium compounds such as selenocysteine. The metabolic pathway with dimethylselenide and trimethylselenonium as known excretory products from inorganic selenium compounds has been worked out (Foster et al. 1986). The p-hydroxylation of metabolite I is most probably catalyzed by cytochrome P-450, and the subsequent O-glucuronidation (metabolite III) is an expected step catalyzed by UDP-glucuronyltransferases. However, the Se-glucuronidation (metabolite IV) is a novel reaction. Apparently, Se-glucuronidation and Se-methylation are competing reactions with 0.75 μmol metabolite (IV) formed versus a total of 2.37 μmol methylated compounds (Müller et al. 1988). Four general categories of glucuronides have so far been established, the O-, N-, C-, and S-glucuronides (Kasper and Henton 1980) and Se-glucuronides thus form a fifth category.

Regarding biological aspects, it is noteworthy that previous work (see Parnham and Graf 1987) has shown that the Se in ebselen is not bioavailable. This conclusion stems from the observation that ^{75}Se from ^{75}Se-ebselen, in contrast to ^{75}Se-selenite, is not available for selenoprotein synthesis in mouse liver (Wendel et al. 1984), and that ebselen given orally for 5 days does not raise GSH peroxidase activity in Se-deficient mouse macrophages (Parnham and Kindt 1984). This would explain the high oral LD_{50} of 6.8 g kg^{-1} (in mouse) and the very low toxicity of ebselen. Therefore, ebselen must be viewed differently from other selenocompounds which readily enter the Se pool of the cell.

3.3 Ebselen Interaction with Cytochrome P-450 Rat Liver Microsomes

The interaction of ebselen with rat liver cytochrome P-450 was used as a model system to quantify possible interactions of this seleno-organic compound with sulfhydryl groups of intracellular membrane-bound proteins. Ebselen induced a unique difference spectrum with a maximum at 405 nm and with minima at 385 and 425 nm after the addition to microsomes in vitro. This spectrum was interpreted to indicate an interaction with the thiolate anion of cytochrome P-450. The formation of the difference spectrum could be blocked by prior addition of dithioerythritol.

The addition of ebselen converted maximally 50% of the cytochrome P-450 to P-420 in a time-dependent fashion, completion being reached within 10 min. The effect was half-maximal at 50 μM using a cytochrome P-450 concentration of 1 nmol ml^{-1} (Fig. 4). It is assumed that only certain isoenzymes of total hepatic cytochrome P-450 are accessible to ebselen.

Bovine serum albumin at physiological concentrations and sulfhydryl compounds such as dithioerythritol or cysteine are effective in preventing this cytochrome P-450 inactivation by ebselen (Table 2). Specificity studies reveal that variation of the N-substituent in the benzisoselenazolone system does not influence cytochrome P-450 inactivation, whereas ebselen derivatives with a methylated or glucuronidated selenium moiety as well as diselenides do not convert cytochrome P-450 to P-420 (Table 3). It is concluded that benzisoselenazolones are able to interact with sulfhydryl groups of membrane-associated proteins

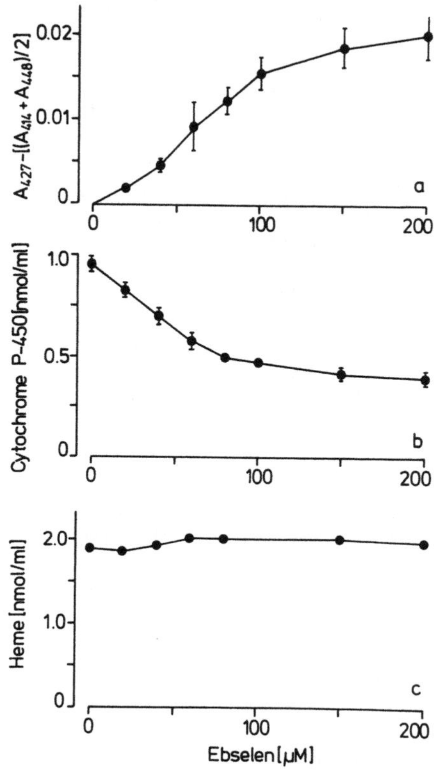

Fig. 4. Comparison of the concentration-dependent development of the ebselen-induced appearance of the reduced-cytochrome P-420 spectrum (**a**), of the ebselen-induced cytochrome P-450 loss (**b**), and of total heme concentrations (Kühn-Velten and Sies 1988)

Table 2. Protective effect of several sulfhydryl compounds (100 µM each) on cytochrome P-450 in the presence of 100 µM ebselen[a] (Kühn-Velten and Sies 1989)

Compounds added	Cytochrome P-450 ($pmol\ ml^{-1}$)	Difference to controls
None	1039 ± 14 (n = 17)	
Ebselen	657 ± 17 (n = 15)	$P<0.0005$
+ Cysteine	940 ± 44 (n = 5)	$P<0.005$
+ Penicillamine	996 ± 9 (n = 3)	NS
+ Mercaptoethanol	1058 ± 32 (n = 3)	NS
+ GSH	1026 ± 46 (n = 5)	NS
+ GSSG	664 ± 27 (n = 5)	$P<0.0005$
+ Dithioerythritol	980 ± 45 (n = 5)	NS
+ Dihydrolipoic acid	909 ± 42 (n = 3)	$P<0.0025$

[a] Sulfhydryl compounds were added to microsomal suspensions (1 mg protein ml^{-1} final concentration). These suspensions were then divided between sample and reference cuvettes. DMSO or ebselen was added to the sample cuvette, and the reference cuvette received DMSO only. Both cuvettes were reduced with dithionite, and the sample gassed with CO. Means ± SEM, number of preparations in parentheses. NS = not significantly different.

Table 3. Effect of ebselen (compound 1) and ebselen derivatives (100 μM each) on hepatic microsomal cytochrome P-450 (Kühn-Velten and Sies 1989)

Compound added (number)	Cytochrome P-450 (pmol ml^{-1})	Difference to control
– None	1080 ± 64	
(1) 2-Phenyl-1,2-benzisoselenazol-3(2H)-one (ebselen)	678 ± 26	$P < 0.0025$
(2) 2-Phenyl-1,2-benzisothiazol-3(2H)-one	841 ± 42	$P < 0.025$
(3) 2-(4-Chlorophenyl)-1,2-benzisoselenazol-3(2H)-one	656 ± 30	$P < 0.0025$
(4) 2-Methyl-1,2-benzisoselenazol-3(2H)-one	859 ± 75	$P < 0.05$
(5) 2-Methylselenobenzanilide	968 ± 31	NS
(6) 4′-Hydroxy-2-methylselenobenzanilide	1031 ± 13	NS
(7) 2-Glucuronylselenobenzanilide	1035 ± 15	NS
(8) S-(2-Phenyl carbamoyl benzeneselenenyl)-glutathione	1142 ± 74	NS
(9) 2,2′-Diselenobis-(N-methyl-benzamide)	1125 ± 62	NS

in vitro. In vivo, however, ebselen is obviously largely bound to plasma proteins. This may contribute to the previously reported low toxicity of this substance (see Parnham and Graf 1987).

4 Concluding Remarks

It is of interest to note that the seleno-organic compound ebselen, exhibiting glutathione peroxidase-like activity (Müller et al. 1984; Wendel et al. 1984) may have a model character with respect to the active site of glutathione peroxidase. Simonarson (1988) has pointed out that this would have corresponded with Willstätter's notions on the active site of an enzyme earlier this century. The very first description of this type of seleno-organic compound was actually published in 1924, during the period of the controversies regarding the nature of proteins (Lesser and Weiß 1924).

The abundance of 13 different selenium-containing proteins in tissue homogenates of the rat (Behne et al. 1988) should be considered in interpreting the functions of seleno-organic compounds as well. Most of these different selenoproteins have not yet been assigned to a known enzyme activity. Therefore, much is to be learned in this respect.

Acknowledgements. Studies in the author's laboratory were supported by the Deutsche Forschungsgemeinschaft and the National Foundation for Cancer Research. The contributions by colleagues as cited in the references and the technical assistance of Ursula Rabe and Bettina Aufmbruch are gratefully acknowledged.

References

Behne D, Hilmert H, Scheid S, Gessner H, Elger W (1988) Evidence for specific selenium target tissues and new biologically important selenoproteins. Biochim Biophys Acta 966:12–21

Cadenas E, Sies H (1985) Oxidative stress: excited oxygen species and enzyme activity. Adv Enzyme Reg 23:217–237

Cadenas E, Wefers H, Müller A, Brigelius R, Sies H (1982) Active oxygen metabolites and their action in the hepatocyte. Studies on chemiluminescence responses and alkane production. Agents Actions Suppl 11:203–216

Cadenas E, Simic MG, Sies H (1989) Antioxidant activity of 5-hydroxytryptophan, 5-hydroxyindole and dopa against microsomal lipid peroxidation and its dependence on vitamin E. Free Radical Res Commun 6:11–17

Cotgreave IA, Sandy MS, Berggren M, Moldeus PW, Smith MT (1987) N-acetylcysteine and glutathione-dependent protective effect of PZ-51 (ebselen) against diquat-induced cytotoxicity in isolated hepatocytes. Biochem Pharmacol 36:2899–2904

Doroshow JH (1986) Role of hydrogen peroxide and hydroxyl radical formation in the killing of Ehrlich tumor cells by anticancer quinones. Proc Natl Acad Sci USA 83:4514–4518

Fischer H, Terlinden R, Löhr JP, Römer A (1988) A novel biologically active selenoorganic compound VIII. Biotransformation of ebselen. Xenobiotica 18:1347–1359

Foster SJ, Kraus RJ, Ganther HE (1986) Formation of dimethyl selenide and trimethylselenonium from selenobetaine in the rat. Arch Biochem Biophys 247:12–19

Hayashi M, Slater TF (1986) Inhibitory effects of ebselen on lipid peroxidation in rat liver microsomes. Free Rad Res Commun 2:179–185

Hemler ME, Cook HW, Lands WEM (1979) Prostaglandin biosynthesis can be triggered by lipid peroxides. Arch Biochem Biophys 193:340–345

Kasper CB, Henton D (1980) Glucuronidation. In: Jakoby WB (ed) Enzymatic basis of detoxication, vol II. Academic Press, London New York, pp 3–36

Kühn-Velten N, Sies H (1989) Optical spectral studies of ebselen interaction with cytochrome P-450 of rat liver microsomes. Biochem Pharmacol 38:619–625

Kuhl P, Borbe HO, Römer A, Fischer H, Parnham MJ (1985) Selective inhibition of leukotriene B_4 formation by ebselen: a novel approach to anti-inflammatory therapy. Agents Actions 17:366–367

Lesser R, Weiß R (1924) Ueber selenhaltige aromatische Verbindungen. Chem Ber 57:1077–1082

Maiorino M, Roveri A, Coassin M, Ursini F (1988) Kinetic mechanism and substrate specificity of glutathione peroxidase activity of ebselen (PZ51). Biochem Pharmacol 37:2267–2271

Müller A, Cadenas E, Graf P, Sies H (1984) A novel biologically active seleno-organic compound. I. Glutathione peroxidase-like activity in vitro and antioxidant capacity of PZ51 (ebselen). Biochem Pharmacol 33:3235–3239

Müller A, Gabriel H, Sies H (1985) A novel biologically active selenoorganic compound. IV. Protective glutathione-dependent effect of PZ51 (ebselen) against ADP-Fe induced lipid peroxidation in isolated hepatocytes. Biochem Pharmacol 34:1185–1189

Müller A, Gabriel H, Sies H, Terlinden R, Fischer H, Römer A (1988) A novel biologically active seleno-organic compound. VII. Biotransformation of ebselen in perfused rat liver. Biochem Pharmacol 37:1103–1109

Parnham MJ, Graf E (1987) Seleno-organic compounds and the therapy of hydroperoxide linked pathological conditions. Biochem Pharmacol 36:3095–3102

Parnham MJ, Kindt S (1984) A novel biologically active seleno-organic compound. III. Effects of PZ 51 (ebselen) on glutathione peroxidase and secretory activities of mouse macrophages. Biochem Pharmacol 33:3247–3250

Parnham MJ, Leyck S, Dereu N, Winkelmann J, Graf E (1986) Ebselen (PZ51): A GSH-peroxidaselike organoselenium compound with anti-inflammatory activity. Adv Inflam Res 10:397–400

Safayhi H, Tiegs G, Wendel A (1985) A novel biologically active seleno-organic compound. V. Inhibition by ebselen (PZ51) of rat peritoneal neutrophil lipoxygenase. Biochem Pharmacol 34:2691–2694

Sies H (1978) The use of perfusion of liver and other organs for the study of microsomal electron transport and cytochrome P-450 systems. Methods Enzymol 52:48–59

Sies H (1986) Biochemistry of oxidative stress. Angew Chem Int Ed Engl 25:1058–1071

Simonarson B (1988) Glutathione peroxidase, selenium and vitamin E in defense against reactive oxygen species. Reactive oxygen species in chemistry, biology and medicine. Plenum, New York London, p 15–36

Wendel A, Tiegs G (1986) A novel biologically active seleno-organic compound. VI. Protection by ebselen (PZ-51) against galactosamine/endotoxin-induced hepatitis in mice. Biochem Pharmacol 35:2115–2118

Wendel A, Fausel M, Safayhi H, Otter R (1984) A novel biologically active seleno-organic compound. II. Activity of PZ-51 in relation to glutathione peroxidase. Biochem Pharmacol 33:3241–3246

The Use of Highly Enriched ^{77}Se in Metabolic Studies of Ebselen in Man – An NMR Investigation

H. FISCHER[1,2], G. HILBOLL[1], A. RÖMER[1], and R. TERLINDEN[1]

There is a lot of prejudice against selenium. The element per se is considered to be toxic, irrespective of its molecular surroundings. This rather uncritical view obviously does not hold true for organoseleno compounds like ***ebselen***. Nevertheless, it is of enormous importance to rule out as much as possible the risk of a metabolic liberation of selenium from the organic carrier in man. Our NMR work with selenium-77 shall contribute to that field.

Selenium provides the advantage of having an NMR-active stable isotope: ^{77}Se. Unfortunately, Se-NMR is rather intensitive due to both the gyro-magnetic

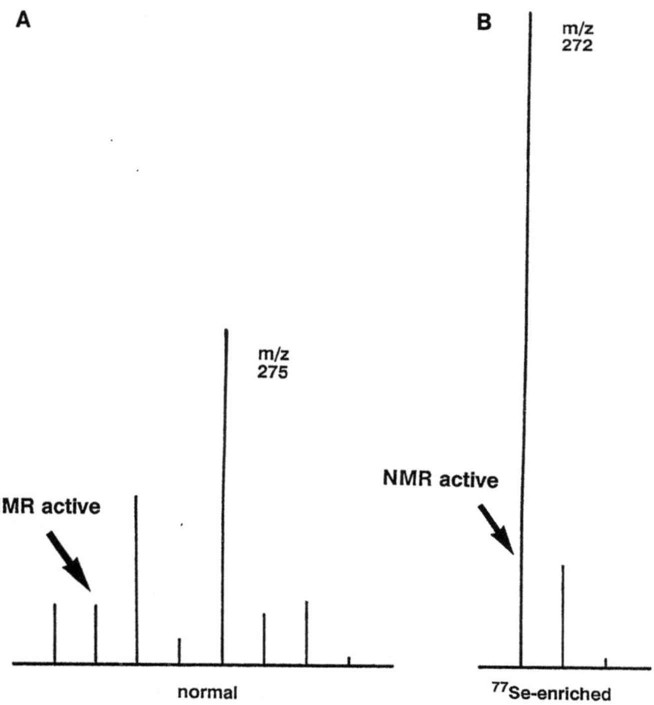

Fig. 1 A, B. Isotopic pattern of the molecular ions of normal ebselen (**A**) and of ebselen which was synthesized using selenium-77 with an isotopic purity of 95% (**B**)

[1] Rhône-Poulenc/Nattermann, Cologne Research Centre, Nattermannallee 1, 5000 Köln 30, FRG
[2] To whom correspondence should be addressed

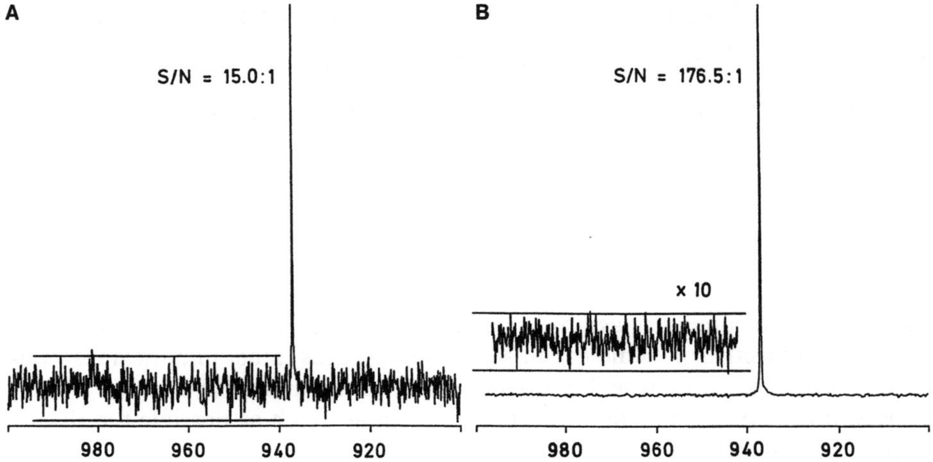

Fig. 2A, B. ^{77}Se NMR spectra of ebselen. **A** With natural selenium; **B** with enriched selenium

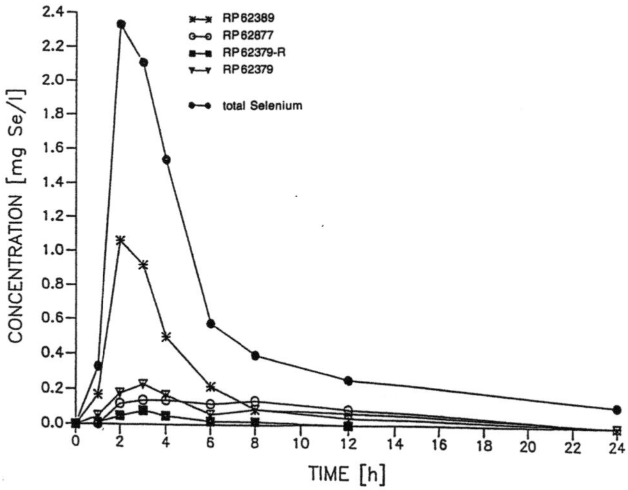

Fig. 3. Plasma concentrations of total selenium and ebselen metabolites after oral administration of 500 mg ^{77}Se-ebselen to a healthy volunteer

ratio and the natural abundance of ^{77}Se, namely 7.6%. By an enrichment of the latter, one should obtain an overall increase in sensitivity by a factor of more than 10. Figure 1 illustrates what that means in our case. The fraction of the NMR-active, ^{77}Se-containing isotopomers is indicated by the arrows in "normal" ebselen and in enriched material.

To verify whether this increase will benefit the actual ^{77}Se-NMR experiment we compared the signal to noise ratio of common ebselen and of such material which was synthesized using selenium highly enriched in the 77 isotope. One can observe the expected increase by a factor of about 12 (Fig. 2).

In this study 500 mg of this ^{77}Se-ebselen were given orally to a volunteer. Figure 3 shows the plasma concentration-time curves of total selenium and the major metabolites of ebselen. This is to prove that ebselen is rapidly and well ab-

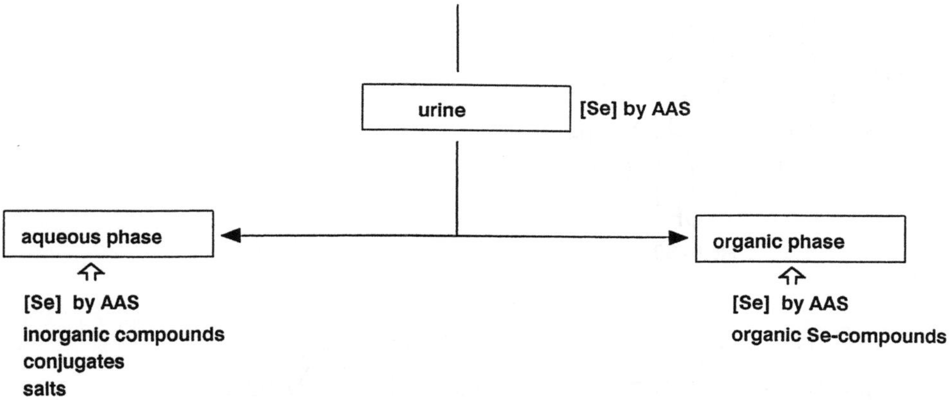

Fig. 4. Monitoring the balance of total selenium. Objective: description of [Se] by chemical entities

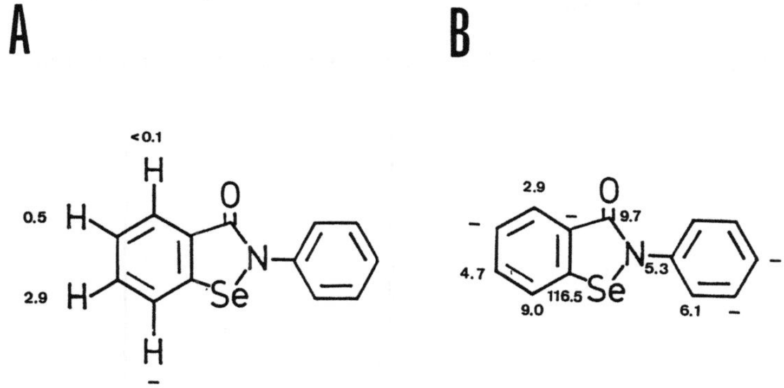

Fig. 5 A, B. Structure of ebselen. **A** With ^{1}H-^{77}Se NMR coupling constants; **B** with ^{13}C-^{77}Se NMR coupling constants

sorbed and that the urine samples are of relevance for our experiments. We chose the 0 to 3 h sample as it contained the highest fraction of the administered dose (18%). The total renal excretion was around 70% of the dose within 72 h. The questions now were:

1. In what form is the selenium excreted?
2. Are there any selenium-containing compounds which are indicative of a metabolic breakdown like trimethylselenonium salts?

The water soluble compounds often escape analysis due to the work-up procedure which is designed to isolate normal lipophilic metabolites (Fig. 4). Therefore, we thought it useful to apply selenium NMR for better coverage.

A major problem is the assignment of Se-NMR signals. One of the best tools for that purpose are the coupling constants to other nuclei (Fig. 5). From the parent ebselen the couplings to hydrogen and carbon are depicted and they are

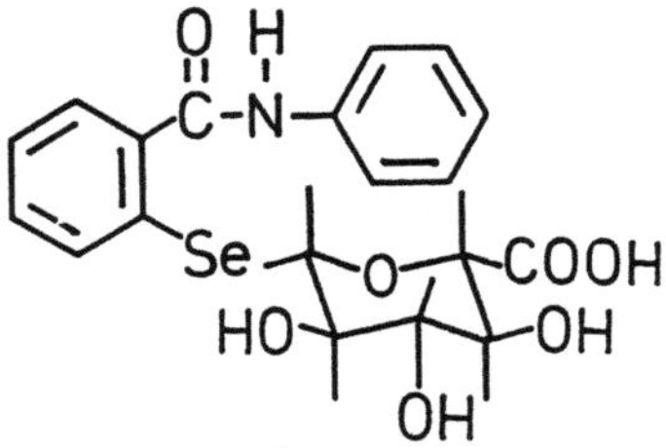

1 RP6289

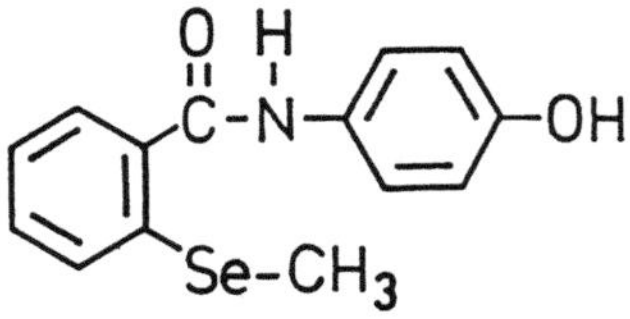

2 RP62379

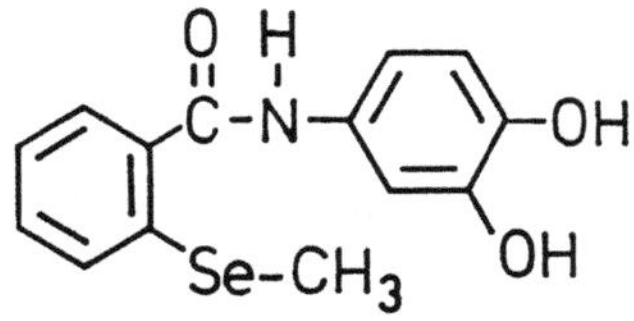

3 RP63897

Fig. 6. Structures of metabolites which were detected in human urine

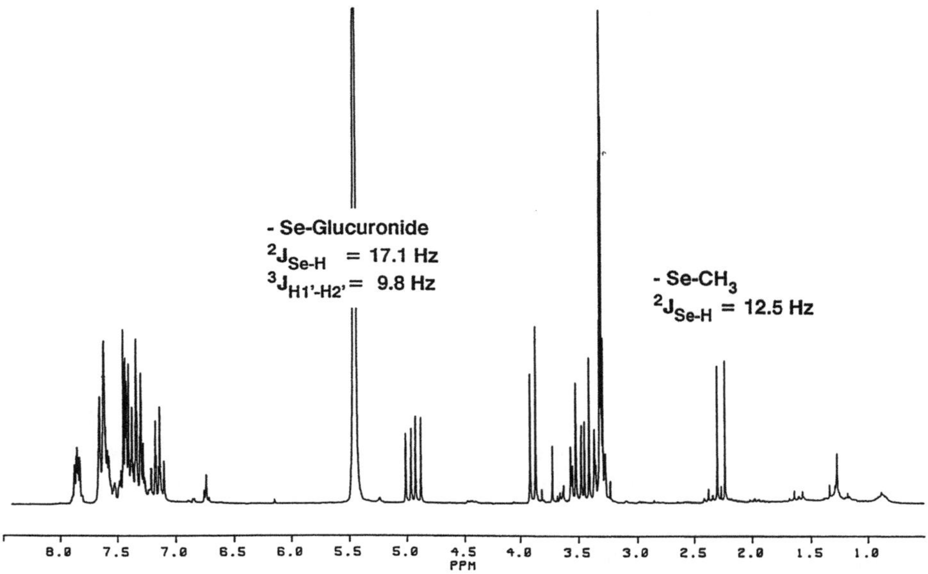

Fig. 7. ^{1}H-NMR spectrum of a HPLC fraction of the urine which was used for the ^{77}Se-NMR spectra, the Se-H couplings of two metabolites are indicated

worthy of interpretation: Note that there is no ^{77}Se-^{1}H coupling to the ortho position. With such a feature the interpretation of spectra of derived metabolites is facilitated.

In Fig. 6 the structures of the metabolites which we found in human urine are shown. The selenium is either glucuronidated or methylated. The respective cou-

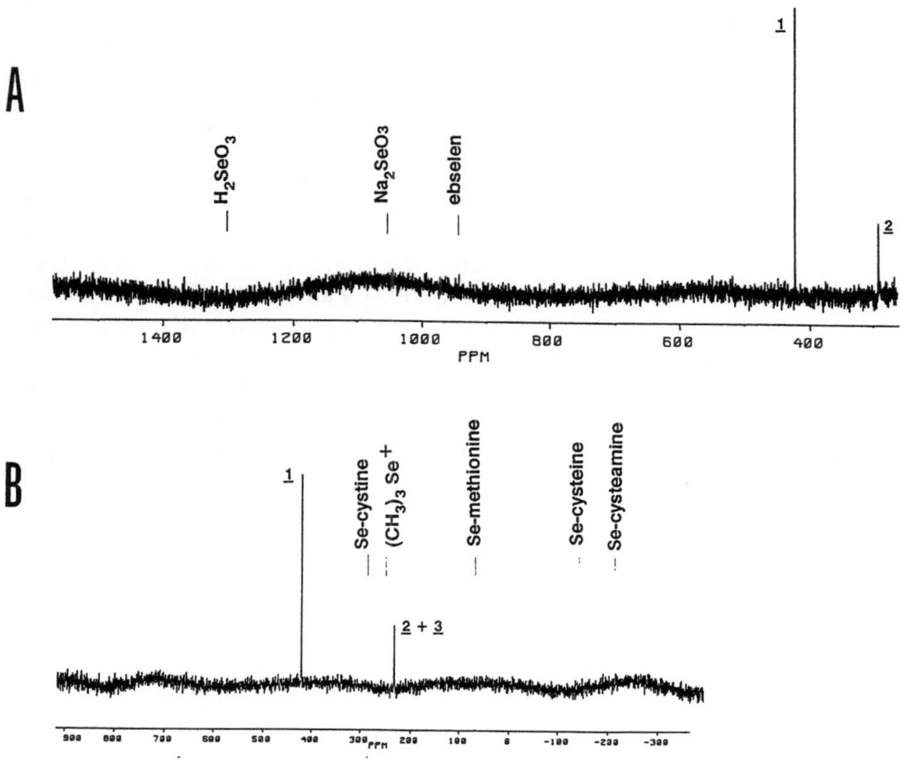

Fig. 8 A, B. ^{77}Se-NMR spectra of total human urine. **A** Low field part; **B** high field part

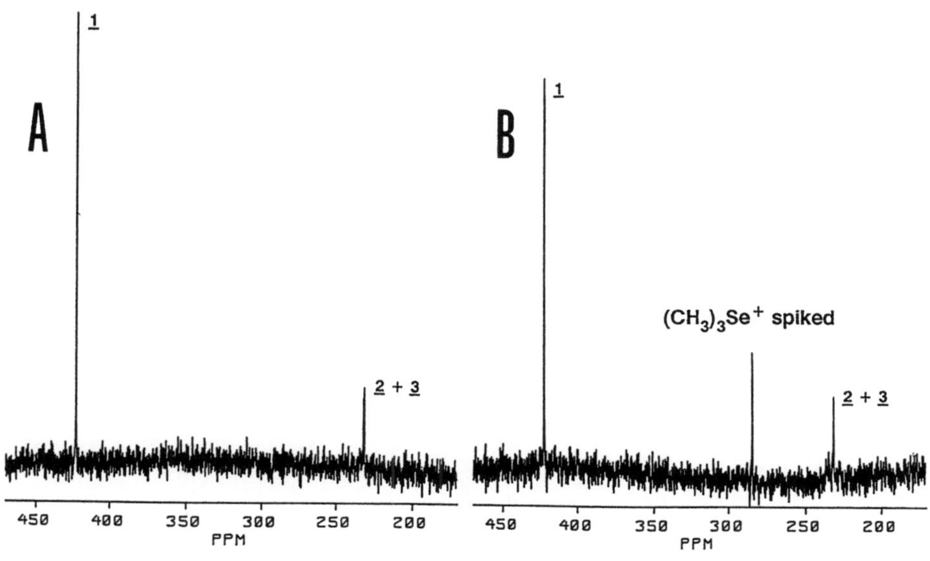

Fig. 9 A, B. Estimation of the detection limit of trimethyl selenonium. **A** Expanded region of the spectrum of Fig. 8; **B** same urine sample, to reach with 500 μg ml^{-1} trimethyl selenonium (non-enriched)

plings to the adjacent protons are easy to detect even in mixtures (Fig. 7). The presence of the glucuronide is proven by the double doublet and the typical splitting of the Se-methyl groups can be observed at high field. This information allows a clear assignment of signals in the selenium NMR.

In the NMR spectrum of the investigated urine sample (Fig. 8) only two signals could be found which are due to the already mentioned metabolites. No further signals were detected. In particular, the areas did not show signals where endogenously formed selenium compounds should turn up. But to make sure that such compounds could be detected by this method, we spiked in this very sample trimethylselenonium with a concentration of 500 $\mu g\ ml^{-1}$. The signal of trimethylselenonium could be detected in the corresponding spectrum (Fig. 9B) with a signal to noise ratio of about 5. This was non-enriched material. From this one can estimate the detection limit in the range of about 10 μg Se ml^{-1} urine without any pretreatment. We hope to decrease the detection limit 10- to 100-fold by preconcentration, thus coming into the range of the natural background.

Effect of Ebselen (PZ-51) in Liver Failure Induced by *Propionibacterium acnes* (*P. acnes*)

M. AKASAKI[1,2], T. IKEDA[1], F. NUMATA[1], Y. KUREBAYASHI[1], and W. TSUKADA[1]

Summary

Intravenous injection of *P. acnes* and lipopolysaccharide (LPS) into BALB/c mice resulted in acute liver failure. Treatment with ebselen (PZ-51), a seleno-organic compound, reduced serum GPT activity, and the number and size of necrotic foci of hepatocytes. These activities are suggested to be due to inhibition of leukotriene (LT) formation by ebselen.

1 Introduction

Ebselen is a seleno-organic compound which exerts novel biological activities, i.e., glutathione peroxidaselike activity (Parnham et al. 1986) and inhibition of leukotriene (LT) formation (Safayhi et al. 1985, 1986, 1987); Tiegs and Wendel (1988) have reported that ebselen (PZ-51) protects mice against galactosamine/endotoxin-induced hepatitis by inhibition of leukotriene (LT) formation. Arthur et al. (1985) have reported that superoxide dismutase (SOD) and allopurinol protected against hepatic necrosis produced by *Corinebacterium parvum*. We have studied the effect of ebselen in acute liver failure induced by *Propionibacterium acnes (P. acnes)*.

2 Methods

BALB/c female mice were purchased from Charles River Japan (Kanagawa, Japan) at 6 weeks of age and conditioned in our facility for 1 week prior to use. Experiments were carried out according to Tsutsui (1985). Briefly, mice were intravenously injected with 0.1 or 1 mg/mouse of heat-killed *P. acnes* (gift from Y. Mizoguchi, The Third Department of Internal Medicine, Osaka City University, School of Medicine). One week later, 1 μg/mouse of bacterial lipopolysaccharide

[1] Exploratory Research Laboratories III, Research Institute, Daiichi Seiyaku Co. Ltd., 1-16-13, Kitakasai, Edogawa-ku, Tokyo, 134, Japan
[2] To whom correspondence should be addressed

(LPS from *Salmonella enteritidis*, DIFCO, Detroit, MI, USA) was given intravenously. The degree of liver injury was assessed by measuring serum GPT activity 24 h after the injection of LPS using a Hitachi 736 Autoanalyzer (Hitachi, Tokyo, Japan), and by histopathological study. PZ-51, synthesized by A. Nattermann & Co., GmbH (Cologne, FRG), was administered orally. Statistical analysis was carried out by Student's t-test.

3 Results and Discussion

Significant elevation of serum GPT activity was observed by the treatment of mice with *P. acnes* and LPS (Table 1). In the case when 1 mg/mouse of *P. acnes* was given, all disease control mice but one died within 24 h after LPS injection. PZ-51 (100, 300 mg kg^{-1}, daily) reduced both the mortality and serum GPT activity (Table 1). PZ-51 was also effective when it was administered only on days 7 and 8 (Table 2). Leukotriene D4 (LTD4, Wako, Osaka, Japan, 100 μg kg^{-1}) was

Table 1. Effect of PZ-51 on *P. acnes*-induced liver failure model[a]

Group	n	n (Alive)	GPT (IU l^{-1}) Mean ± SE
Normal control	11	11	41.4 ± 2.2
Disease control	11	1	1161
PZ-51 30 mg kg	11	4	913 ± 115
100	11	10	543 ± 142
300	11	11	282 ± 36

[a] 1 mg kg^{-1} of *P. acnes* was intravenously injected into BALB/c mice. One week later, LPS (1 μg kg^{-1}) was intravenously injected. Serum GPT activity was measured 24 h after LPS injection. $p<0.001$; PZ-51 (30 mg kg^{-1}) vs. PZ-51 (300 mg kg^{-1})

Table 2. Involvement of LTD_4 in *P. acnes*-induced liver failure model[a]

Group	n	n (Alive)	GPT (IU l^{-1}) Mean ± SE
Normal control	7	7	28.7 ± 1.6
Disease control	7	0	–
PZ-51 (100 mg kg^{-1}) daily	8	8	276 ± 45**
PZ-51 (100 mg kg^{-1}) days 7, 8	7	7	207 ± 37**
LTD_4 (100 μg kg^{-1})	7	7	90.1 ± 2.5*

[a] The experiment was carried out as described in the legend to Table 1. LTD_4 was intravenously injected instead of LPS. * = $p<0.005$, ** = $p<0.001$ vs. normal control.

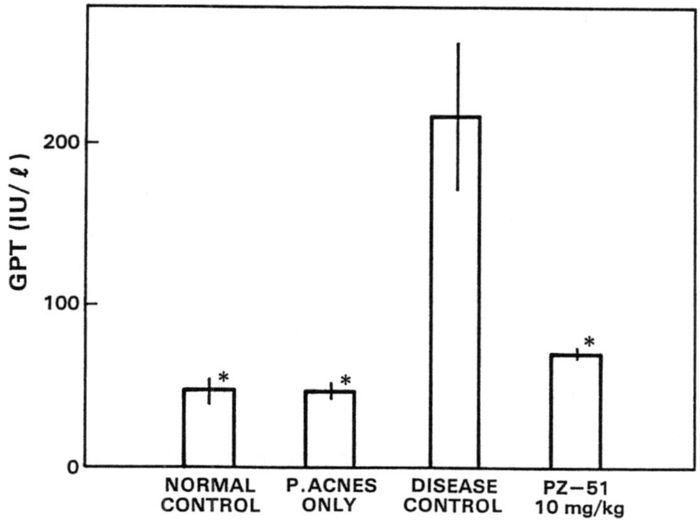

Fig. 1. Liver failure induced by smaller amount of *P. acnes*. 0.1 mg kg^{-1} of *P. acnes* was injected in this experiment. * = $p<0.05$ vs. disease control

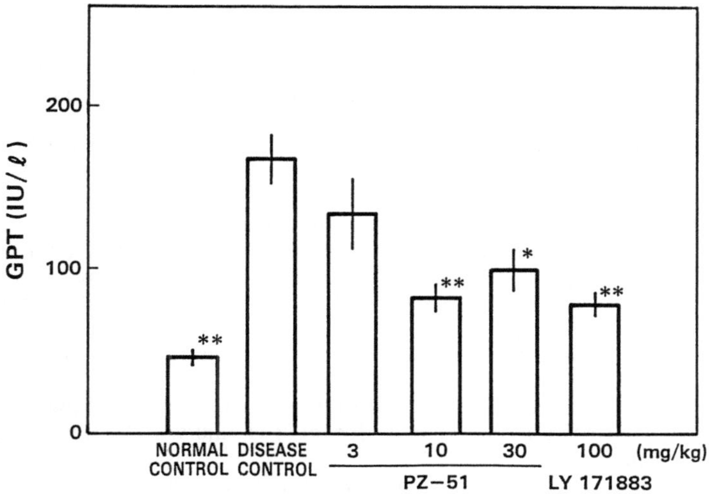

Fig. 2. Liver failure induced by smaller amount of *P. acnes*. See the legend to Fig. 1. * = $p<0.05$; ** = $p<0.001$ vs. disease control

intravenously injected on day 7 instead of LPS in this experiment, and LTD4 was confirmed to be involved in the mechanism of this liver failure model. When the amount of *P. acnes* was reduced to one-tenth (0.1 mg/mouse), no death of disease control mice was observed, and PZ-51, also under this experimental condition, decreased serum GPT activity at a lower dose (10 mg kg^{-1}) only on day 7 (Fig. 1). In this experiment, no liver injury was seen without LPS injection, showing that a combination of *P. acnes* and LPS was essential for induction of liver injury.

These results were reproducible and LY171883, an LTD4 antagonist, was also effective (Fig. 2). In histopathological studies, degeneration and necrosis of liver cells, in association with mononuclear cell infiltration, were found in disease control mice. Both the number and size of necrotic foci were markedly reduced in PZ-51-treated mice (photographs not shown).

As described above, PZ-51 was found to have a potent protective activity against *P. acnes*-induced liver injury. Mizoguchi et al. (1987) have reported that LTs play an important role in this liver failure model, since some 5-lipoxygenase inhibitors like azelastine and AA-861 showed protective activity. As shown in Table 2, LTD4 increased the serum GPT level in our experiment as well, and LY171883 prevented the increase in serum GPT activity (Fig. 2). It has been reported that PZ-51 has 5-lipoxygenase inhibitory activity and exerts protective activity against galactosamine/endotoxin-induced hepatitis (Wendel and Tiegs 1986, 1987; Tiegs and Wendel 1988). Taken together, these results suggest that PZ-51 protects mice against *P. acnes*-induced liver injury by inhibition of LT formation.

Acknowledgements. We thank Miss N. Teshima for excellent technical assistance.

References

Arthur MJP, Bentley IS, Tanner AR, Kowalski Saunders P, Millward-Sadler GH, Wright R (1985) Oxygen-derived free radicals promote hepatic injury in the rat. Gastroenterology 89:1114–1122

Mizoguchi Y, Sakagami Y, Kuboi H, Yamamoto S, Morisawa S (1987) Effect of azelastine on an experimentally-induced acute hepatic failure model. Jpn J Allergol 36:108–111 (in Japanese)

Parnham MJ, Leyck S, Dereu N, Winkelmann J, Graf E (1986) A GSH-peroxydase-like organoselenium compound with anti-inflammatory activity. Adv Inflam Res 10:397–400

Safayhi H, Tiegs G, Wendel A (1985) A novel biologically active seleno-organic compound – V. Inhibition by ebselen (PZ-51) of rat peritoneal neutrophil lipoxygenase. Biochem Pharmacol 34:2691–2694

Tiegs G, Wendel A (1988) Leukotriene-mediated liver injury. Biochem Pharmacol 37:2569–2573

Tsutsui H (1985) Studies on the experimentally induced acute hepatic failure in mice: the role of adherent cells in the induction of liver injury. Acta Hepatol Jpn 26:1438–1444 (in Japanese)

Wendel A, Tiegs G (1986) A novel biologically active seleno-organic compound – VI. Protection by ebselen (PZ-51) against galactosamine/endotoxin-induced hepatitis in mice. Biochem Pharmacol 35:2115–2118

Wendel A, Tiegs G (1987) Leukotriene D4 mediates galactosamine/endotoxin-induced hepatitis in mice. Biochem Pharmacol 36:1867

Ebselen (PZ-51) Inhibits the Formation of Ischemic Brain Edema

J. TANAKA [1,2] and F. YAMADA [1]

1 Introduction

Recent evidence has suggested the involvement of lipoxygenase metabolites (Asano et al. 1985; Dempsey et al. 1986) and/or active oxygen such as superoxide anion (Chan et al. 1985) in the development of ischemic brain edema. Ebselen (PZ-51; Fig. 1) is a novel organoselenium compound which catalytically inactivates peroxides in vitro in a manner similar to that of glutathione peroxides (Müller et al. 1984). In addition, this compound has an inhibitory effect on lipoxygenase and reduces leukotriene B_4 formation with a concomitant decrease in 5-HETE production (Kuhl et al. 1986).

We have examined the effect of ebselen on ischemic brain edema in the rat embolic stroke model.

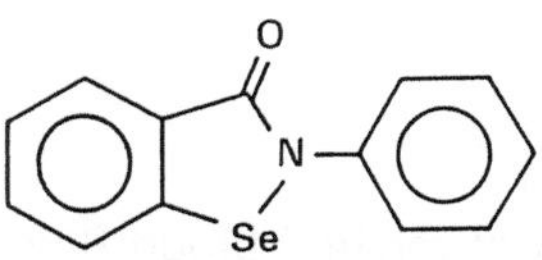

Fig. 1. The structure of ebselen (PZ-51)

2 Experiments

Ebselen was provided by A. Nattermann & Co. GmbH (Cologne, FRG). Male Wistar rats (8 weeks old, Funabashi Farm Co. Ltd., Funabashi, Japan) were used in the experiments. They were allowed free access to water until the experiments, and they were fasted overnight before use.

Ebselen in a 0.5% carboxymethyl cellulose solution was administered orally 30 min before the induction of focal cerebral ischemia. Furthermore, ebselen in plasma containing 5% dimethylformadehylde was injected into the tail vein continuously (0.5 mg ml^{-1} h^{-1} $body^{-1}$) from 30 min before until 3 h after the microsphere injection.

[1] Research Institute, Daiichi Seiyaku Co., Ltd., 1-16-13, Kitakasai, Edogawa-ku, Tokyo 134, Japan
[2] To whom correspondence should be addressed

A focal ischemia was induced in the same way previously reported (Kogure et al. 1974) except that degradable microspheres (Daiichi Radioisotope Institutes, Tokyo, Japan), which gradually dissolve in blood and produce the recovery of cerebral blood flow, were utilized. Namely, 50 μg of the degradable microspheres, whose diameter was 10–60 μm, were injected into the right internal carotid artery under GOF anesthesia (N_2O/O_2/halothane = 70% vol./30% vol./1.5%). Immediately after the microsphere injection, the anesthesia was stopped, and the animals were allowed to move freely or were placed individually in Bormann cages.

Brain edema was determined by comparison of wet and dry weight. The rats were decapitated 3 h after the induction of cerebral ischemia. Then the brains were quickly removed from the skulls and divided into hemispheres under 100% humidity atmosphere. The wet weight of each hemisphere was measured according to chemical balance as soon as possible. After drying the brain in an oven at 130 °C overnight, the dry weight was obtained. The water contents of each hemisphere were calculated as follows:

water contents (%) = (wet weight − dry weight) × 100/wet weight.

After the solubilization the dried brain with 1 ml of 60% nitric acid at 80 °C, the solution was neutralized with Tris solution. Then, sodium and potassium ions were measured by flame photometry and expressed as mEq kg^{-1} dry weight. Welhi's t-test was used for statistical analysis.

3 Results

The character of this model was preliminarily studied. Local cerebral blood flow (l-CBF) was observed by autoradiography using radiolabeled iodoantipyrine. L-CBF of the microsphere-injected hemisphere was prominently reduced immediately after the microsphere injection, and the recovery of l-CBF caused by the degradation of microspheres and the formation of collateral circulation was observed heterogeneously for several hours after the microsphere injection. Recovery of l-CBF was observed except in the infarcted area 24 h after the microsphere injection. The brain water content of the microsphere-injected hemisphere increased prominently soon after the microsphere injection and lasted up to 24 h. Moreover, an increase of the sodium concentration and a decrease of the potassium concentration in the microsphere-injected hemisphere were also observed with a concomitant increase of the brain water contents. However, the water contents and sodium-potassium concentration of the contralateral hemisphere did not change significantly. One week after the microsphere injection, the necrotic regions were observed by microscopy. Small, but many and widely spread, infarcted or gliosis areas were observed in the microsphere-injected hemisphere.

The effect of ebselen on ischemic brain edema was examined in this rat embolic stroke model. Ebselen or its vehicle was administered orally 30 min before the microsphere injection, or injected continuously into the tail vein. In the ebselen-treated group, there was a significant reduction in the brain water con-

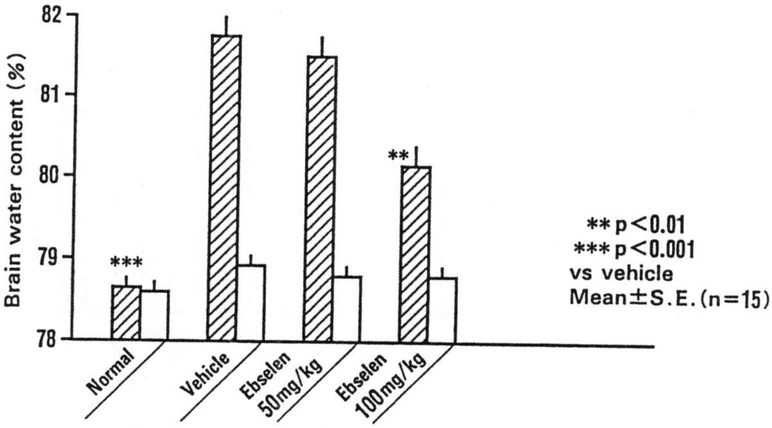

Fig. 2. The effect of ebselen on brain water content when ebselen was administered orally. *hatched column*: Ischemic hemisphere; *open column*: contralateral hemisphere

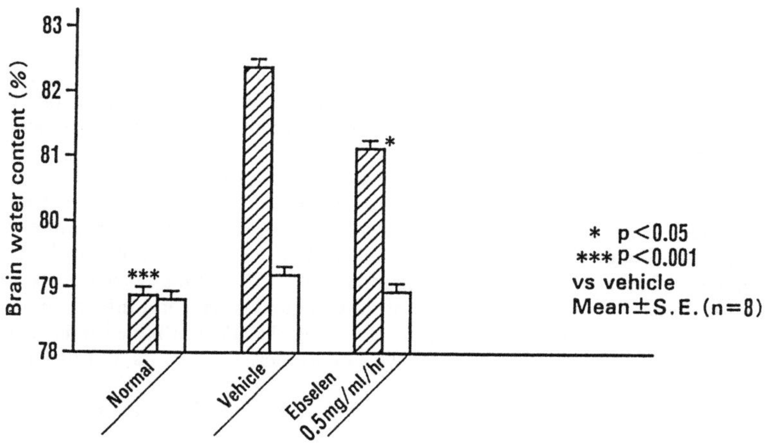

Fig. 3. The effect of ebselen on brain water content when ebselen was administered intravenously. *hatched column*: Ischemic hemisphere; *open column*: contralateral hemisphere

tents of the microsphere-injected hemisphere, when ebselen was orally administered (Fig. 2) and also intravenously injected (Fig. 3). Furthermore, the tendency to ameliorate the sodium concentration was also observed in the ebselen-treated group (Fig. 4).

4 Discussion

The reduction of ischemic brain edema by ebselen was observed. As mentioned above, ebselen possesses a glutathione peroxidaselike activity and a lipoxygenase inhibitory effect. These results support the involvement of lipoxygenase metabolites and/or active oxygen in the development of ischemic brain edema, and also suggest that ebselen is a useful compound for the therapy of cerebrovascular diseases.

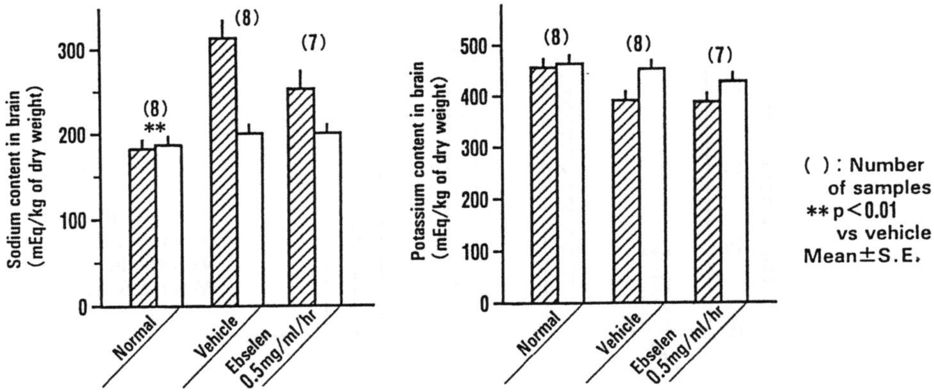

Fig. 4. The effect of ebselen on brain sodium-potassium concentration. *hatched column*: Ischemic hemisphere; *open column*: contralateral hemisphere

References

Asano T, Gotoh O, Koide T, Takakura K (1985) Ischemic brain edema following occlusion of the middle cerebral artery in the rat. II: Alteration of the eicosanoid synthesis profile of brain microvessels. Stroke 16:110–113

Chan PH, Fishman RA, Longar S, Chen S, Yu A (1985) Cellular and molecular effects of polyunsaturated fatty acids in brain ischemia and injury. Prog Brain Res 63:227–235

Dempsey RJ, Roy MW, Cowen DE, Maley ME (1986) Lipoxygenase metabolites of arachidonic acid and the development of ischaemic cerebral oedema. Neurol Res 8:53–56

Kogure K, Busto B, Scheinberg P, Reinmuth OM (1974) Energy metabolites and water content in rat brain during the early stage of development of cerebral infarction. Brain 97:103–114

Kuhl P, Borbe HO, Fisher H, Romer A, Safayhi H (1986) Ebselen reduces the formation of LTB4 in human and porcine leukocytes by isomerisation to its 5S,12R-6-trans-isomer. Prostaglandines 31:1029–1048

Müller A, Cadenas E, Graf P, Sies H (1984) A novel biologically active seleno-organic compound – I. Biochem Pharmacol 33:3235–3239

Effects of Ebselen (PZ-51) on Experimental Cerebral Vasospasm

T. WATANABE[1,5], M. NISHIYAMA[1], H. OKAMOTO[1], T. HORI[1], T. ASANO[2], T. SHIMIZU[3], and H. MASAYASU[4]

1 Introduction

1.1 Delayed Cerebral Vasospasm

The delayed cerebral vasospasm occurring approximately from the fourth to the fourteenth day after subarachnoid hemorrhage (SAH) is a serious problem in neurosurgery. The delayed vasospasm is manifested clinically in about 40% of all SAH patients. Its delayed onset and long-lasting contraction of vessels are very interesting, but incomprehensible phenomena. Recently, computerized tomography has verified the close correlation between the presence of a subarachnoid clot and the occurrence of delayed vasospasm. However, its cause is still unknown and its treatment has not yet been established. So even at present, its occurrence affects the outcome of SAH seriously. So neurosurgeons should be aware of the pathogenesis of delayed cerebral vasospasm. We proposed that lipid peroxidation induced by the lysis of a subarachnoid clot could be the causative agent of delayed cerebral vasospasm after SAH (Asano et al. 1980).

1.2 Lipid Peroxides and Glutathione Peroxidase in Serum of SAH Patients

Since thiobarbituric acid reactive substances (TRS) are regarded as indicators of lipid peroxides, we observed sequential changes of TRS in the serum of SAH patients (Fig. 1). The TRS value in the serum of the patients was elevated and reached a plateau on day 4 or 5 and remained unchanged until day 10 or 11, when it started to decrease gradually. This time course matched well that of delayed vasospasm. The shaded area shows the normal values of middle-aged patient who had no organic diseases. Glutathione peroxidase is well known to be the most effective scavenger of lipid peroxides. Figure 2 shows the sequential changes of glutathione peroxidase activity in the serum of the SAH patients. The increase of glutathione peroxidase activity in the serum was observed during the period when

[1] Department of Neurosurgery, Faculty of Medicine, University of Tottori, 86 Nishimachi, Yonago Tottori 683, Japan
[2] Department of Neurosurgery, Saitama Medical School, Saitama, Japan
[3] Department of Physiological Chemistry and Nutrition, Faculty of Medicine, University of Tokyo, Tokyo 134, Japan
[4] Daiichi Seiyaku Co. Ltd., Tokyo 134, Japan
[5] To whom correspondence should be addressed

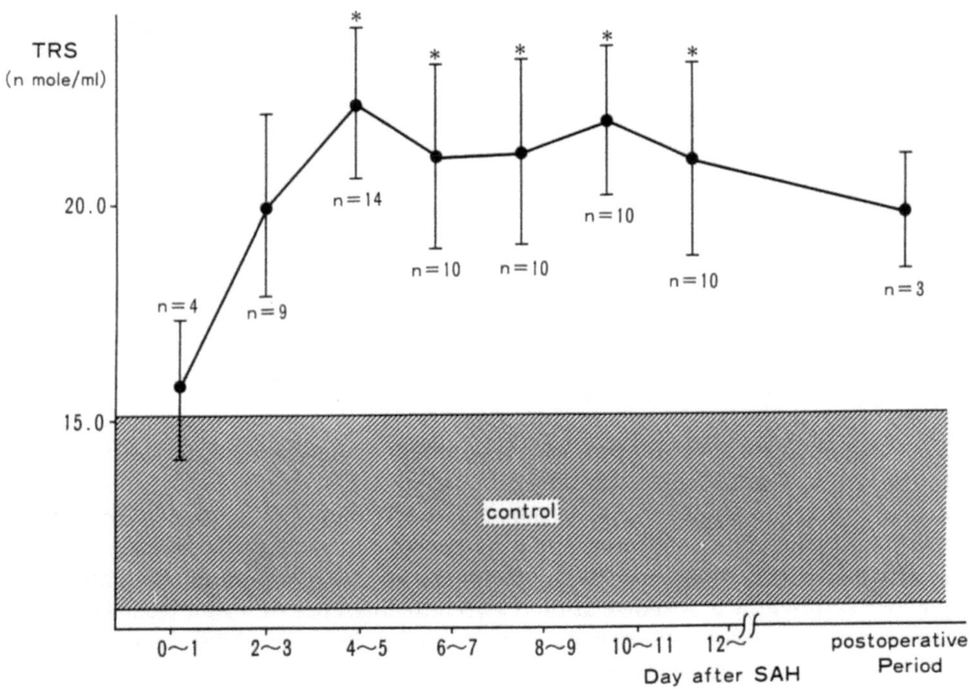

Fig. 1. Sequential changes in serum TRS of SAH patients. Each *dot* and *bar* represent the mean value ±SE. The *asterists* denote significantly different from the control value ($P<0.02$)

delayed vasospasm should be occurring as well as the TRS. These data were obtained from the SAH patients whether delayed vasospasm occurred or not. There is no significant difference between values of the patients who showed delayed vasospasm clinically and the patients without vasospasm (Watanabe et al. 1988a). Case 1 represents a patient who did not indicate delayed vasospasm clinically (Fig. 3). The changes of glutathione peroxidase activities in the serum corresponded to that of TRS. Case 2 represents a patients who revealed vasospasm clinically from day 9 (Fig. 4). During clinical manifestations of delayed vasospasm, an increase of the TRS value was observed with a decrease of glutathione peroxidase activity. This tendency was observed for nearly all patients. This suggested that if glutathione peroxidase activity did not respond well to an increased level of lipid peroxides, a critical increase of lipid peroxides wouls cause severe delayed vasospasm.

1.3 Arachidonic Acid Lipoxygenase Products in Clot and Basilar Artery of Experimental SAH Dogs

In order to examine the involvement of lipoxygenase products from arachidonic acid in the pathogenesis of delayed vasospasm, we measured the contents of hydroxyeicosatetraenoic acids (HETE) in the subarachnoid clot and lipoxygenase

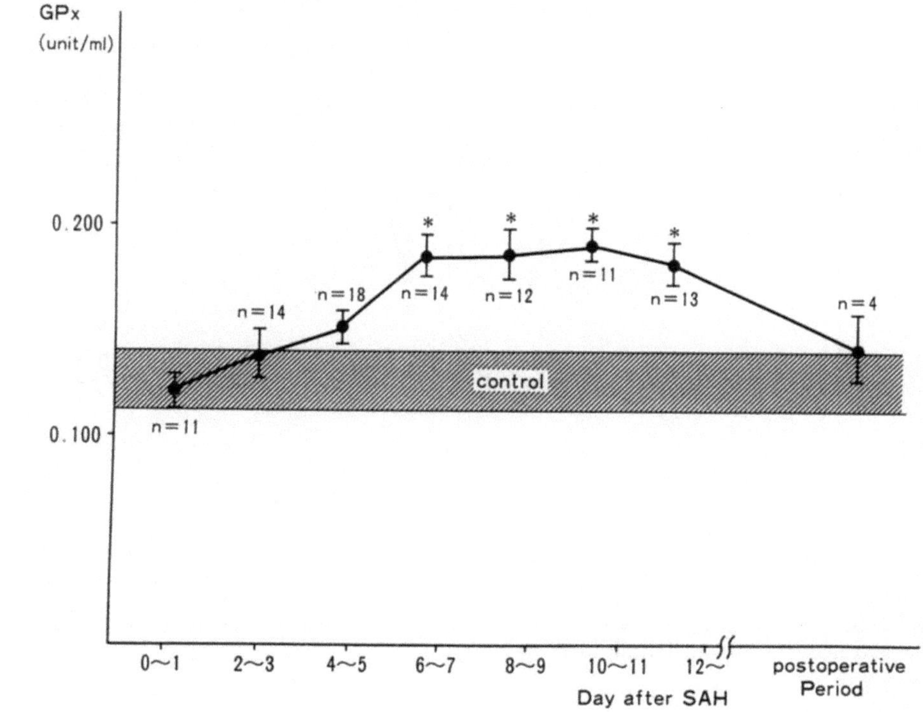

Fig. 2. Sequential changes in serum glutathione peroxidase activity of SAH patients. Each *dot* and *bar* represent the mean value ±SE. The *asterisks* denote significantly different from the control value ($P<0.01$)

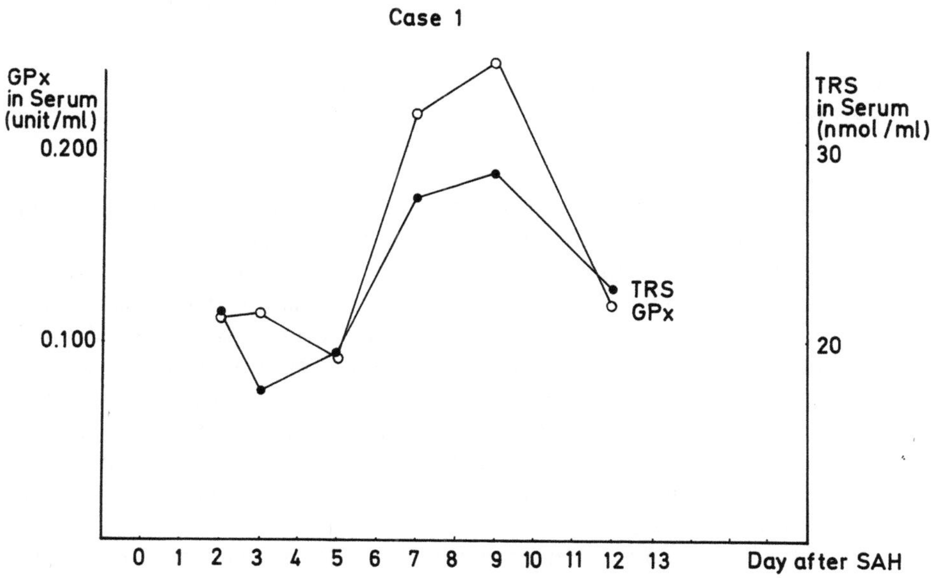

Fig. 3. Sequential changes in serum TRS and glutathione peroxidase activity of SAH patient who did not indicate delayed vasospasm clinically

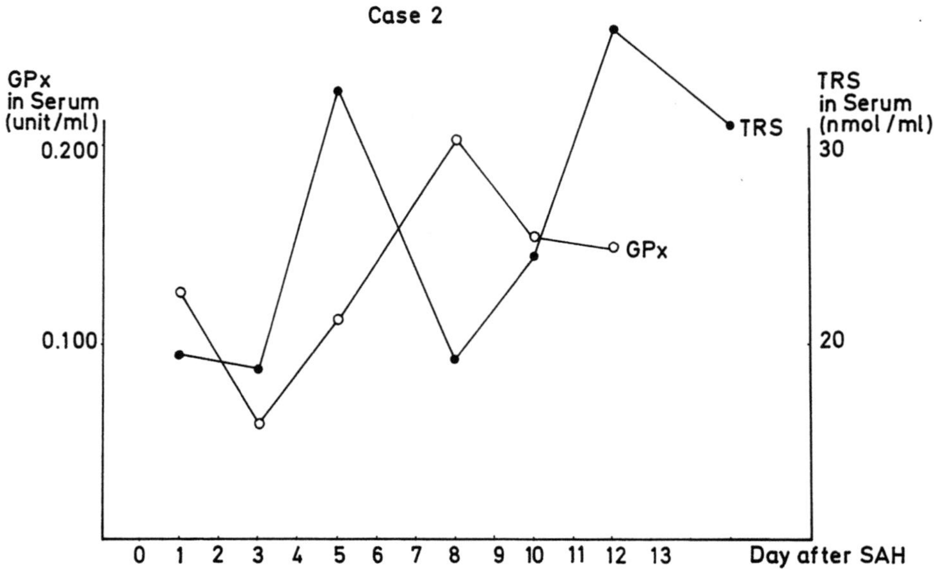

Fig. 4. Sequential changes in serum TRS and glutathione peroxidase activity of SAH patient who revealed vasospasm clinically from days 9–15

Table 1. HETE contents of subarachnoid clot[a]

	Day 2	Day 3	Day 5	Day 8
Subarachnoid clot (12-HETE, nmol g^{-1} tissue)	1.88 ± 2.00 (n = 5)	1.29 ± 0.60 (n = 5)	1.76 ± 2.43 (n = 5)	1.89 ± 2.16 (n = 6)

[a] These data were from samples homogenized with methanol.
Values are expressed as mean ± SD

activity in the basilar artery, using the canine two-hemorrhage model. Adult mongrel dogs weighing from 7 to 14 kg were anesthetized and the right vertebral artery was exposed and cannulated. Angiograms were taken on days 1 and 8. Autologous blood (0.4 ml kg^{-1}) was injected into cisterna magna on days 1 and 3. The diameter of the basilar artery, visualized on the angiogram, was measured at the midpoint of the artery. The magnitude of vasospasm in each dog was expressed as the diameter ratio to the control angiogram. On day 8, each animal was exsanguinated and the brain was perfused with ice-cold saline containing heparin via the vertebral artery. An opening hole was made in the clivus. The subarachnoid clot and the basilar artery were obtained for high performance liquid chromatography (HPLC) analysis.

The subarachnoid clot was homogenized with methanol and centrifuged. The supernatant was dried and the residue was applied to reverse-phase HPLC. In the subarachnoid clot, only 12-HETE was detected from day 1–8 (Table 1). The mass spectrum of 12-HETE obtained from the subarachnoid clot is compatible with

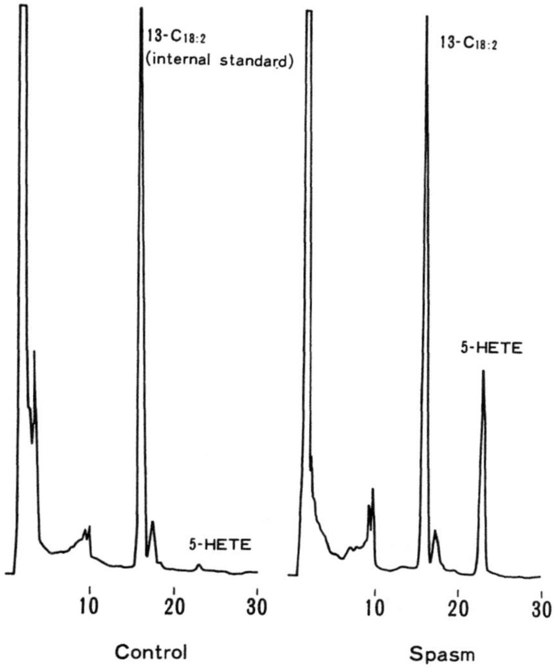

Fig. 5. Reverse-phase HPLC analysis of products after incubation of the basilar artery with arachidonic acid and A23187. The absorbance at 234 nm was monitored

Table 2. Stimulation of basilar artery with A23 187[a]

	Day 8 (nmol of 5-HETE mg^{-1} protein)
Two-hemorrhage	1.34 ± 0.52 (n = 5)
Control	0.01 ± 0.01 (n = 5)

[a] These data indicate the quantity of 5-HETE formed from endogenous and/or exogenous arachidonic acid when the basilar artery was incubated with arachidonic acid and A23 187. Values are expressed as mean ± SD. There is a statistical significance ($p<0.01$ t-test)

that of the authentic 12-HETE. The basilar artery segments obtained from normal dogs and two-hemorrhage models were cut into small pieces with scissors and incubated for 5 min at 37 °C in 0.5 ml Tris-Tyrode solution with 2 μM calcium ionophore and 80 μM arachidonic acid. The products were also analyzed by HPLC. A representative HPLC chart shows that the canine basilar artery contained only 5-lipoxygenase and that the vessel with spasm produced a larger quantity of 5-HETE than the vessel from the control animal (Fig. 5). The quantity of 5-HETE was measured by the peak area ratio to the internal standard. Table 2 shows that 5-lipoxygenase activity of the basilar artery segment was significantly increased in the two-hemorrhage group as compared to the control group. This spectrum of 5-HETE obtained from the basilar artery is practically identical to that of the authentic 5-HETE. As for the same samples of which we measured 5-lipoxygenase activities, we also measured leukotrienes using reverse-phase

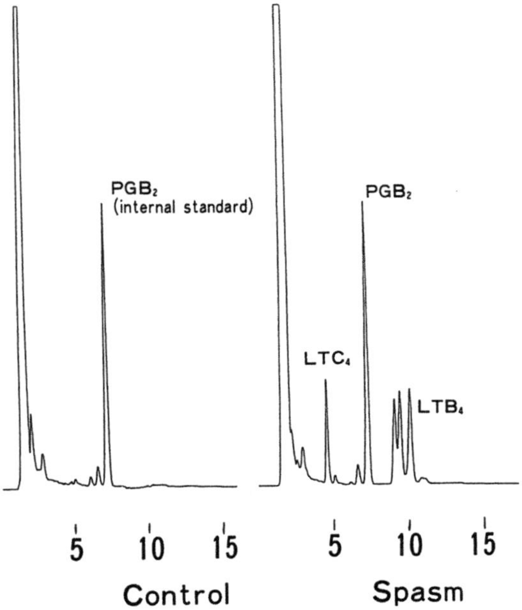

Fig. 6. Reverse-phase HPLC analysis of products after incubation of the basilar artery with arachidonic acid and A23187. The absorbance at 280 nm was monitored

HPLC (Fig. 6). The left side is the representative HPLC chart obtained from the basilar artery with vasospasm. The right side is the control. PGB_2 is an internal standard. The leukotrienes B_4 and C_4 are produced in the basilar artery with vasospasm but are not produced by the normal vessel. Thus, the 5-lipoxygenase pathway was remarkably activated in the basilar artery exposed to SAH compared with the normal basilar artery (Watanabe et al. 1988b).

Thus, PZ-51 is a seleno-organic compound which exerts novel biological activities, i.e., glutathione peroxidaselike activity and lipoxygenase inhibitory activity. PZ-51 was the very agent we were looking for as a treatment of delayed cerebral vasospasm by controlling the lipid peroxides. So, we examined the effects of administration of PZ-51 in the canine two-hemorrhage model.

2 Experimental Results

As a non-treated group, 19 dogs were subjected to two-hemorrhage SAH. Thus, 100 mg kg^{-1} PZ-51 was administered per os to 12 dogs exposed to the two-hemorrhage procedure.

The administration of PZ-51 was started on day 2, once a day (six dogs) and twice a day (six dogs), and was maintained for 7 days. The diameter ratio on day 8 was compared to that of the control two-hemorrhage group. The diameter ratios of the groups treated with PZ-51 once a day and twice a day were 64.8 and 68.1% respectively, a significantly greater ratio than the 41.3% obtained in the non-treated group (Table 3). So angiographically, the administration of PZ-51

Table 3. Diameter ratio of the basilar artery

	Day 8[a] (%)
Untreated	41.3 ± 4.59*, ** (n = 19)
PZ-51 treated (P.O.)	
100 mg kg^{-1} day^{-1} (once a day)	64.8 ± 17.4* (n = 6)
100 mg kg^{-1} day^{-1} (twice a day)	68.1 ± 6.41** (n = 6)

[a] Values are expressed as the means ± SD
* = $P<0.05$, ** = $P<0.001$ (t-test)

Table 4. 5-Lipoxygenase activity in the basilar artery

	Day 8 (nmol of 5-HETE mg^{-1} protein)
Untreated	6.73 ± 0.95 (n = 3)
PZ-51 treated (P.O.) 100 mg kg^{-1} day^{-1} (twice a day)	3.32 ± 1.82 (n = 5)

[a] Values are expressed as mean ± SD. There is a statistical significance ($p<0.01$ t-test)

ameliorated delayed vasospasm in the two-hemorrhage model. Moreover, PZ-51 is the most effective agent we have ever examined in the two-hemorrhage model. The basilar artery segments obtained from three non-treated dogs and five dogs treated with PZ-51 twice a day were cut into small pieces with scissors. These samples were homogenized with 0.5 ml of 20 mM phosphate buffer (pH 7.4) containing 2 mM EDTA and 0.5 mM DTT and centrifuged at 10000 rpm for 10 min. The supernatant, i.e., the enzyme extract, was mixed with 0.1 M Tris-HCl buffer containing 5 mM $CaCl_2$ and 2 mM ATP. 100 µM arachidonic acid and 4 µM 15-hydroperoxyeicosatetraenoic acid were added and the samples were incubated for 5 min at 37 °C. The products were analyzed by reverse-phase HPLC. The basilar artery segment of the PZ-51-treated group produced 3.32 nmol of 5-HETE mg^{-1} protein, significantly less than the 6.73 nmol of 5-HETE mg^{-1} protein of the non-treated group (Table 4). Thus, the administration of PZ-51 inhibited the increase of 5-lipoxygenase activity in the two-hemorrhage model.

3 Conclusion

PZ-51, an anti-inflammatory agent, significantly suppressed vasospasm angiographically together with 5-lipoxygenase activity in the basilar artery, so PZ-51 represents a promising agent in the treatment of delayed vasospasm. In addition, these findings indicate that inflammatory changes are involved in the pathogenesis of cerebral vasospasm.

References

Asano T, Tanishima T, Sasaki T, Sano K (1980) Possible participation of free radical reactions initiated by clot lysis in the pathogenesis of vasospasm after subarachnoid hemorrhage. In: Wilkins RH (ed) Cerebral arterial spasm. Proc 2nd Int Workshop. Williams & Wilkins, Baltimore, pp 190–201

Watanabe T, Sasaki T, Asano T, Takakura K, Sano K, Fuchinoue T, Watanabe K, Yosimura S, Abe K (1988a) Changes in glutathione peroxidase and lipid peroxides in cerebrospinal fluid and serum after subarachnoid hemorrhage – with special reference to the occurrence of cerebral vasospasm. Neurol Med Chir 28:645–649

Watanabe T, Asano T, Shimizu T, Seyama Y, Takakura K (1988b) Participation of lipoxygenase products from arachidonic acid in the pathogenesis of cerebral vasopasm. J Neurochem 50:1145–1150

Ebselen, a Seleno-Organic Compound, Inhibits Leukocyte Aggregation and Reduces the Plasma Leukotriene B_4 Level in Humans and Rats

K. SAKUMA[1,3], H. MASAYASU[1], H. SHIBATA[2], and S. ASHIDA[1]

1 Introduction

Recent evidence suggests that intravascular activation of leukocytes may be implicated in the pathogenesis of inflammation, postischemic injury and atherothrombotic diseases (Harlan 1985). Thus, the prevention of leukocyte activation represents a possible target for therapeutic intervention in these diseases. Ebselen [2-phenyl-1,2-benzisoselenazol-3(2H)-one] is a seleno-organic compound which catalyzes the degradation of hydroperoxides (Müller et al. 1984), converts leukotriene B_4 (LTB_4) into its biologically inactive 6-trans isomer (Kuhl et al. 1986) in the presence of glutathione, and inhibits arachidonate 5-lipoxygenase (Safayhi et al. 1985) and superoxide anion production (Ichikawa et al. 1987) in leukocytes. We have, therefore, studied the effect of ebselen on leukocyte activation in rats and humans after oral administration. This was done by assessing its inhibition of LTB_4-induced whole blood aggregation ex vivo. In addition, we have investigated possible effects of ebselen on platelet aggregation and the concentrations of LTB_4, thromboxane B (TXB_2) and 6-keto-prostaglandin $F_{1\alpha}$ (6-keto-$PGF_{1\alpha}$) in plasma after oral administration.

2 Materials and Methods

Ebselen was supplied by Nattermann & Co. GmbH (Cologne, FRG). Nafazatrom was synthesized at Daiichi Seiyaku Research Institute (Tokyo, Japan). In an animal experiment, ebselen and nafazatrom were given orally to male Wistar-Imamichi rats weighing 230–260 g (Institute of Animal Reproduction, Saitama, Japan) fasted overnight. Blood was collected by a clean heart puncture under pentobarbital anesthesia 1 h after drug administration and anticoagulated with 0.1 vol of citrate (3.13%) or heparin (40 u ml^{-1}). In a study in humans, ebselen was given orally to five healthy, fasting, male volunteers, aged 23 to 37 years, weighing between 60 and 86 kg who denied taking any medications for at least 10 days before the initiation of the study. Blood was taken by clean venipuncture

[1] Research Institute, Daiichi Seiyaku Co. Ltd., Edogawa-ku, Tokyo 134, Japan
[2] Department of Internal Medicine, Kitasato University School of Medicine, Kitasato, Sagamihara-shi, Kanagawa 228, Japan
[3] To whom correspondence should be addressed

into 0.1 vol of citrate, heparin and indomethacine (2 mM)-containing citrate. Subjects with little aggregability in the predrug, whole blood aggregation test were excluded from the protocol. Whole blood aggregation was measured by use of a Chlono-Log impedance aggregrometer, model 540 (Chlono-Log Co., USA). Aggregation was induced by adding LTB_4 (Pacesel Co., FRG) or N-formyl-methionyl-leucyl-phenylalanine (fMLP, Sigma Co., USA) at a final concentration of 0.1 μM to 1 ml aliquots of the heparinized blood preincubated with 5 μg ml^{-1} cytochalasin B (Sigma) for 1 min. The extent of aggregation was measured as the increase in impedance after 4 min. Platelet aggregration was assessed turbidometrically with the Chlono-Log aggregometer by adding ADP (2.10 μM) or collagen suspension (1 μg ml^{-1}) (Hormon Chemie, FRG) to platelet-rich plasma (PRP). PRP was prepared by centrifugation of the citrated blood for 7 min at 270×g (rat blood) or 190×g (human blood). The concentrations of LTB_4, TXB_2 and 6-keto-$PGF_{1\alpha}$ in plasma were determined by radioimmunoassay after extraction of the samples with SEP-PACK C18 cartridges (Waters, USA).

3 Results

Oral administration of ebselen at 0.3 – 300 mg kg^{-1} to rats resulted in a dose-dependent inhibition of LTB_4-induced whole blood aggregation, while a reference drug, nafazatrom, which was shown to inhibit LTB_4 production in neutrophils (Strasser et al. 1985), failed to produce a significant inhibition at 10 – 30 mg kg^{-1} (Fig. 1). Platelet aggregation in PRP induced by ADP was not influenced by oral administration of ebselen or nafazatrom at 30 mg kg^{-1}. In the human study, LTB_4-induced whole blood aggregation was suppressed again 2 h after ingestion of a 200-mg dose of ebselen but fMLP-induced aggregation was

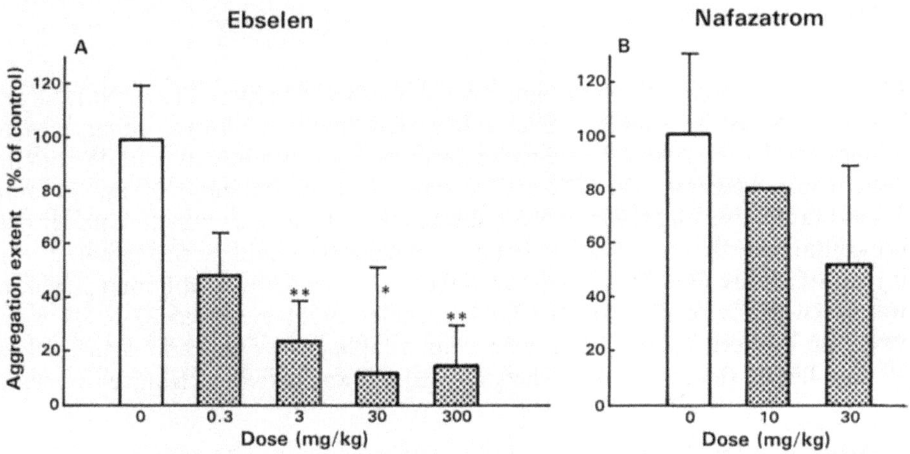

Fig. 1A,B. Whole blood aggregation induced by 0.1 μM LTB_4 ex vivo 1 h after oral administration of ebselen and nafazatrom to rats. Mean±SE; n = 5; *$p<0.05$, **$p<0.01$ vs. control

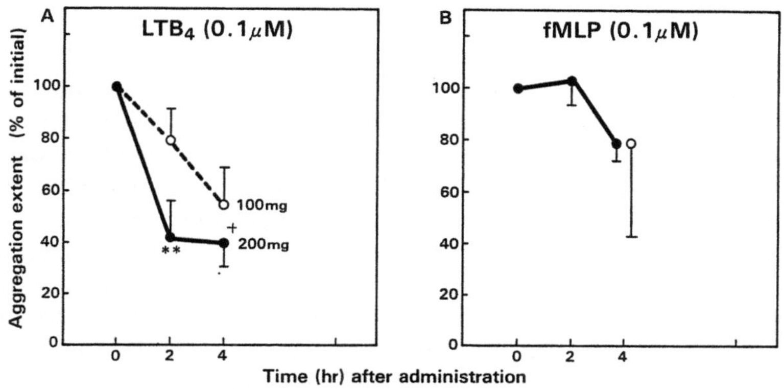

Fig. 2A, B. Whole blood aggregation induced by LTB_4 and fMLP ex vivo after oral administration of ebselen to healthy volunteers. Mean ± SE; n = 3; $+p<0.1$, $^{**}p<0.01$ vs. initial value

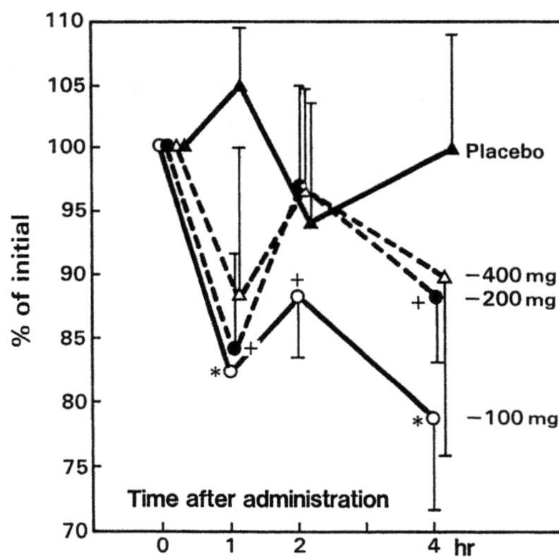

Fig. 3. Plasma LTB_4 level after oral administration of ebselen to male volunteers. Mean ± SE; n = 5; $+p<0.1$, $^{*}p<0.05$ vs. initial value

not significantly affected (Fig. 2). Platelet aggregation in PRP in response to ADP or collagen was not affected by an oral dose of 400 mg of ebselen. The plasma LTB_4 level, which was 121.3 ± 14.9 pg ml^{-1} before ebselen, was reduced to about 80% of the initial level 1 and 4 h after 100 mg of ebselen (Fig. 3). A similar reduction in LTB_4 level was also produced after 200 and 400 mg dosing although it was not statistically significant. Plasma TXB_2 and 6-keto-$PGF_{1\alpha}$ levels were not affected significantly by ebselen.

4 Discussion

Leukotriene B_4 is a potent mediator of chemotaxis and aggregation of neutrophils. LTB_4-induced whole blood aggregation, however, involved not only neutrophils but also platelets (Sakuma and Ashida, unpublished data). In the present study, ebselen inhibited LTB_4-induced whole blood aggregation ex vivo without affecting ADP-induced platelet aggregation in PRP, suggesting that this drug selectively attenuates leukocyte aggregation. The precise mechanism for this attenuation is not clear. It seems, however, reasonable to assume that the effect is mediated by inactivation of LTB_4 (Kuhl et al. 1986), since ebselen had a greater inhibitory effect on LTB_4-induced aggregation than on an fMLP-induced one (Fig. 1) and a reduced plasma LTB_4 level (Fig. 3).

In conclusion, ebselen inhibited LTB_4-induced leukocyte aggregation and reduced the plasma LTB_4 level in healthy volunteers. These effects of ebselen suggest a therapeutic potential of this drug in inflammation and ischemic injury, where leukocytes may play a pathogenic role.

References

Harlan JM (1985) Leukocytes-endothelial interactions. Blood 65:513–525

Ichikawa S, Omura K, Katayama T, Okamura N, Ohtsuka T, Ishibashi S, Masayasu H (1987) Inhibition of superoxide anion production in guinea pig polymorphonuclear leukocytes by seleno-organic compound, ebselen. J Pharmacobio-Dyn 10:595–597

Kuhl P, Borbe OH, Fischer H, Romer A, Safayhi H (1986) Ebselen reduces the formation of LTB_4 in human and porcine leukocytes by isomerization to its 5S,12R-6-trans-isomer. Prostaglandins 31:1029–1048

Müller A, Cadenas E, Graf P, Sies H (1984) A novel biologically active seleno-organic compound – I; glutathione peroxidase-like activity in vitro and antioxidant capacity of PZ-51 (ebselen). Biochem Pharmacol 33:3235–3239

Safayhi H, Tiegs G, Wendel A (1985) A novel biologically active seleno-organic compound – V; inhibition by ebselen (PZ-51) of rat peritoneal neutrophil lipoxygenase. Biochem Pharmacol 34:2691–2694

Strasser Th, Fischer S, Weber PC (1985) Inhibition of leukotriene B_4 formation in human neutrophils after oral nafazatrom (BAYG 6575). Biochem Pharmacol 34:1981–1984

Binding of Ebselen to Plasma Protein

H. Nomura[1,2], H. Hakusui[1], and T. Takegoshi[1]

1 Introduction

Ebselen (2-phenyl-1,2-benzisoselenazol-3(2H)-one) is a selenium-containing heterocyclic compound, which exhibits GSH peroxidaselike activity in vitro and anti-inflammatory activity in vivo (Müller et al. 1984; Kuhl et al. 1985).

The metabolism of ebselen has been studied in various animals and man (Fischer et al. 1988; Müller et al. 1988). Benzisoselenazolones are known to react very rapidly with thiols. It can be assumed that the isoselenazolone ring of ebselen is reductively opened, and reacts with sulfhydryl groups of proteins in plasma.

In this study, ebselen binding to bovine serum albumin (BSA), SH-modified BSA and plasma proteins in rat and man was examined using ^{14}C-ebselen in vitro to elucidate the nature of ebselen binding to protein.

2 Material and Methods

Material. Ebselen uniformly labeled with ^{14}C in an aniline moiety (9.9 mCi $mmol^{-1}$) and unlabeled ebselen were obtained from Nattermann & Cie. GmbH (Cologne, FRG). Warfarin labeled with ^{14}C was purchased from Amersham (England). Bovine serum albumin (BSA) was obtained from Organon Teknika (Holland) and SH-modified BSA, which has less than 1% of the sulfhydryl group found in normal BSA, from Miles Scientific (Naperville, IL, USA). Other reagents were all commercially available and of reagent grade.

Preparation of Rat and Human Plasma. Male Sprague Dawley rats (SD rats) were used to obtain the normal rat plasma. Male Nagase Analbuminemic rats (NAR) kindly donated from S. Nagase (Sasaki Institute, Tokyo) were used to obtain the rat plasma which is deficient in albumin (Sugiyama et al. 1980). Human plasma was freshly prepared just before use.

[1] Drug Metabolism Research Center, Research Institute, Daiichi Seiyaku Co., Ltd., 16–13, Kitakasai 1-Chome, Edogawa-ku, Tokyo 134, Japan
[2] To whom correspondence should be addressed

Determination of Ebselen Binding to Protein. Binding studies were performed at room temperature. The binding reaction was initiated by adding 20 µl of ^{14}C-ebselen in dimethylformamide with or without unlabeled ebselen to 4% BSA or SH-modified BSA solution or plasma. The unbound ebselen was extracted with chloroform quickly after incubation for 10 min. For determination of the amount of ebselen covalently bound to proteins, radioactivity in an aliquot of the aqueous layer was counted by a liquid scintillation counter. An ultrafiltration method using MPS-1 (Amicon) equipped with a YMB membrane (14 µm, Amicon) was also performed to determine the amount of ebselen bound to proteins.

Measurement of SH Group Concentration. The SH group concentration in proteins was determined by the sulfhydryl-disulfide exchange reaction with 5,5′-dithiobis(2-nitrobenzoic acid) (DTNB, Ellman's reagent) followed by measurement of visible absorption at 410 nm (Simplicio et al. 1985). A calibration curve was prepared by using known amounts of a reduced type of glutathione (GSH) as a standard.

3 Results and Discussion

Ebselen was bound to BSA immediately at 4°, 25° and 37°C (data not shown). To determine the capacity of ebselen binding to proteins, the binding of ^{14}C-ebselen to BSA or SH-modified BSA was studied by adding increasing amounts of unlabeled ebselen to a fixed amount of ^{14}C-ebselen (Fig. 1). This study demonstrated that the binding of ebselen to BSA was saturable and the maximum binding capacity was 0.4 mM in 4% BSA solution. However, the binding capacity

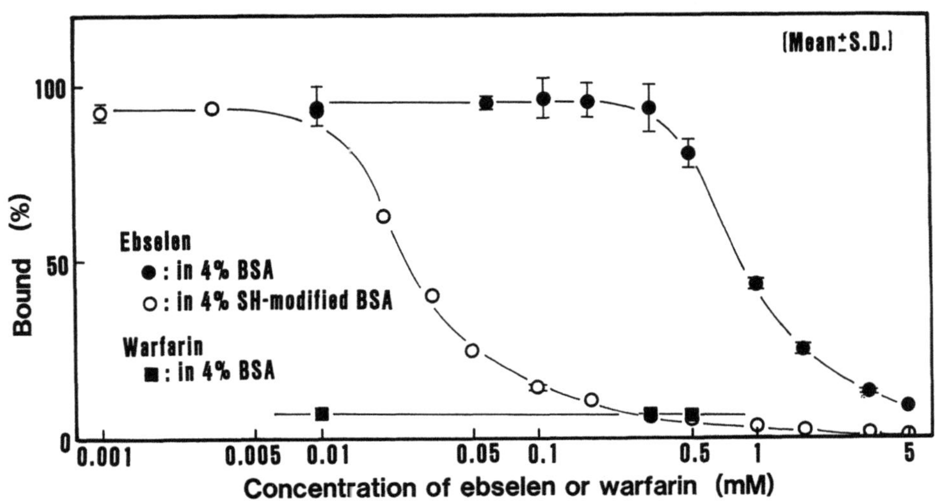

Fig. 1. Binding of ^{14}C-ebselen and ^{14}C-warfarin to BSA or SH-modified BSA. The amount of binding was determined by chloroform extraction

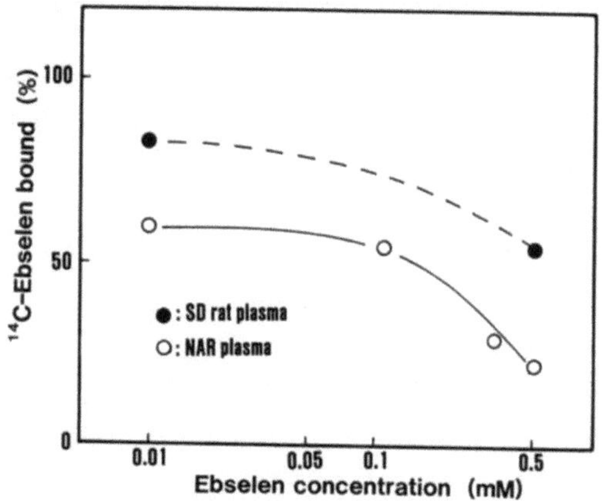

Fig. 2. Binding of ^{14}C-ebselen to plasma in SD rat and NAR. The amount of ebselen binding was determined by chloroform extraction

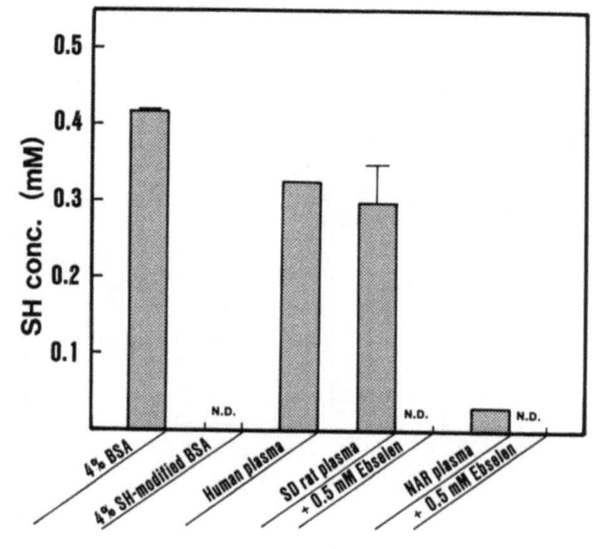

Fig. 3. Concentration of the SH group in 4% BSA, 4% SH-modified BSA and various plasmas measured with Ellman's reagent. ND = not determined

of ebselen in 4% SH-modified BSA was only 0.01 mM. The ^{14}C-warfarin binding to BSA was investigated as a reference compound which has no sulfhydryl group in the molecule. The binding of warfarin to BSA as detected by this extraction method was negligible (Fig. 1).

The binding of ^{14}C-ebselen to NAR plasma was less than that found in normal rat plasma (Fig. 2). Because NARs possess a trace amount of albumin and a larger amount of cholesterol and globulin in their serum, ebselen may interact with these substances.

Figure 3 gives the values of the SH group concentration in 4% BSA, 4% SH-modified BSA and various plasmas measured with DTNB. While the concentration of the SH group in 4% BSA was 0.4 mM, that in 4% SH-modified BSA was undetectable. The plasma SH group concentration in NAR was significantly lower than that found in normal rats. The SH groups found in rat plasma disappeared

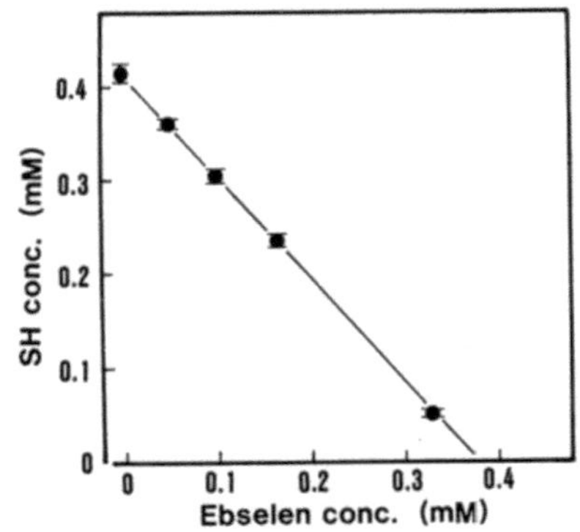

Fig. 4. Effect of ebselen on SH group concentration in 4% BSA solution. The concentration of the SH group was measured with Ellman's reagent

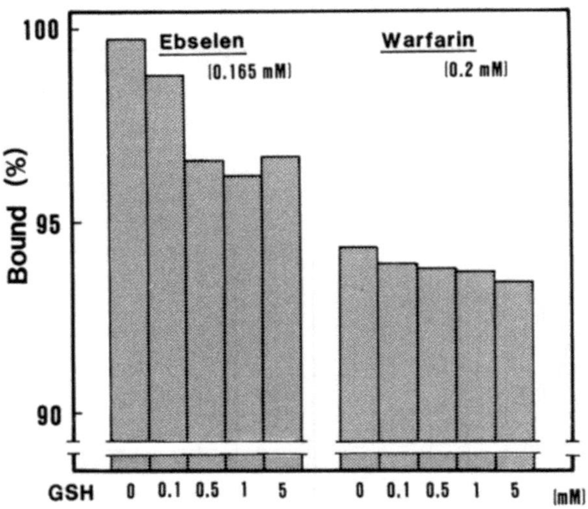

Fig. 5. Effect of GSH on binding of ebselen and warfarin to BSA. The amount of binding was determined by an ultrafiltration technique

completely with 0.5 mM ebselen. The SH group concentration in 4% BSA, 0.4 mM, which corresponds to the capacity of ebselen binding in 4% BSA, decreased proportionally with an increase of the amount of ebselen added (Fig. 4). These results suggested that the SH group in plasma was blocked directly with added ebselen.

The effect of the reduced type of glutathione (GSH) on binding of ebselen and warfarin is shown in Fig. 5. As bound percents shown in Fig. 5 were determined by an ultrafiltration method, the values represent not only the covalently bound ratio but also the non-covalently bound one. However, it is clear that the binding of ebselen significantly decreased with more than 0.5 mM of GSH. On the other hand, the binding of warfarin was not influenced by GSH. 2-Mercaptoethanol (2-ME), which is a strong agent for cleaving the disulfide bond, remarkably decreased the binding of ebselen to BSA (Table 1). From these results, it was suggested that these agents, i.e. GSH and 2-ME, cleaved the selenium-sulfur bond involved in the binding of ebselen to plasma proteins.

Table 1. Effect of 2-mercaptoethanol (2-ME) on binding of ebselen to BSA[a]

Volume of 2-ME added (μl)	Bound (%)
0	94.8
10	22.7
20	15.2
50	5.19

[a] After the indicated volume of 2-ME was added to 0.5 ml of 4% BSA solution with 7 μM of ^{14}C-ebselen, the extraction was performed with 3 ml chloroform (see Sect. 2)

In conclusion, we have found that ebselen forms covalent bonds, a selenium-sulfur linkage, with plasma proteins in vitro. Judging from the results of the present study and the pharmacokinetic properties of ebselen, it can be considered that the covalent bonds formed may readily dissociate in vivo and the reaction may be affected by endogenous thiols.

References

Fischer H, Terlinden R, Löhr JP, Römer A (1988) A novel biologically active selenoorganic compound. VIII. Biotransformation of ebselen. Xenobiotica 18:1347–1359

Kuhl P, Borbe HO, Römer A, Fischer H, Parnham MJ (1985) Selective inhibition of leukotriene B_4, formation by ebselen: a novel approach to anti-inflammatory therapy. Agents Actions 17:366–367

Müller A, Cadenas E, Graf P, Sies H (1984) A novel biologically active seleno-organic compound – I. Glutathione peroxidaselike activity in vitro and antioxidant capacity of PZ-51 (ebselen). Biochem Pharmacol 33:3235–3239

Müller A, Gabriel H, Sies H, Terlinden R, Fischer H, Römer A (1988) A novel biologically active seleno-organic compound – VII. Biotransformation of ebselen in perfused rat liver. Biochem Pharmacol 37:1103–1109

Simplicio PD, Ignesti G, Pirisino R (1985) Relationship between SH group reactivity and concentration of bovine serum albumin and rat plasma. Pharmacol Res Commun 17:913–925

Sugiyama K, Emori T, Nagase S (1980) Absence of albumin in tissues of analbuminemic rats. J Biochem 88:1413–1417

Protection of Experimental Cobalt Cardiomyopathy in the Rat by Selenium Pretreatment

JIAN-QI CUI[1] and GUANG-LU XU[1,2]

1 Introduction

Clinical evidence has implicated cobalt (Co) in the etiology of a primary myocardial disease occurring among certain heavy beer drinkers when Co was added to the beer (McDermott et al. 1966; Bonenfant et al. 1967; Morin et al. 1967; Kesteloot et al. 1968; Alexander 1972). Investigators have reported in experimental animals Co-induced myocardial damage similar to the pathologic manifestations of the beer drinker's heart disease (Sandusky et al. 1981). It has been reported that Co acts to accelerate the in vivo formation of lipid peroxides (LPO) in the blood of rabbits and causes a decrease in superoxide dismutase (SOD) activity, but fails to alter glutathione peroxidase (GSH-Px) in erythrocytes (Kuno et al. 1980; Morita et al. 1982). Selenium (Se)-vitamin E protection against acute Co cardiotoxicity has been observed in swine (Van Vleet et al. 1977).

The present studies were designed (1) to investigate the cardiomyopathy produced in rats by the administration of Co in terms of its relation to lipid peroxidation and protective agents involved in peroxidative pathways, and (2) to evaluate the ability of prior treatment with sodium selenite to provide protection against myocardial lesion.

2 Materials and Methods

In the present study, 72 Wistar rats weighing 100 g were divided into Co, Co with Se pretreatment (Co+Se) and control groups of 24 rats each, and supplied with a stock diet ad libitum. The Co group received i.p. injections of 5 mg Co^{2+} kg^{-1} body wt daily for 3 days, and then every 2 days four times. The Co+Se group was given i.p. seven doses of 50 µg Se kg^{-1} body wt prior to Co injection, and then received Se together with Co at the same interval. The control was given the same volume of saline injection. Each group was subdivided into three subgroups of eight rats each, and rats were killed after 3, 5 and 7 injections of Co respectively.

[1] Research Laboratory of Keshan Disease, Xi'an Medical University, Xi'an 710061, Shaanxi, The People's Republic of China
[2] To whom correspondence should be addressed

Samples of blood and heart were collected for histopathologic studies with hematoxylin and eosin staining and histochemical examinations of succinic dehydrogenase (SDH) by nitroneotetrazolium chloride and of lipid by bromine sudan black, and for determinations of the following substances: creatine phosphokinase (CPK) and glutamic oxaloacetic transaminase (GOT) by automated analyzer, free fatty acid (FFA) colorimetrically with diphenylcarbazide, Se by 2,3-DAN fluorometry, glutathione peroxidase (GSH-Px) by NADPH-coupled assay, superoxide dismutase (SOD) by ferricytochrome C assay and lipid peroxide (LPO) by thiobarbituric acid reaction.

3 Results

The Co-treated rats, as compared to the control, showed significant increases in plasma CPK and GOT after injections of Co (Tables 1 and 2). Under the light microscope, the hearts in part of the rats (4/7) after seven doses had remarkable cardiac lesions: interstitial edema, dissolution of myofibres and inflammatory cell infiltration. Histochemically, SDH was less intense and lipid staining was increased in intensity in myocardium of Co-treated rats. A marked increase in cardiac

Table 1. Plasma CPK (IU l^{-1})

Group	Doses of Co injection*		
	3	5	7
Control	107 ± 9 (7)^a	182 ± 20 (7)^a	98 ± 13 (8)^a
Co	196 ± 41 (6)^b	405 ± 77 (8)^b	163 ± 58 (8)^a
Co + Se	140 ± 10 (6)^b	142 ± 7 (8)^a	132 ± 23 (8)^a

* $\bar{x}$ + SE (N)
Means with different superscripts in the same vertical column are significantly different (p < 0.05 or 0.01)

Table 2. Plasma GOT (IU l^{-1})

Group	Doses of Co injection*		
	3	5	7
Control	91 ± 6 (7)^a	93 ± 4 (7)^a	71 ± 8 (8)^a
Co	204 ± 65 (6)^a	237 ± 43 (8)^b	218 ± 61 (8)^b
Co + Se	136 ± 25 (6)^a	119 ± 33 (8)^a	59 ± 9 (8)^a

* $\bar{x}$ + SE (N)
Means with different superscripts in the same vertical column are significantly different (p < 0.05 or 0.01)

Table 3. Cardiac FFA

Group	mEq g^{-1} Protein*
Control	10.4±2.3 (6)[a]
Co	26.7±5.6 (6)[b]
Co+Se	8.8±0.4 (6)[a]

* $\bar{x}$+SE (N)
Means with different superscripts in the same vertical column are significantly different ($p<0.05$ or 0.01)

Table 4. Cardiac Se contents (ng g^{-1})

Group	Doses of Co injection*		
	3	5	7
Control	133±18 (7)[a]	124±12 (7)[a]	121± 9 (8)[a]
Co	118±18 (6)[a]	121±22 (8)[a]	110±16 (8)[a]
Co+Se	245±11 (6)[b]	256±19 (8)[b]	255±24 (8)[b]

* $\bar{x}$+SE (N)
Means with different superscripts in the same vertical column are significantly different ($p<0.05$ or 0.01)

Table 5. Cardiac SOD (U mg^{-1} protein)

Group	Doses of Co injection[a]		
	3	5	7
Control	18.2±0.6 (7)[a]	17.9±0.7 (7)[a]	17.6±1.0 (8)[a]
Co	17.5±1.0 (6)[a]	18.4±1.0 (7)[a]	17.8±0.6 (8)[a]
Co+Se	17.7±0.5 (6)[a]	17.4±0.4 (8)[a]	17.4±1.1 (8)[a]

[a] $\bar{x}$+SE (N)
Means with different superscripts in the same vertical column are significantly different ($p<0.05$ or 0.01)

FFA content was seen (Table 3). No significant changes for cardiac Se content (Table 4) and SOD activity were observed (Table 5). Cardiac GSH-Px activities were unchanged after three and five doses (Table 6). Although the activity decreased after seven injections (Table 6), there was no change in cardiac LPO level (Table 7).

Pretreatment of Se resulted in protection against Co cardiomyopathy as demonstrated by the decrease in plasma CPK and GOT (Tables 1 and 2), absence of histopathologic myocardial damage and lack of histochemical abnormalities in

Table 6. Cardiac GSH-Px (mU mg^{-1} protein)

Group	Doses of Co injection*		
	3	5	7
Control	77 ± 9 (6)[a]	73 ± 10 (7)[a]	85 ± 14 (7)[a]
Co	73 ± 11 (6)[a]	69 ± 10 (7)[a]	64 ± 10 (8)[b]
Co + Se	142 ± 10 (6)[b]	204 ± 38 (7)[b]	258 ± 16 (8)[c]

* $\bar{x} \pm SE$ (N)
Means with different superscripts in the same vertical column are significantly different ($p < 0.05$ or 0.01)

Table 7. Cardiac LPO

Group	nmol MDA mg^{-1} Protein*
Control	2.22 ± 0.37 (7)[a]
Co	2.18 ± 0.35 (7)[a]
Co + Se	2.97 ± 0.40 (6)[a]

* $\bar{x} + SE$ (N)
Means with different superscripts in the same vertical column are significantly different ($p < 0.05$ or 0.01)

SDH reaction and lipid staining as well as elevated cardiac FFA levels (Table 3). Se acted to elevate the Se contents in hearts along with an increase of cardiac GSH-Px activity (Tables 4 and 6), but failed to alter the SOD activity and formation of LPO in the heart (Tables 5 and 7).

4 Conclusion

The results demonstrate that first, the Co-induced cardiomyopathy in rats might not be caused by an elevation of LPO production, secondly, Se pretreatment provides protection against Co-induced cardiomyopathy and thirdly, the protective effect of Se appears to be related to the reversal of the inhibition of SDH and deposition of lipid in hearts rather than the detoxification of LPO through the increased GSH-Px activity.

Acknowledgements. The authors thank N.-Z. Chen, W.-L. Xue, Y.-Q. Xue, C.-H. Tang and J.-P. Tian for their valuable advice and technical assistance.

References

Alexander CS (1972) Cobalt beer cardiomyopathy: a clinical and pathologic study of 28 cases. Am J Med 53:395–417

Bonenfant JL, Miller G, Roy PE (1967) Quebec beer-drinkers' cardiomyopathy: pathologic studies. Can Med Assoc J 97:910–916

Kesteloot H, Roelandt J, Williams J, Claes JH, Joossens JV (1968) An inquiry about the role of cobalt in the heart disease of chronic beer drinkers. Circulation 37:854–864

Kuno Y, Tochihara N, Koike S (1980) The effects of cobalt chloride on the formation of blood lipid peroxide related to glutathione peroxidase in the erythrocytes of rabbits. Jpn J Hyg 35:665–669

McDermott PH, Delaney RL, Egan JD, Sullivan JF (1966) Myocardosis and cardiac failure in men. JAMA 198:253–256

Morin Y, Daniel P (1967) Quebec beer drinkers' cardiomyopathy: etiological consideration. Can Med Assoc J 97:926–929

Morita H, Kuno Y, Koike S (1982) The effects of cobalt on superoxide dismutase activity, methemoglobin formation and lipid peroxide in rabbit. Jpn J Hyg 37:597–600

Sandusky GE, Henk WG, Roberts ED (1981) Histochemistry and ultrastructure of the heart in experimental cobalt cardiomyopathy in the dog. Toxicol Appl Pharmacol 61:89–95

Van Vleet JF, Rebar AH, Ferrans VJ (1977) Acute cobalt and isoproterenol cardiotoxicity in swine: protection by selenium-vitamin E supplementation and potentiation by stress-susceptible phenotype. Am J Vet Res 38:991–1002

Correlation of Levels of Platinum and Selenium in Rats Treated with Cisplatin and Sodium Selenite

H. SAKURAI[1,4], K. TSUCHIYA[1], S. HIRAI[2], Y. OKADA[2], and H. HARAGUCHI[3]

The ability of selenium compounds to reduce the toxicity of mercury compounds has been demonstrated in animals (Parizek and Ostadalove 1967; Hill 1974) and humans (Kosta et al. 1975), the molar ratio of mercury to selenium being approximately 1:1 in certain organs after this combination therapy. Among selenium compounds, sodium selenite was found by us (Komiya et al. 1977) to have the most powerful protective effect against toxicity of mercuric chloride.

Cisplatin [cis-dichlorodiamine platinum(II)], currently used therapeutically (Rosenberg 1977), has antitumor activity against a number of human neoplasms (Einhorn and Donohue 1977; Merrin 1979; Yagoda et al. 1976; Young et al. 1979). However, it cannot be used for long periods or at high doses because it has adverse effects on renal functions. Recently, Naganuma et al. (1983) reported that co-administration of a selenium compound, sodium selenite, protected experimental animals against the toxicity of cisplatin without influencing its therapeutic effect. Thus, we expected that there would be some correlation between the levels of platinum and selenium in animals treated with cisplatin plus a selenium compound.

Male Wistar rats were divided into four groups of two or three animals each: a control group and groups administered with sodium selenite (Komiya et al. 1977), cisplatin and sodium selenite plus cisplatin. Sodium selenite and cisplatin were given s.c. at 11.6 and 16.7 μmol kg^{-1} body weight, respectively, and the animals were killed 24 h later. Portions of various organs and subcellular fractions of the liver and kidney were freeze-dried. The platinum and selenium contents of the samples were determined by neutron activation analysis on the nuclear reactor of the Atomic Energy Research Laboratory of Musashi Institute of Technology, using the following nuclear reactions: ${}^{198}_{78}Pt(n,\gamma,\beta){}^{199}_{79}Au$ (half-life 3.1 days, 158 keV) (Heydorn 1983) and ${}^{76}_{34}Se(n,\gamma){}^{77m}_{34}Se$ (half-life 17.5 s, 162 keV) (Hirai et al. 1978). Protein was measured by the method of Lowry et al. (1951).

The results are shown in Fig. 1, in which the levels of both elements are represented as mean values for two or three rats. The results can be summarized: (1)

[1] Faculty of Pharmaceutical Sciences, University of Tokushima, Sho-machi 1, Tokushima 770, Japan
[2] Atomic Energy Research Laboratory, Musashi Institute of Technology, Ozenji 971, Tama-ku, Kawasaki-shi, Kanagawa 215, Japan
[3] Department of Chemistry, Faculty of Science, University of Tokyo, Hongo 7-3-1, Bukyo-ku, Tokyo 113, Japan
[4] To whom correspondence should be addressed

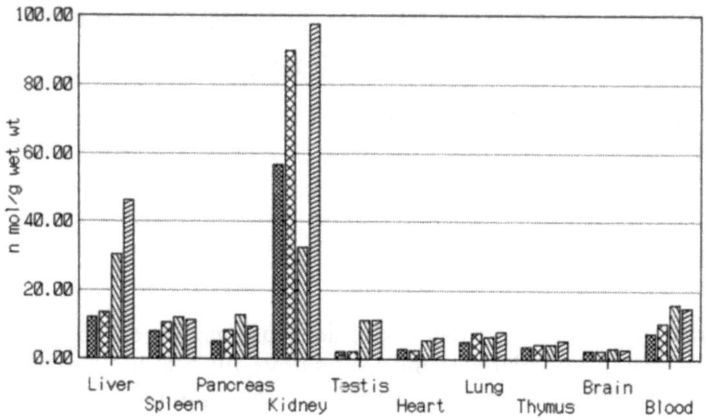

Fig. 1. Distributions of platinum and selenium in various organs of rats treated with cisplatin, sodium selenite or cisplatin plus sodium selenite. ▩ Platinum level in rats treated with cisplatin; ▩ platinum level in rats treated with cisplatin plus sodium selenite; ▧ selenium level in rats treated with sodium selenite; ▨ selenium level in rats treated with cisplatin plus sodium selenite

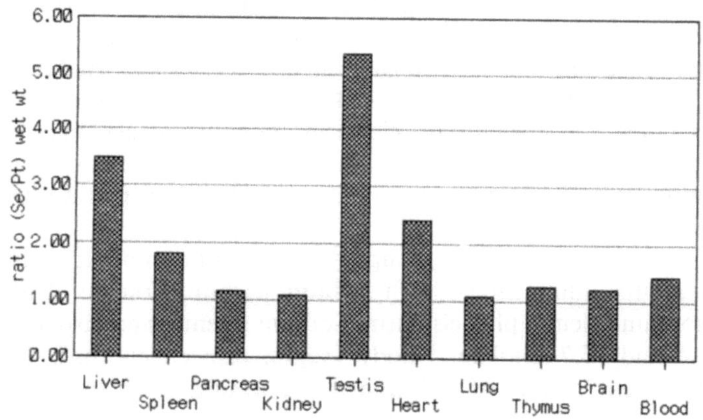

Fig. 2. Se/Pt ratios in various organs of rats treated with cisplatin plus sodium selenite

The accumulations of both elements were highest in the kidney. (2) Retention of selenium in the kidney and liver was enhanced by co-administration of cisplatin. (3) The accumulation of platinum in the kidney, spleen, pancreas, lung and blood was increased by selenium administration. (4) The total accumulated levels of platinum and selenium were about 35 and 24% higher, respectively, after administration of cisplatin plus sodium selenite than after injection of each compound singly.

The molar ratio of selenium to platinum (Se/Pt) ranged from about 1:1 to 5:1, depending on organs, as shown in Fig. 2. It is interesting to note that the ratio (Se/Pt) in the liver and kidney was about 3:1 and 1:1, respectively.

Since the accumulations of both elements were greatest in the liver and kidney, the levels and ratio of the two elements in subcellular fractions of these organs

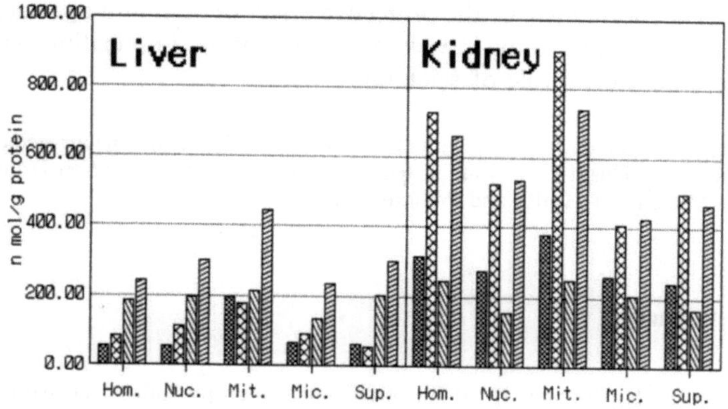

Fig. 3. Levels of platinum and selenium in subcellular fractions of liver and kidney of rats treated with cisplatin, sodium selenite or cisplatin plus sodium selenite. *Symbols* are the same as for Fig. 1

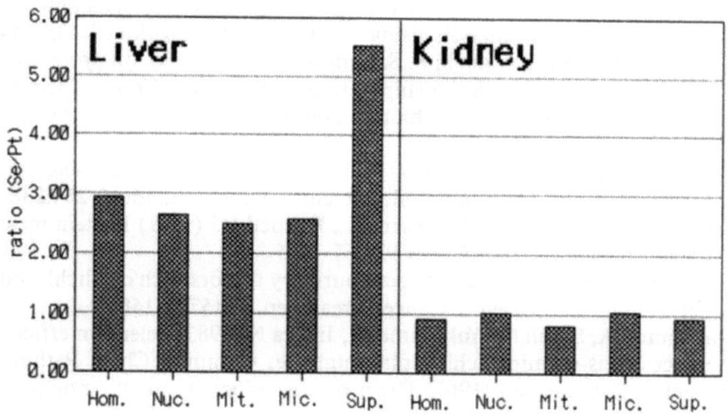

Fig. 4. Se/Pt ratios in subcellular fractions of liver and kidney of rats treated with cisplatin plus sodium selenite

were studied. The accumulation of both platinum and selenium was observed in all fractions of the kidney and the nuclear and microsomal fractions of the liver (Fig. 3). The ratio of Se/Pt in the subcellular fractions was nearly 3 : 1 in all subcellular fractions of the liver except the supernatant and was 1 : 1 in all subcellular fractions of the kidney (Fig. 4).

Selenium is normally present in all organs (data not shown), but the level of platinum in various organs was not influenced by the naturally occurring selenium. However, a significant accumulation of platinum in the kidney, liver, spleen, pancreas and lung was found after administration of sodium selenite with cisplatin. This observation differs from findings on mercury detoxification by selenium, which indicated that co-administration of naturally occurring selenium was autoprotective (Kosta et al. 1975).

The mechanism of the protective effect of selenium against platinum toxicity requires further study, in view of its therapeutic implications and its relevance to the biochemical role of selenium.

Acknowledgements. We are grateful to Prof. K. Fuwa for encouragement and constructive discussion. This study was supported in part by a grant-in-aid for New Drug Developments from the Ministry of Health and Welfare of Japan.

References

Einhorn LH, Donohue JP (1977) cis-Daminedichloroplatinum, vinblastine and bleomycin combination chemotherapy in disseminated testicular cancer. Ann Int Med 87:293–298

Heydorn K (1983) Trace elements in clinical chemistry determined by neutron activation analysis. In: Sigel H (ed) Metal ions in biological systems vol 16. Dekker, New York, Basel, p 123

Hill CH (1974) Reversal of selenium toxicity in chicks by mercury, copper and cadmium. J Nutr 104:593–598

Hirai S, Suzuki S, Okamoto M (1978) Determination of selenium in biological materials by neutron activation analysis using ^{77m}Se. Bunseki Kagaku 27:435–440

Komiya K, Koike I, Kawauchi S, Sakurai H (1977) Studies on protective effect of mercurial toxicity by selenium I: relationship between protective effects of various selenium compounds on the toxicity of mercuric chloride and distributions of mercury and selenium in rat tissues. Eisei Kagaku 23:235–243

Kosta L, Byrne AR, Zelenko V (1975) Correlation between selenium and mercury in man following exposure to inorganic mercury. Nature (London) 254:238–239

Lowry OH, Rosenbrough NJ, Farr AL, Randell RJ (1951) Protein measurement with the Folin phenol reagent. J Biol Chem 93:265–275

Merrin CE (1979) Treatment of genitourinary tumors with cis-dichlorodiamineplatinum(II): experience in 250 patients. Cancer Treat Rep 63:1579–1584

Naganuma A, Satoh M, Yokoyama M, Imura N (1983) Selenium efficiently depressed toxic side effect of cis-diaminedichloroplatinum. Res Commun Chem Pathol Pharmacol 42:127–134

Parizek J, Ostadalove I (1967) The protective effect of small amounts of selenium in sublimate intoxication. Experientia 23:142–144

Rosenberg B (1977) Noble meal complexes in cancer chemotherapy. Adv Exp Med Biol 91:129–150

Yagoda A, Watson RC, Gonalez-Vitale JC, Grabstald H, Whitmore WF (1976) cis-Dichlorodiamineplatinum(II) in advanced bladder cancer. Cancer Treat Rep 60:917–923

Young RC, Von Hoff DD, Gormley P, Makuch R, Cassidy J, Howser D, Bull JM (1979) cis-Dichlorodiamineplatinum(II) for the treatment of advanced ovarian cancer. Cancer Treat Rep 63:1539–1544

Part IV Selenium and Human Health

Progress in Establishing Human Nutritional Requirements and Dietary Recommendations for Selenium

O. A. LEVANDER [1]

1 Introduction

In 1980, the Committee on Dietary Allowances of the U.S. National Research Council established the first human dietary standard for selenium (National Research Council 1980). This standard, called an "Estimated Safe and Adequate Daily Dietary Intake", was presented as a range to reflect the uncertainty in this value. Since that time, several studies have been published which increase greatly our understanding of human selenium requirements (Levander 1987). This new information enabled the 1985 Committee to propose the first Recommended Dietary Allowance (RDA) for selenium.

2 Selenium Requirements and Dietary Standards for Adults

Early attempts to estimate human selenium requirements were based on extrapolation from animal studies but such estimates could vary widely depending upon the species and deficiency criterion selected (Levander 1986). Likewise, the method of metabolic balance has not been very helpful in delineating human selenium requirements. Studies carried out in New Zealand, the USA, and China have shown that people can adapt to a wide range of dietary selenium intakes and can maintain balance even at intakes associated with Keshan disease (Levander 1986). Dietary surveys conducted in China have provided information regarding minimum daily requirements for selenium. Comparison of dietary intakes in areas with and without Keshan disease indicated that Chinese men and women require 19 and 13 μg day^{-1}, respectively, to prevent the disease (Yang et al. 1987).

Perhaps the most promising approach to the determination of human selenium requirements is that of Yang et al. (1987) who measured the amount of dietary selenium needed to saturate plasma glutathione peroxidase activity. In this study, adult men living in a low selenium area (average dietary intake only 10 μg day^{-1}) were supplemented with graded doses of selenomethionine for several months. It was found that the activity of the plasma enzyme plateaued at selenium intakes of 40 μg or more. This figure (with appropriate adjustments) was used to calculate a RDA for North Americans by the 1985 RDA Committee.

---

[1] Vitamin and Mineral Nutrition Laboratory, US Department of Agriculture, Agricultural Research Service, Beltsville Human Nutrition Research Center, Beltsville, MD 20705, USA

First, Yang's value had to be adjusted for body weight (the men in Yang's study weighed 60 kg) since this variable is thought to influence the selenium requirement (Levander 1987). RDAs also generally include some kind of safety factor, often based on the experimental variance observed in the estimate of the requirement. The 1985 RDA Committee used a safety factor of 1.3 based on an arbitrarily assumed coefficient of variation of 15% (doubling the coefficient of variation would allow the proposed recommendation to encompass, at least on theoretical statistical grounds, 97.5% of a given population group). This estimate of variation was considered reasonable since several studies of human protein requirements have similar coefficients of variation (National Research Council 1980) and the metabolism of selenium is closely linked to that of protein. Multiplying the 40 μg estimated requirement figure by 79/60 or 62/60 to adjust for body weights and 1.3 to provide a safety factor yields a dietary recommendation after rounding of 70 and 55 μg of selenium per day for the standard North American 79-kg male and 62-kg female, respectively.

3 Selenium Requirements and Dietary Standards for Children and Infants

Yang et al. (1987) also compared the selenium intakes of children and infants living in areas of China with and without Keshan disease. The mean dietary selenium intakes for 3- to 17-year-old age groups were 5 to 14 μg day^{-1} in nonendemic areas vs 3 to 6 μg day^{-1} in endemic areas. The 1985 RDA Committee knew of no other data regarding selenium requirements during this stage of the life cycle so the recommendations for children were extrapolated from the US adult recommendation based on the factor body $weight^{0.75}$, an index of metabolic body size. Such calculations yielded recommended intakes of 20 to 50 μg day^{-1} for the 2- to 18-year-old age groups.

As in the case of children, there are no direct experimental determinations of infant selenium requirements. However, a dietary survey conducted by Yang et al. (1987) of breast-fed infants 1 year of age or younger showed that the selenium intake of Chinese babies was 1.0 to 2.8 and 2.0 to 3.4 μg day^{-1} in areas endemic and nonendemic for Keshan disease, respectively. In North America, the average selenium content of breast milk ranges between 15 and 20 μg l^{-1} (Mannon and Picciano 1987). The 1985 RDA Committee assumed that breast-fed infants consume about 750 ml of milk daily and proposed a selenium recommendation for infants of 13 μg day^{-1}.

4 Selenium Requirements and Dietary Standards for the Elderly

Dietary surveys carried out in countries with soils relatively rich in selenium (e.g., Canada, USA) indicate that elderly people have an adequate selenium status when measured by the typical criteria (dietary intakes and blood selenium levels and glutathione peroxidase activities) (Mertz et al. 1988). In the absence of any evidence indicating an increased need for selenium in the elderly, the 1985 RDA Committee recommended intakes of 70 and 55 μg day^{-1} for men and women over the age of 70, respectively, the same recommendation made for young and middle-aged adults.

5 Selenium Requirements and Dietary Standards for Pregnancy and Lactation

Swanson et al. (1983) found that women retained 10 and 23 μg of selenium per day during the second and fourth quarters of pregnancy. However, the amount of selenium fed to these subjects during that balance study (150 μg day^{-1}) was high compared to their usual intakes and this may have contributed to misleadingly high retention values. Fetal demand can also be estimated by the factorial technique. For example, if it is assumed that 5 kg of lean tissue containing an average selenium concentration of 0.25 μg g^{-1} (Schroeder et al. 1970) are laid down during gestation, a total of 1.25 mg of selenium would have to be retained over the course of the pregnancy. Since most fetal growth occurs during the last trimester, fetal demand during this time would be 14 μg day^{-1}. The 1985 RDA Committee assumed an absorption rate of 80% (Levander 1983) and recommended an increase in dietary selenium of 18 μg day^{-1} during the last trimester of pregnancy.

Increased selenium needs during lactation can be estimated from the selenium lost in breast milk. For North American women, the 1985 RDA Committee calculated a lacteal loss of 13 μg Se day^{-1} by assuming an average selenium concentration of 17.5 μg l^{-1} milk and a mean milk production of 750 ml day^{-1}. Assuming again an absorption rate of 80%, an additional selenium intake of 16 μg day^{-1} was recommended during lactation to prevent selenium depletion of the mother and to maintain the level of selenium in the breast milk.

6 The Appropriateness of the Proposed US Dietary Standard for Selenium

Two types of information were used by the 1985 RDA Committee to justify their dietary selenium recommendations. The recommendations for adults, children, and the elderly were based on the metabolic experiment of Yang et al. (1987). On the other hand, the recommendations for infants, pregnant women, and lactating women were based on the selenium content of human milk and tissue in North America.

With regard to the recommendations based on the Yang study, it could rightly be asked whether maximization of glutathione peroxidase activity is really necessary for optimal human health. The utilization of glutathione to reduce hydroperoxides via glutathione peroxidase depends on a complex series of metabolic reactions for the regeneration of reduced glutathione including those catalyzed by glutathione reductase and glucose-6-phosphate dehydrogenase. Should glutathione peroxidase not be the kinetically rate-limiting step in this overall sequence, then maximization of its activity may not be all that physiologically relevant and dietary selenium standards could be set correspondingly lower.

In the case of the recommendations based on North American human milk and tissue selenium concentrations, it should be understood that those apply specifically to North Americans and probably represent luxus levels of intake. For example, the recommended selenium intake for infants based on breast milk selenium content presumably incorporates a substantial safety margin since no untoward effects of low selenium intake have been seen in Finland or New Zealand where infants are fed breast milk that contains less than half the selenium content of North American breast milk. Likewise, the *additional* selenium intake recommended during the last trimester of pregnancy approaches the *total* selenium intake reported in some healthy New Zealand women, so again the North American standard probably includes a sizeable safety factor.

The 1985 RDA Committee concerned itself solely with dietary recommendations for North Americans and, given the lack of evidence for any selenium deficiency in the healthy US population, recognized the need to propose standards that could readily be met by a typical well-balanced North American diet. Extensive surveys of US diets found mean selenium intakes of 18, 52, and 108 $\mu g\ day^{-1}$ for infants, toddlers (1- to 3-year age group), and adults, respectively (Pennington et al. 1984), enough to meet the proposed RDA. National diets in certain other countries do not provide such generous amounts of selenium (World Health Organization 1987) and in fact would not meet the 1985 Committee's recommendation without some sort of supplementation or fortification program. Supplements in the form of pills containing sodium selenite have been used extensively in China to protect children against Keshan disease. Primarily on the basis of epidemiological evidence, authorities in Finland initiated a national program to increase the selenium content of their food supply through the addition of selenate to fertilizers (Varo et al. 1988). Scientists in New Zealand, however, have uncovered no evidence of any human health problem due to the low selenium status of their population and no steps are being taken to alter that status. Each country will have to develop its own national policy on selenium nutrition appropriate to its own circumstances.

One could perhaps criticize the US dietary selenium standards as being unrealistically high and difficult for certain other nations to meet. However, when the 1985 RDA Committee proposed its dietary selenium recommendations, it did not consider the putative beneficial effects of high levels of selenium against certain degenerative human diseases, such as cancer. Some advocates of the use of selenium in the chemoprevention of cancer have recommended intakes in excess of the upper limit of the safe and adequate range (i.e., $>200\ \mu g\ day^{-1}$) (National Research Council 1980). Several epidemiological and experimental animal studies have now been published which are inconsistent with a chemopreventive role for selenium in cancer (Levander 1987). Moreover, Birt et al. (1988) have recently observed that selenium may have cancer-promoting activity under certain conditions. Therefore, it would seem prudent not to recommend excessive levels of selenium supplementation at this time.

References

Birt DF, Julius AD, Runice CE, White LT, Lawson T, Pour PM (1988) Enhancement of BOP-induced pancreatic carcinogenesis in selenium-fed Syrian golden hamsters under specific dietary conditions. Nutr Cancer 11:21–33

Levander OA (1983) Considerations in the design of selenium bioavailability studies. Fed Proc 42:1721–1725

Levander OA (1986) Selenium. In: Mertz W (ed) Trace elements in human and animal nutrition, 5th edn, vol 2. Academic Press, London New York, p 209

Levander OA (1987) A global view of human selenium nutrition. Annu Rev Nutr 7:227–250

Mannon S, Picciano MF (1987) Influence of maternal selenium status on human milk selenium concentration and glutathione peroxidase activity. Am J Clin Nutr 46:95–100

Mertz W, Morris ER, Smith JC, Udomkesmalee E, Fields M, Levander OA, Anderson RA (1989) Trace elements in the elderly: metabolism, requirements and recommendations for intakes. In: Munro HN (ed) Nutrition and aging, a volume in human nutrition: a comprehensive treatise. Plenum, New York London, p 195

National Research Council (1980) Recommended dietary allowances, 9th ed. Natl Acad Sci, Washington DC

Pennington JAT, Wilson DB, Newell RF, Harland BF, Johnson RD, Vanderveen JE (1984) Selected minerals in food surveys, 1974 to 1981/82. J Am Diet Assoc 84:771–780

Schroeder HA, Frost DV, Balassa JJ (1970) Essential trace metals in man: selenium. J Chron Dis 23:227–243

Swanson CA, Reamer DC, Veillon C, King JC, Levander OA (1983) Quantitative and qualitative aspects of selenium utilization in pregnant and nonpregnant women: an application of stable isotope methodology. Am J Clin Nutr 38:169–180

Varo P, Alfthan G, Ekholm P, Aro A, Koivistoinen P (1988) Selenium intake and serum selenium in Finland: effects of soil fertilization with selenium. Am J Clin Nutr 48:324–329

World Health Organization (1987) Environmental Health Criteria 58: Selenium. A report of the International Programme on Chemical Safety. World Health Organization, Geneva

Yang GQ, Zhu LZ, Liu SJ, Gu LZ, Qian PC, Huang JH, Lu MO (1987) Human selenium requirements in China. In: Combs GF, Spallholz JE, Levander OA, Oldfield JE (eds) Proc 3rd Int Symp Selenium Biol Med. AVI Press, Westport, Connecticut, p 589

Supplementation of a Population with Selenium – Reasons for Action and Initial Experience

G. ALFTHAN[1], A. ARO[1], and J.K. HUTTUNEN[1,2]

1 Introduction

The Ice Age deprived Finland of much of its selenium (Se) through leaching. Over 60% of Finnish rock is igneous with a very low selenium concentration. As a result, the Se concentration of cultivated soils is low, $0.2-0.3\ mg\ kg^{-1}$. This factor combined with climatic conditions, and a low pH and high iron content of soil, provides poor availability of Se for plants (Yläranta 1985). Because of financial losses caused by Se-deficiency diseases in domestic animals, Finland was among the first countries to initiate supplementation of animal feeds with selenite in 1969. This measure was successful, and Se deficiency diseases of animals are currently rare in Finland.

2 Selenium in Food and Human Serum in Finland

In the mid-1970s, a survey of 30 mineral elements in several 100 food items revealed the Se content of food produced in Finland to be exceptionally low (yielding an intake of $20-30\ \mu g$ Se day^{-1}), as compared to that of other nations (Koivistoinen 1980). Analysis of human blood and serum samples confirmed that the Se status of the population in Finland was similar to that reported in New Zealand, another country with exceptionally low dietary Se intake (Westermarck et al. 1977; Thomson et al. 1985).

It has been estimated that during the years 1941 and 1981 the mean Se intake of the Finnish population varied between 20 and $55\ \mu g\ day^{-1}$, depending on the ratio of the imported high-Se wheat and rye to the low-Se domestic grain in the diet (Mutanen and Koivistoinen 1983). In the 1970s, when data for several epidemiological studies were collected, grain imports were limited. Thus, the mean intake of Se was low throughout that decade. During the first half of the 1980s, however, Finland experienced crop failures and was forced to import Se-rich grain from North America. Between 1975 and 1984 the mean serum Se concentration of healthy adults in southern Finland varied annually between 50 and $100\ \mu g\ l^{-1}$ (Alfthan 1988). The mean serum Se concentration correlated well ($r = 0.89$) with the estimated national intake on an annual basis.

---

[1] National Public Health Institute, Mannerheimintie 166, 00280 Helsinki, Finland
[2] To whom correspondence should be addressed

3 Clinical and Epidemiological Studies

In 1982 a bioavailability trial was carried out among healthy, middle-aged men in central Finland (Levander et al. 1983) to determine whether the Se intake characteristic of the population in the early 1980s (50–60 μg day^{-1}) was sufficient to maintain maximal activity of platelet glutathione peroxidase (GSH-Px). Platelet GSH-Px activity increased in 4 weeks by 70% in subjects receiving 200 μg Se day^{-1} from selenate and Se-rich wheat, and remained stable during the remainder of the supplementation period. Similar responses have been reported in other populations with low intakes, but not in those with high Se intakes (cf. Thomson et al. 1985). The substantial increase in GSH-Px activity during Se supplementation suggested, therefore, that dietary intake was suboptimal.

Several longitudinal, epidemiological investigations were conducted in cohorts examined in the 1970s and 1980s to test the hypothesis that a low Se intake is linked to cardiovascular disease and cancer. Many of these studies suggested that a very low level of serum Se (below 45 μg l^{-1}) is predictive of the development of coronary heart disease, stroke or cancer (cf. Salonen and Huttunen 1986; Salonen 1987). Similar findings have been reported from other populations, although the data are still conflicting (for review, see Stampfer et al. 1987). A recent clinical trial report from Finland (Salonen et al. 1988) suggests that low Se levels may increase the aggregability of platelets, which could explain the association between Se deficiency and atherosclerotic diseases. Other studies have shown that Se supplements prevent, at least in pharmacological doses, cancer in experimental animals (Ip 1986).

4 Measures to Improve Se Status in Finland

As it became evident that the mean dietary intake of Se in Finland was exceptionally low, concern was expressed about the nutritional adequacy of Se in agricultural products (Koivistoinen and Huttunen 1986). An expert group appointed by the Ministry of Agriculture and Forestry evaluated the role of Se in human nutrition and health (Expert Group on Selenium 1983) and recommended that the Se content of feeds and basic foods should be increased by adding sodium selenate to multimineral fertilizers. The proposed measure was economical on a national scale and likely to be comprehensive in its effects on animal and human nutrition. It was thought that the experimental, epidemiological and clinical data provided suggestive evidence for the role of inadequate selenium nutrition in the etiology of human diseases. Supplementation via fertilizers was deemed to be safe, since the plants would act as a natural buffer and supplementation could be discontinued rapidly if considered necessary.

The Se concentration of domestically produced spring wheat increased from 0.01 mg kg^{-1} before to over 0.20 mg kg^{-1} after the fertilization program (Varo et al. 1988). The Se levels of beef, pork and milk increased threefold. Concentrations in some other foods of animal origin, such as eggs and liver, increased less because of the earlier selenite supplementation of animal feeds.

Over a period of about 6 months the serum Se concentration of both urban and rural subjects in the follow-up groups increased from 70 to 100 $\mu g\,l^{-1}$ (Varo et al. 1988). The rate of increase in the Se level of foodstuffs has levelled off after 3 years of Se fertilization, and serum Se has stabilized at the level of 110 $\mu g\,l^{-1}$ (Aro et al. 1989). The mean national daily Se intake is now calculated to be 100–110 $\mu g\,day^{-1}$ at an energy intake of 10 MJ.

A unique opportunity was afforded when 45 of the original 50 subjects in the Se bioavailability trial (Levander et al. 1983) 6 years earlier were found willing to participate in a similar study in 1987 (Alfthan, Aro and Arvilommi unpublished results).

Three groups of ten men were supplemented for 16 weeks with 200 µg daily of selenium in various supplements. The mean prestudy plasma Se concentration was 110 $\mu g\,l^{-1}$. Both selenate and selenite increased platelet GSH-Px activity after 4 weeks by 25% compared to placebo, an effect distinctly smaller than the 70% increase observed in the same subjects following the earlier basal intake of 55 $\mu g\,day^{-1}$ (Levander et al. 1983). Thus, the present Se intake of 100 $\mu g\,day^{-1}$ in Finland approaches the level needed to maintain maximum activity of thrombocyte GSH-Px, which has been estimated to occur at a plasma Se concentration of 100–135 $\mu g\,l^{-1}$ (Levander et al. 1983).

5 Conclusions

For geochemical reasons Finland is a low-Se area. Supplementation of animal feeds with Se in the 1960s successfully eliminated Se-deficiency diseases in domestic animals. The decision to add Se to multimineral fertilizers, beginning in 1984, was based on studies of the Se content of human foods, epidemiological investigations and clinical trials of Se supplements. As a result, the mean Se intake in Finland has increased from about 20–50 to over 100 $\mu g\,day^{-1}$ in 1986. At the same time the mean serum Se concentration has increased from 70 to 110 $\mu g\,l^{-1}$. The Se intake of the population is now within the limits considered to be safe and adequate.

References

Alfthan G (1988) Longitudinal study on the selenium status of healthy adults in Finland during 1975–1984. Nutr Res 8:467–476

Aro A, Alfthan G, Varo P (1989) Se supplementation of fertilizers has increased the Se intake and serum Se concentration of Finnish people. Proc 4th Int Symp Selenium Biol Med. Springer, Berlin Heidelberg New York

Expert Group on Selenium (1983) Addition of selenium to fertilizers – review and proposal for actions. Rep Minist Agric For, Finland (in Finnish)

Ip C (1986) The chemopreventive role of selenium in carcinogenesis. J Am Coll Toxicol 5:7–20

Koivistoinen P (ed) (1980) Mineral element composition of Finnish foods I–XII. Acta Agric Scand Suppl 22

Koivistoinen P, Huttunen JK (1986) Selenium in food and nutrition in Finland. An overview on research and action. Ann Clin Res 18:13–17

Levander OA, Alfthan G, Arvilommi H, Gref CG, Huttunen JK, Kataja M, Koivistoinen P, Pikkarainen J (1983) Bioavailability of selenium to Finnish men as assessed by platelet glutathione peroxidase activity and other blood parameters. Am J Clin Nutr 37:887–897

Mutanen M, Koivistoinen P (1983) The role of imported grain on the selenium intake of Finnish population in 1941–1981. Int J Vitam Nutr Res 53:102–108

Salonen JT (1987) Selenium in ischaemic heart disease. Int J Epid (Suppl) 16:323–328

Salonen JT, Huttunen JK (1986) Selenium in cardiovascular diseases. Ann Clin Res 18:30–35

Salonen JT, Salonen R, Seppänen K, Kantola M, Parviainen M, Alfthan G, Mäenpää PH, Taskinen E, Rauramaa R (1988) Relationship of serum selenium and antioxidants to plasma lipoproteins, platelet aggregability and prevalent ischaemic heart disease in Eastern Finnish men. Atherosclerosis 70:155–160

Stampfer MJ, Colditz GA, Willett WC (1987) The epidemiology of selenium and cancer. Cancer Surv 6:623–633

Thomson CD, Ong LK, Robinson MF (1985) Effects of supplementation with high-selenium wheat bread on selenium, glutathione peroxidase and related enzymes in blood components of New Zealand residents. Am J Clin Nutr 41:1015–1022

Varo P, Alfthan G, Ekholm P, Aro A, Koivistoinen P (1988) Selenium intake and serum selenium in Finland: effects of soil fertilization with selenium. Am J Clin Nutr 48:324–329

Westermarck T, Raunu P, Kirjarinta M, Lappalainen L (1977) Selenium content of whole blood and serum in adults and children of different ages from different parts of Finland. Acta Pharmacol Toxicol 40:465–475

Yläranta T (1985) Increasing the selenium content of cereal and grass crops in Finland. Academic diss, Yliopistopaino, Helsinki

Selenium Status and Cardiovascular Disease: Dutch Epidemiologic Data

F. J. KOK[1,2,3] and A. HOFMAN[1]

1 Introduction

Several mechanisms have been suggested which link selenium to myocardial ischemia (Salonen 1987). The antioxidant hypothesis is most popular and quite well established by experimental studies. A decreased glutathione peroxidase (GSH-Px) activity due to selenium deficiency may result in an accumulation of hydroperoxides. These peroxides may damage arterial endothelium and thus accelerate coronary atherosclerosis. On the other hand, platelet aggregation may be increased, resulting in an increased risk of coronary thrombosis. Other mechanisms probably not mediated by GSH-Px such as increased sensitivity to catecholamines, lower protection against heavy metals, and lower calcium transport, and their influence on blood pressure, thrombosis and atherosclerosis are more speculative.

Epidemiologic data do not clearly suggest an association between low selenium and atherosclerosis (Moore et al. 1984; Aro et al. 1986) or its clinical manifestations (Wang et al. 1981; Oster et al. 1986; Salonen and Huttunen 1986). Motivated by previous findings, we were wondering whether there would be a relation between low selenium status and the risk of cardiovascular disease in the Netherlands with intermediate selenium intake but relatively large variability (Kok et al. 1987a). Therefore, three epidemiologic studies were conducted to investigate the association between selenium and cardiovascular morbidity and mortality.

2 Selenium and Mortality from Cardiovascular Disease

In a case-control analysis of data obtained in a Dutch prospective follow-up study, we investigated the association between low serum selenium and cardiovascular mortality. Results, which have been published previously (Kok et al. 1987b), suggest no increase in the risk of cardiovascular mortality for low levels

[1] Department of Epidemiology, Erasmus University Medical School, Rotterdam, The Netherlands
[2] Department of Human Nutrition, TNO-CIVO Toxicology and Nutrition Institute, P.O. Box 360, 3700 AJ Zeist, The Netherlands
[3] To whom correspondence should be addressed

of serum selenium ($< 105.0\,ng\,ml^{-1}$). So far, two of six prospective studies (four Finnish, one Norwegian, one Dutch) reported a positive association, the other four showed no clear relationship. Possible explanations for the different findings have been discussed by Salonen (1987).

3 Selenium and Incidence of Acute Myocardial Infarction

Recently we completed a case-control study in which we compared plasma, erythrocyte and toenail selenium levels, and the activity of erythrocyte GSH-Px activity of 84 acute myocardial infarction patients and 84 healthy population controls (Kok et al. 1989; Table 1). Patients clinically diagnosed with a first myocardial infarction had no cardiac complaints longer than 6 months prior to the event and no clinical evidence for chronic heart failure. Blood was taken within 4 h of the first symptoms. Cases and controls were on their regular diet (i.e., no dietary regimen) and had not been treated for at least 5 years for kidney or lung disease, alcohol or drug abuse. Moreover, controls should have no history of any cardiovascular event and/or heart surgery. Controls were matched to the cases for age (mean age, 59.4 years) and gender (70% males) and were residing in the hospital's catchment area. Although selenium status was assessed comprehensively, our main focus was on toenail selenium, because it reflects long-term status and thus may be etiologically interpreted. GSH-Px activity was measured in erythrocytes, because over 90% of GSH-Px activity in whole blood is in the erythrocytes, with less than 5% in plasma and 2% in leucocytes and platelets. Furthermore, erythrocytes are more stable, have a longer biological half-life and are probably less susceptible to acute effects of myocardial infarction as compared to platelets. Selenium levels were measured using instrumental neutron activation analysis. GSH-Px activity was measured according to the coupled assay of Paglia and Valentine. Non-specific peroxidase activity of hemoglobin was suppressed by adding Drabkin's reagent.

Results for selenium are presented in Table 1. All selenium measurements were lower in the cases than controls. In contrast, erythrocyte GSH-Px activity in the patients was significantly higher. For an unbiased comparison of the selenium levels it is important to take into account other cardiovascular risk factors which are associated with selenium status, because they may explain the observed differences. Multivariate analysis showed a more than four-fold higher risk of acute myocardial infarction for subjects in the lowest quartile of toenail selenium ($< 0.66\,mg\,kg^{-1}$) compared to those in the highest quartile ($> 0.90\,mg\,kg^{-1}$). The relative odds was 4.5 (95% confidence interval: CI = 1.3 – 15.7). This risk estimate was adjusted for age, gender, serum LDL- and HDL-cholesterol, systolic and diastolic blood pressure, cigarette smoking, body mass index, alcohol intake, family history of myocardial infarction and antihypertensive medications. Serum (n-3) and (n-6) fatty acids, serum α-tocopherol and retinol, use of analgesic drugs and use of other medications had no further effect on this association. Moreover, no modifying effect of serum α-tocopherol, retinol and analgesic medication was observed.

Table 1. Selenium status of acute myocardial infarction (AMI) patients and population controls

Status parameter[a]	AMI cases (n = 84[c])	Population controls (n = 84)[c]	p-Value[b]
Selenium in			
Plasma ($ng\ ml^{-1}$)	100.8 ± 3.0	106.8 ± 2.6	0.34
Erythrocytes ($\mu g\ gHb^{-1}$)	0.54 ± 0.01	0.59 ± 0.02	0.02
Toenails ($mg\ kg^{-1}$)	0.70 ± 0.02	0.78 ± 0.02	0.01
Erythrocyte GSH-Px activity ($U\ gHb^{-1}$)	31.3 ± 0.9	28.0 ± 0.9	0.01

[a] Mean ± standard error
[b] Tested by t-test for paired observations
[c] Some of the individual parameters have missing data

Table 2. Selenium status of patients with angiographically defined coronary atherosclerosis

Status parameter[a]	Cases[b] (n = 91)	Controls[b] (n = 72)	p-Value[c]
Selenium in			
Plasma ($ng\ ml^{-1}$)	95.1 ± 2.2	108.8 ± 3.4	0.001
Erythrocytes ($\mu g\ gHb^{-1}$)	0.53 ± 0.02	0.52 ± 0.02	0.78
Toenails ($mg\ kg^{-1}$)	0.78 ± 0.02	0.79 ± 0.02	0.77
Erythrocyte GSH-Px activity ($U\ gHb^{-1}$)	32.4 ± 0.9	29.7 ± 0.9	0.03

[a] Mean ± standard error
[b] Cases had 85% or more stenosis in at least one coronary vessel; controls had less than 50% stenosis in all three coronary vessels
[c] Tested by t-test for paired observations

In conclusion, low selenium status was present in the patients before the infarction and thus may be associated with increased risk of acute myocardial infarction. The higher erythrocyte GSH-Px activity we observed in the patients may be interpreted as a defense against increased oxidant stress either preceding or following the acute event.

4 Selenium and Coronary Atherosclerosis

To provide further insight in the possible role of selenium in atherosclerosis, we compared the selenium status and GSH-Px activity of patients with varying degrees of angiographically determined coronary atherosclerosis. Cases (n = 91; mean age, 55.4 years; 70% males) had 85% or more stenosis in at least one coronary vessel Controls (n = 72; mean age, 53.7 years; 60% males) has less than 50% stenosis in all three coronary vessels. Not eligible were patients with a history of a myocardial infarction longer than 3 months before angiography (dietary influences) and patients who were under medical treatment for longer than 2 years (drug influences).

Erythrocyte and toenail selenium levels were similar in cases and controls, whereas plasma selenium levels were significantly lower in cases (Table 2). Erythrocyte GSH-Px activity was significantly higher in cases than controls. Multivariate analysis based on quartiles of plasma selenium showed a significant inverse trend between severity of coronary sclerosis and plasma selenium level. The multivariate relative odds for subjects with plasma selenium levels below 83 $ng\ ml^{-1}$ (lowest quarile) was 5.5 (95%, CI = 1.7–17.9) compared to those in the highest quartile (>128 $ng\ ml^{-1}$).

In conclusion, no consistent association between selenium status and coronary atherosclerosis was observed. However, since arterial endothelial cells are involved in atherogenesis and are undergoing constant, rapid turnover, plasma selenium may be the best measure of selenium most available for incorporation into these cells (Moore et al. 1984). The higher GSH-Px activity in the cases is in accord with the findings in the myocardial infarction patients.

5 General Conclusions

1. None of the different types of epidemiologic studies give convincing support to an etiologic role of selenium in atherosclerosis and its clinical manifestations.

2. The case-control studies among myocardial infarction patients, however, suggest that selenium and/or GSH-Px are implicated in the disease. The oxidant stress hypothesis in myocardial infarction, i.e., elevated levels of circulating free radicals and lipid peroxides which induce synthesis of detoxifying enzymes, like GSH-Px, needs further study. An alternative possibility which deserves attention is that the lower selenium levels in our patients may have an etiologic role other than mediated by the GSH-Px enzyme.

3. Future studies require careful consideration of the relevant selenium parameters to be measured, i.e., tissues, selenium species and enzyme activities.

Acknowledgement. This research was supported by the Netherlands Heart Foundation (grant 34.013).

References

Aro A, Alfthan G, Soimakallio S, Voutilainen E (1986) Se concentrations in serum and angiographically defined coronary artery disease are uncorrelated. Clin Chem 32:911–912

Kok FJ, Bruijn AM de, Hofman A, Valkenburg HA (1987a) Selenium status and chronic disease mortality: Dutch epidemiological findings. Int J Epidemiol (Suppl)16:329–332

Kok FJ, Bruijn AM de, Vermeeren R, Hofman A, van Laar A, Bruin M de, Hermus RJJ, Valkenburg HA (1987b) Serum selenium, vitamin antioxidants and cardiovascular mortality: a 9-year follow-up study in the Netherlands. Am J Clin Nutr 45:462–468

Kok FJ, Hofman A, Witteman JCM, Bruijn AM de, Kruyssen DHCM, Bruin M de, Valkenburg HA (1989) Decreased selenium levels in acute myocardial infarction. JAMA (in press)

Moore JA, Noiva R, Wells IC (1984) Selenium concentrations in plasma of patients with arteriographically defined coronary atherosclerosis. Clin Chem 30:1171–1173
Oster O, Drexler M, Schenk J, Meinertz T, Kasper W, Schuster CJ, Prellwitz W (1986) The serum selenium concentration of patients with acute myocardial infarction. Ann Clin Res 18:36–42
Salonen JT (1987) Selenium in ischaemic heart disease. Int J Epidemiol (Suppl)16:323–328
Salonen JT, Huttunen JK (1986) Selenium in cardiovascular diseases. Ann Clin Res 18:30–35
Wang YX, Böcker K, Reuter H, Kiem J, Kasperek K, Iyengar GV, Loogen F, Gross R, Feinendegen LE (1981) Selenium and myocardial infarction: glutathione peroxidase in platelets. Klin Wochenschr 59:817–818

Bioavailability of Selenium in a Selected Dutch Population

H. W. VAN DER TORRE[1,2], H. VAN STEENBRUGGE[1], S. PELUPESSY[1], J. A. WESSTRA[1], W. VAN DOKKUM[1], G. SCHAAFSMA[1] and TH. OCKHUIZEN[1]

1 Introduction

It is suggested that selenium intake is marginal for a large part of the Dutch population as compared with the safe and adequate daily dietary intake of 50–200 μg day^{-1}. In order to evaluate the selenium intake, knowledge is required regarding the bioavailability of selenium in the main sources of the average daily selenium supply. The bioavailability of selenium in wheat has been studied (Levander et al. 1983; Thomson et al. 1985), but as far as we know no information is available on the bioavailability of selenium in meat. The aim of this study was to evaluate the influence of a variable selenium intake on various blood parameters and to establish the bioavailability of selenium in wheat products in comparison with that in meat products.

2 Methods

Twenty-three healthy male subjects, 20–30 years of age, participated in the study. After a 5-week period of low Se intake (30–40 μg Se day^{-1}; 23 subjects), a 9-week period was programmed during which the selenium intake was either 40–70 μg day^{-1} (eight subjects), 130–140 μg day^{-1} (seven subjects) or 205–225 mg day^{-1} (8 subjects). Finally, the subjects were on a self-selected diet for a 10-week period. The study was executed in the Institute's metabolic ward. The diet was constant during the mentioned periods of 5 and 9 weeks. Selenium was supplemented through naturally selenium-enriched bread or meat. The wheat and meat were imported from South Dakota, a seleniferous area of the USA. The wheat selenium source was given daily as six rolls, the meat selenium source as daily portions of 150 g minced meat. Fasting blood samples were taken between 8.00 and 9.00 a.m. at the beginning of the experiment (week 0) and after 3, 5 (start supplementation period), 6, 9 and 14 weeks from the beginning. Finally, a blood sample was taken at the end of the 10-week post-dosing period. The blood was collected into polycarbonate tubes containing NaEDTA as anticoagulant. The

[1] Department of Nutrition, TNO-CIVO Toxicology and Nutrition Institute, P.O. Box 360, 3700 AJ Zeist, The Netherlands
[2] To whom correspondence should be addressed

blood samples were analyzed for the selenium concentrations in plasma and red blood cells (RBC) and for the glutathione peroxidase activities in plasma, RBC and platelets. Selenium was measured by the fluorimetric method of Koh and Benson (1983). The glutathione peroxidase activities were assayed by the method of Pleban et al. (1982), using H_2O_2 (8.8 mM) as substrate and 2 mmol GSH at 37 °C, pH 7.6. The glutathione peroxidase activity was expressed as mU mg^{-1} protein. Protein was measured by the method of Lowry et al. (1951).

3 Results and Discussion

The plasma and RBC selenium levels at week 0 did not differ significantly among the six treatment groups (Figs. 1 and 2). During the 5-week period of low selenium intake there was a slight decrease in plasma and RBC selenium levels. After a 9-week supplementation period the plasma and RBC selenium concentrations increased depending on the level of selenium intake (21–132 and 0–63% respectively). The activity of glutathione peroxidase in plasma and RBC did not show a significant change during the period of low selenium intake, the plasma enzyme

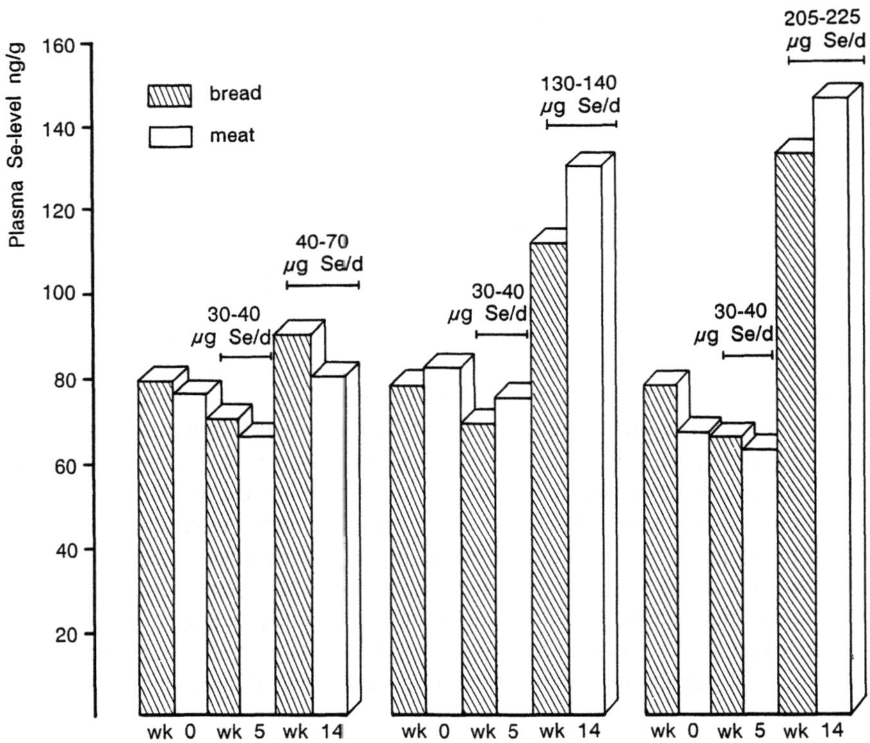

Fig. 1. Plasma selenium levels at the beginning of the intervention trial (week 0), at the end of the period of low selenium intake (week 5) and at the end of the supplementation period (week 14). Forms of selenium given were selenium-rich bread (▧) and selenium-rich meat (□)

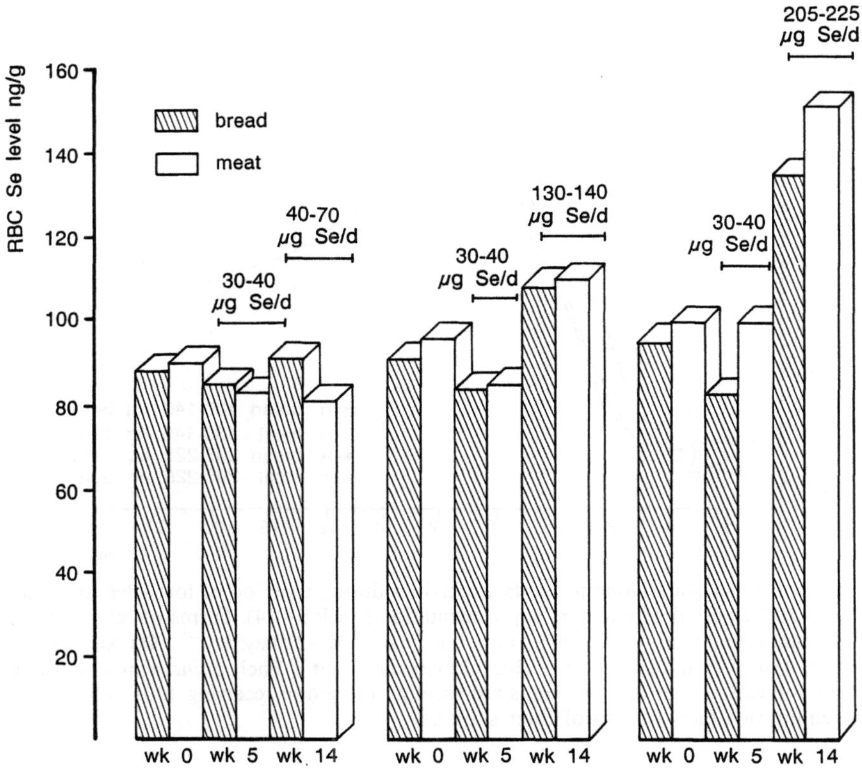

Fig. 2. RBC selenium levels at the beginning of the intervention trial (week 0), at the end of the period of low selenium intake (week 5) and at the end of the supplementation period (week 14). Forms of selenium given were selenium-rich bread (▨) and selenium-rich meat (▭)

increased slightly during the selenium supplementation. Platelet glutathione peroxidase activities were similar during the period of low selenium intake. In contrast, there was a rapid and significant increase in the platelet glutathione peroxidase activity to a maximum of 159% of the activity at week 5 after a supplementation with 205−225 µg Se day^{-1} for 9 weeks (Fig. 3). At the end of the post-dosing period the glutathione peroxidase activities were still above baseline values. These results suggest that both sources of selenium (meat and bread) were stored in the tissues in a form that could be mobilized and used for platelet glutathione peroxidase. The bread and meat supplements caused similar increases in plasma and RBC selenium levels and in platelet glutathione peroxidase activities. Measurement of glutathione peroxidase activity appears to be the method of choice for estimating selenium bioavailability (Levander 1983). Platelet glutathione peroxidase activity seems to be particularly promising for bioavailability studies because the enzyme is sensitive to changes in dietary selenium intake (Levander 1983). Unfortunately, there was no information available on the response of platelet glutathione peroxidase activity to a selenium supplementation of 40−70 µg day^{-1}. It is possible that the dosis of 130−140 µg Se day^{-1} is too

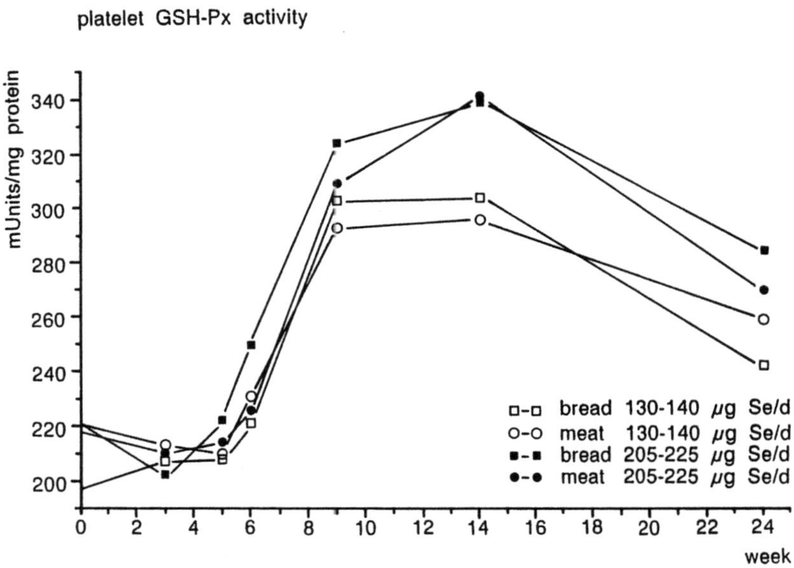

Fig. 3. Platelet glutathione peroxidase activities during a period of low selenium intake (week 0–5) and a period of selenium supplementation (week 5–14). Forms of selenium given were selenium-rich bread, 130–140 μg day^{-1} (□) or 205–220 μg day^{-1} (■) and selenium-rich meat, 130–140 μg day^{-1} (○) or 205–220 μg day^{-1} (●). Each *point* represents the mean of four subjects, except for the points representing the group receiving 130–140 μg selenium as selenium-rich meat (means of three subjects)

high to distinguish between the bioavailability of the two selenium sources. So we suggest that a selenium intake of less than 130 μg day^{-1} is required for comparing the bioavailability of selenium in wheat and that in meat, using platelet glutathione peroxidase activity as a parameter.

It is concluded that both sources of selenium (bread and meat) were capable of increasing different selenium status parameters in a dose-dependent way.

References

Koh TS, Benson TH (1983) J Assoc Off Anal Chem 66:918–926

Levander OA, Alfthan G, Arvilommi H, Gref CG, Huttunen JK, Kataja M, Koivistoinen P, Pakkarainen J (1983) Am J Clin Nutr 37:887–897

Levander OA (1983) Fed Proc 42:1721–1725

Lowry OH, Roseborough NJ, Farr AC, Randall RJ (1951) J Biol Chem 193:265–275

Pleban PA, Muyani A, Beachum J (1982) Clin Chem 28:311–316

Thomson CD, Ong LK, Robinson MF (1985) Am J Clin Nutr 41:1015–1022

Studies of Human Maximal and Minimal Safe Intake and Requirement of Selenium

G. YANG[1,2], L. GU[1], R. ZHOU[1], and S. YIN[1]

1 Introduction

Although the geographical relationship between human cancer incidence and local environmental Se status has been repeatedly confirmed since 1969 (Shamberger and Frost 1969), the levels of Se used in animal studies, which support the role of Se as inhibiting cancer development, are potentially toxic. It was proposed that the justification for carrying out human intervention trials with Se to prevent cancer should be at best marginal or in the pharmacologic range. A promising approach to resolve the dose problem was the suggested use of levels similar to the intakes of populations residing in regions naturally high in Se (Clark and Combs 1986). Such an investigation has been launched since 1985 in a seleniferous area in Enshi (Yang et al. 1988a, b). A daily Se intake of 50 μg by extrapolation from animals was proposed to meet the human requirement. Since extremely low Se areas with Se-deficiency-related endemic diseases appear in China, the study of the human requirement has been conducted since 1983, and part of the work was recently confirmed in non-affected areas (Yang et al. 1988c). The results mentioned above are presented here.

2 Human Marginal and Maximal Safe Selenium Intakes

2.1 Study Design

Three sites with low, medium and high Se levels in the seleniferous region in Enshi county were chosen for this study. Two hundred to 300 subjects of different ages and sexes were included from each site. The chemical form of Se ingested was mainly in a natural food form. The subjects are life-long residents and cases of selenosis appeared in the high Se site. Field studies conducted in April and September 1986 followed the pilot study of November 1985. Foods consumed during the 3-day survey were collected for Se analysis. Blood, 24-h urine, breast milk and tissue samples were collected for Se, Hg, Cd, Zn, Cu or As determination. In order to determine the expected level of Se intake, individual Se intakes or tissue-

[1] Institute of Nutrition and Food Hygiene, Chinese Academy of Preventive Medicine, 29 Nan Wei Road, Beijing, Peoples' Republic of China
[2] To whom correspondence should be addressed

Table 1. Average daily Se intake (μg) of adults[a]

Se status of site	Male	Female	Average	Ratio
Low	70.5 ± 4.8 (51)	62.0 ± 3.6 (51)	66.3	1
Medium	194.7 ± 22.9 (76)	198.1 ± 23.8 (82)	196.4	3
High	1438.2 ± 76.3 (58)	1238.5 ± 64.6 (61)	1338.4	20

[a] Results are expressed as mean ± SE.
Numbers in parentheses indicate the number of subjects examined.

Se levels were correlated with either biochemical parameters or clinical signs. If the tissue-Se level was known, it could be transferred to the respective Se intakes through regression equations.

2.2 Se-Intake and Tissue-Se Level

The average daily Se intakes in medium and high Se sites were around 3- to 20-fold that of the low Se site (Table 1). No significant differences were found between male and female adults. Se intake and tissue or body fluid Se concentration were found to be significantly correlated, and similar relationships were also found between Se concentrations of tissue.

Of the metals analyzed, only Cd and As were considered high either in the blood and hair or hair respectively of high Se site residents, but their effects on Se metabolism remain to be studied.

2.3 Clinical Diagnosis

Diagnosis of selenosis was based mainly upon morphological alterations of the fingernails. Results indicated that the lowest blood-Se level of selenosis diagnosed among 349 residents in the sites with different levels of blood-Se, had a blood-Se level of 1.02 $\mu g\,ml^{-1}$ (875 μg daily Se intake). More definite evidence obtained from five patients with long, persisting, distinct clinical signs showed a blood-Se range of 1.054 – 1.854 $\mu g\,ml^{-1}$. The minimal blood-Se concentration of 1.054 $\mu g\,ml^{-1}$ (909 μg daily Se intake) confirmed the previous result and would approximate the marginal level of Se toxicity which cannot be tolerated by the susceptible cases.

2.4 Percentage of Se Excretion from Se Intake

The percentage varied within 40 – 45 of a wide range of Se intakes until approximately 2000 $\mu g\,day^{-1}$ (10 – 2000 μg) was attained. However, such a homeostatic regulation weakened and the excretion percentage began to plateau as Se intake increased to near 800 $\mu g\,day^{-1}$ (blood-Se level 0.95 $\mu g\,ml^{-1}$). As the Se intake further increased to over 1000 μg, the percentage excretion increases nearly to a

Table 2. Prothrombin time of residents in relation to concentration of blood-Se

Whole blood Se level ($\mu g\ ml^{-1}$)	Number of subjects	Average time[a] (s)	Samples with prolonged prothrombin time (≥ 14 s)	
			Cases	(%)
<0.20	17	12.2 ± 0.2^a	0	–
0.2–0.99	20	12.2 ± 0.2^a	1	5.0
1.0–1.49	29	13.4 ± 0.3^b	13	44.8
1.5–	18	13.5 ± 0.3^b	8	44.4

[a] Results are expressed as mean ± SE.
Means not sharing a common superscript letter are significantly different ($p < 0.01$).

constant. This seems to indicate that as Se intake increases to around 800 $\mu g\ day^{-1}$, the kidney would begin failing to regulate the excretion of Se in the urine. No notable change of the $(CH_3)_3Se^+$ content in urine was observed.

2.5 Plasma Prothrombin Time

Excessive ingestion of Se may damage the parenchymal cells of the liver and impair the hepatic synthesis of clotting factor(s). Blood samples of inhabitants from different sites were collected for measurement of plasma prothrombin time. It was found that as blood-Se increased to a level above 1 $\mu g\ ml^{-1}$ (850 μg Se intake), cases with a prolonged prothrombin time increased abruptly (Table 2).

2.6 Other Biochemical Parameters

As Se intake increased to a high level, changes of other biochemical parameters were observed: (1) decrease of the plasma-Se to erythrocyte-Se ratio may indicate an increased burden of Se to the body. (2) Decrease of whole blood GSH concentration may indicate the shifting of blood GSH metabolism towards a higher oxidation state. (3) No indication of anemia was observed from hemoglobin, hematocrit or erythrocyte-free proporphyrin content determination. (4) Severe liver damage was again not detected either from analyses of biochemical indicators such as serum GPT, ZnTT, TTT, icteric index and alkaline phosphatase activity or through supersonic B morphologic examination. (5) White cell count increased stepwise from the low to high Se site; its significance is not clear.

2.7 Summary of the Results

Results are summarized in Table 3. It shows that clinical signs may occur at a marginal blood-Se level of approximately $\geq 1\ \mu g\ ml^{-1}$ or a corresponding marginal daily Se-intake of 850 $\mu g\ day^{-1}$, while biochemical alterations suggested a mar-

Table 3. Evidence obtained for the safe level of Se intake[a]

Evidence	Se intake ($\mu g\ day^{-1}$)	Blood Se level ($\mu g\ ml^{-1}$)
Manifestation of clinical signs	≤(875)	≤ 1.02
Long persistence of clinical signs	≤(909)	≤ 1.05
Delay of prothrombin time	≤(853)	≤ 1.00
Reduction of whole blood GSH content	~(853)	~ 1.0
Plateau of percentage of Se excretion to Se intake	≤ 800	≤(0.95)
Remarkable reduction of plasma-Se to erythrocyte-Se ratio	~ 743	~(0.9)

[a] Numbers in parentheses are calculated from regression equations.

ginal blood-Se level of around ≤ 1.0 $\mu g\ ml^{-1}$ or an approximate, marginal, daily safe Se intake of 750 $\mu g\ day^{-1}$. As a safety factor of 1.3 is used, a maximal daily safe Se intake of 550 $\mu g\ day^{-1}$ is suggested for inhabitants in seleniferous area.

3 The Selenium Requirement of Humans (Yang et al. 1987, 1988c)

3.1 Minimal Se Requirement

It is estimated as the approximate, minimal Se intake required to protect the inhabitants in low Se areas against KD (Keshan disease). Such a Se intake was estimated to average 17 $\mu g\ day^{-1}$ for adults (Yang et al. 1987). This result is in agreement with the suggestion from a previous balance study on New Zealand women indicating that the minimal Se requirement is probably not more than 20 $\mu g\ day^{-1}$ (Stewart et al. 1975). This is also in agreement with the statement that Keshan disease appeared only under extremely poor Se status in the presence of environmental or host stress (Levander 1987).

3.2 Physiological Se Requirement

It is estimated as the level of Se intake sufficient to maintain plasma GSH-Px activity at a maximal, constant level. A previous study indicated that as graded levels of Se as selenomethionine was supplemented to male adults living in a low Se area, the plasma GSH-Px activity plateaued at a daily Se intake of ≤ 40.8 μg (Yang et al. 1987). Since more than 70% of the Se intake is quantitatively supplemented, individual variations in Se intakes cannot be obtained in this study. Levander (1987) used a safety factor of 1.3 as the coefficient of variation. Thus, a daily Se intake of 53 μg was obtained for male adults in this study.

It was accidentally found recently that the blood-Se level of the institute workers in Beijing were marginal. In view of the variation of values in food availability and the complicated factors shown to influence the GSH-Px activities in tissues, observation was thus made on such a population. The 12 subjects were

Table 4. Influence of Se supplementation as Na_2SeO_3 on Se levels and GSH-Px activities of blood, erythrocytes and plasma of male and female adults

Week	Se supplemented[a] (μg day^{-1})	Blood Se (μg ml^{-1})	Erythrocyte Se (μg ml^{-1})	Plasma Se (μg ml^{-1})	Plasma GSH-Px activity[b]
1	0	0.127±0.012	0.162±0.035	0.097±0.012	0.40±0.05
2	300	0.135±0.017	0.155±0.029	0.128±0.011	0.40±0.04
3	700	0.144±0.016	0.173±0.033	0.128±0.025	0.39±0.06
4	900	0.160±0.020	0.197±0.054	0.132±0.017	0.41±0.03

[a] Results are expressed as mean±SE. Supplemented to a basal diet providing an average dietary Se intake of 45.5 μg day^{-1} for both sexes.
[b] nmol NADPH min^{-1} ml^{-1}.

Table 5. Influence of Se supplementation as Se-methionine on Se levels and GSH-Px activities of blood, erythrocytes and plasma of male and female adults

Week	Se supplemented[a] (μg day^{-1})	Blood Se (μg ml^{-1})	Erythrocyte Se (μg ml^{-1})	Plasma Se (μg ml^{-1})	Plasma GSH-Px activity[b]
1	0	0.126±0.016	0.158±0.039	0.102±0.025	0.40±0.01
2	300	0.151±0.007	0.154±0.010	0.152±0.012	0.42±0.08
3	700	0.218±0.020	0.204±0.022	0.207±0.015	0.41±0.08
4	900	0.294±0.036	0.285±0.042	0.297±0.063	0.42±0.07

[a] Results are expressed as mean±SE. Supplemented to a basal diet providing an average dietary Se intake of 45.4 μg day^{-1} for both sexes.
[b] nmole NADPH min^{-1} ml^{-1}.

middle-aged and young laboratory workers with equal numbers of both sexes. They were divided into two groups of eight and four each. The average body weight was 60 kg for males and 56 kg for females. Short-term supplementation of higher doses of Se either as sodium selenite or as selenomethionine was made. Blood samples were collected for Se and enzyme activity estimation. The results showed that plasma GSH-Px activities of both groups remained unchanged during the period of Se supplementation (Tables 4 and 5). The amount of average Se intake provided from the natural diet was 48.3 μg, which is within the variation range of previous studies on male adults in a low Se area in which a safe daily Se intake of 53 μg was able to maintain the plasma GSH-Px activity at plateau. This study confirmed the previous result. However, as a safety factor of 1.3 was used in this study, the safety intake was 62 μg for males and 55 μg for females consuming a natural diet.

Another approach is to determine the amount of supplemented Se as sodium selenite necessary to maintain the plasma Se level at plateau (Yang et al. 1988c). A preliminary experiment indicated that a 20 μg Se supplementation daily (or a daily Se intake of 68.3 μg) could raise the plasma Se to a constant level. As the safety factor 1.3 was used, the daily Se intake should approximate 90 μg day^{-1} for 60 kg male adults. The results are summarized in Table 6.

Table 6. Daily selenium requirement and suggested dietary allowance for humans

Se intake	Method of estimation	Experimental value (μg day^{-1})	Safe intake[a] (μg day^{-1})
Minimal requirement	1. Dietary survey in KD area	19.1 ± 5.5 (M)	–
		13.3 ± 4.1 (F)	–
	2. Calculated from marginal Se level of cereal	19.2 (M)	–
		16.0 (F)	–
Physiological requirement	1. Suppl. Se-Met. to max. plasma GSH-Px activity	≤ 40.9 ± 0.6 (M)	50 (M)
		–	–
	2. Natural dietary Se intake to max. plasma GSH-Px activity	≤ 48.3 ± 20.2 (M)	60 (M)
		≤ 42.8 ± 20.4 (F)	55 (F)
	3. Suppl. selenite to max. plasma Se level	~ 68.3 ± 20.2 (M)	90 (M)
		~ 62.8 ± 20.4 (F)	80 (F)
Suggested optimum dietary allowance range	Summation of the data obtained		50 – 250

[a] Results are expressed as mean ± SD. A safety factor of 1.3 is used after Levander, OA [21]. M, male; F, female.

Acknowledgements. The Enshi work was funded by the US National Cancer Institute and supported by the Research Agreement from the US Department of Agriculture.

References

Clark LC, Combs CF Jr (1986) Selenium compounds and the prevention of cancer: research needs and public health implications. J Nutr 116:170–173

Levander OA (1987) A global view of human selenium nutrition. Annu Rev Nutr 7:225–250

Shamberger RJ, Frost DV (1969) Possible protective effect of selenium against human cancer. Can Med J 100:682

Stewart RDH, Griffiths NM, Thomson CD, Robinson MF (1978) Quantitative selenium metabolism in normal New Zealand women. Br J Nutr 40:45–54

Yang GQ, Zhu LZ, Liu SJ, Gu LZ, Qian PC, Huang JH, Lu MD (1987) Human selenium requirements in China. In: Combs GF Jr, Spallholz JE, Levander OA, Oldfield JE (eds) 3rd Int Symp Selenium Biol Med, part B. Van Nostrand Reinhold, New York, pp 589–607

Yang GQ, Zhou RH, Yin SA, Gu LZ, Yan BW, Liu YQ (1988a) Studies of maximal daily selenium intake of safety in seleniferous area in China. 1. Selenium intake and tissue selenium levels of the inhabitants. J Trace Element Electrolyt Health Dis (submitted)

Yang GQ, Zhou RH, Yin SA, Gu LZ, Yan BW, Liu YQ (1988b) Studies of maximal daily selenium intake of seleniferous area in China. 2. Relation between selenium intake and the manifestation of clinical signs and some biochemical alterations. J Trace Element Electrolyt Health Dis (submitted)

Yang GQ, Zhou RH, Yin SA, Gu LZ (1988c) Further studies on the physiological selenium requirement of humans in China, and a brief discussion of the results. J Hyg Res 1988 (in press) (in Chinese)

Are Germans Selenium-Deficient?

O. Oster and W. Prellwitz[1]

1 Introduction

The title of this study deserves attention for the following reasons:

1. Hartfiel and Bahners (1988) have measured the selenium content of soil and found for West Germany a lower selenium content in soil than e.g. Finland, a country which was considered low in dietary daily selenium supply in the past.
2. Oster and Prellwitz (1982) and Oster et al. (1988a) reported serum, whole blood and organ selenium concentrations of healthy populations of Germany, which were only slightly higher than those reported from Finland (before Se fertilization of soil) (Westermarck 1977) and New Zealand (Thomson and Robinson 1980), also a country which is considered to have a low dietary selenium supply.
3. Schrauzer and White (1978) presented a formula correlating whole blood selenium with the daily selenium intake. From this formula a daily dietary selenium intake of ca. 40 µg is calculated for West Germans.

To describe the selenium status of a population many selenium-related parameters must be considered, in particular, whether clinical symptoms correlate with some of the parameters. One important parameter to evaluate the selenium status of a population is the knowledge of the selenium content of the daily consumed food. This chapter concentrates on the daily dietary selenium intake of West German adults.

2 Material and Methods

Equipment. A model 5000 atomic absorption spectrometer, equipped with the Mercury Hydride system (MHS 20), and an electrodeless selenium discharge lamp, both from Perkin Elmer, were used.

Reagents. The following reagents were used: selenium standard (Merck Titrisol), HNO_3, $HClO_4$ and H_2SO_4 (Merck suprapure reagents), sodium borohydride p.a. (Merck), sodium hydroxide (Merck), phenantroline (Merck), antifoam Dow Corning DB 110 A. Bidistilled H_2O was obtained from Millipore equipment.

[1] Institut für Klinische Chemie und Laboratoriumsmedizin, Universitätsklinik Mainz, 6500 Mainz, FRG

The Wet Digestion of Food. 0.3 to 1.0 g of food were treated at room temperature overnight with 8 ml of a mixture of $HNO_3/HClO_4$ (5 : 1). In the morning 2.5 ml H_2SO_4 were added. The digestion proceeded under reflux with a Büchi Digestor 450 as described previously (Oster et al. 1983). Reduction of hexavalent selenium to selenium in the oxidation state 4^+ was done with HCl, also as described previously (Oster et al. 1983, 1986b).

The Determination of Selenium with the Hydride AAS System. Selenium was determined with the MHS 20 System as described previously (Oster et al. 1986b) using phenantroline as complexing agent.

3 Results and Discussion

The food in which the selenium content was determined was purchased in shops and supermarkets in the city of Mainz, Germany in November 1987.

Pork and beef (from 12 different butchers) were obtained from the back of the animal and were free of fat. Fish was purchased on two different days within two different weeks from the largest supplier of fish in Germany, the Nordseehalle. Eggs were obtained from five different suppliers. Bread was from three different bakeries and cow milk from 15 different suppliers.

From Table 1 it can be seen that egg yolk (cooked) has the highest selenium content, about 2.5 to 4 times more than egg white. Seafish has a nearly similar selenium content as egg yolk and distinctly more than freshwater fish (trout).

Pork has twice the selenium content of beef. This may be due to the difference in the feeding habits of pigs and cattle. Pigs are raised mostly with a relatively standardized diet to which selenium in the form of selenite is added. Cattle feed mostly on grassland. Also, the high selenium content of eggs might be due to selenium supplementation of food.

From the selenium content of food the daily selenium intake of Germans is calculated from the corrected data for the consumption of food from the German Nutrition Report 1984 (Deutscher Ernährungsbericht 1984; see Table 2). In Table 2 approximations are made. For sausage an average selenium content of 48 μg was assumed. For food which is cooked (meat, eggs, potatoes) a correction is made (10% reduction). In Table 2 the daily dietary intake of selenium is calculated for men and women. Men consume about 20% more selenium per day by food than women.

In Fig. 1 the distribution of the total daily dietary selenium intake among the single nutrients is graphically presented. From Fig. 1 it can be seen that pork with approximately 20% is the main source of the daily dietary selenium for Germans followed by egg and fish. Meat, including sausage, participates with approximately 33% of the daily dietary selenium intake. From Fig. 1 it is also clear that most of the selenium intake of Germans is through consumption of animal protein (pork and beef, eggs, fish, cheese). This means that the selenium species is mostly selenocysteine and selenomethionine.

Table 1. Selenium content of German food

Food	N	Selenium content (ng g^{-1})	Range
Pork	12	118 ± 20	89 – 146
Beef	12	58 ± 22	20 – 83
Sausage from			
Liver	2	54/60	11 – 81
Blood	2	11/15	
Meat	2	44/49	
Bacon	2	74/81	
Eggs (boiled)			
Egg white	5	102 ± 35	58 – 138
Egg yolk	5	368 ± 88	233 – 475
Bread			
White bread	8	19 ± 4	13 – 25
Rye bread	5	14 ± 2	13 – 17
Whole rye bread		18/74/84	
Flour	2	23/25	
Fish			
Cod fish	2	168/368	
Red perch	2	368/433	
Flounder	2	171/187	
Haddock	2	225/313	
Trout	2	115/125	
Potato	2	5/5	
Spaghetti/Noodle	3	45/48/54	
Cheese (fresh cheese)	3	35/37/41	
Milk (fresh milk)	15	7.2 ± 2.8	
Water (Mainz)	5	<2	

Coming back to the title of this work, the daily dietary selenium intake of Germans is, compared to other industrialized countries, e.g., the USA (which is considered to have an adequate selenium supply), Canada and Japan (the dietary selenium intake is considered to be high), low and only slightly higher than that of Finland (before fertilization of soil) (Koivistoinen and Huttunen 1986) and New Zealand (Thomson and Robinson 1980). The daily dietary selenium intake of Germans is distinctly lower than the safe and adequate daily dietary selenium intake of 50 – 200 μg day^{-1} recommended by the US National Research Council (US National Research Council 1980). No clinical symptoms with epidemiological character are, however, observed in Germany, which might be due to a general selenium deficiency in food. However it has to be considered that there might be risk groups due to an inadequate selenium supply by food e.g., pregnant women, lactating women, alcoholics, persons on parenteral nutrition, vegetarians, persons with malnutrition and malabsorption. It has further to be considered that in certain diseases in Germany subnormal serum selenium concentrations are found e.g., patients with congestive cardiomyopathy (Oster etal. 1983), alcoholics,

Table 2. Daily dietary selenium intake of Germans by the following food

Food	Food consumed per day (g)		Selenium content of the food (ng g^{-1})	Daily intake of selenium (µg)	
	Men	Women		Men	Women
Pork	76.3	63.3	118	9.0	7.5
Beef	38.2	31.7	58	2.2	1.8
Sausage + bacon	80.6	54.2	48	3.8	2.6
Bread	175.0	150.0	18	3.2	2.7
Fish	17.8	13.9	247	4.4	3.4
Egg (whole)	36.9	35.4		7.6	6.7
Egg white	26.6	23.6	101	2.7	2.4
Egg yolk	13.3	11.8	368	4.9	4.3
Milk	251.4	240.0	7	1.8	1.7
Cheese	39.8	34.8	38	1.5	1.3
Potato	147.7	116.0	5	0.7	0.6
Flour	35.8	33.0	24	0.9	0.8
Noodle	10.4	8.2	48	0.5	0.4
Potato products	11.4	5.9	8	0.1	0.1
Flour + potato products (others)	36.3	27.8	33	1.2	0.9
Vegetables	279.1	266.7	8	2.2	2.2
Beverages + soups	1400	1000	2	2.8	2.0
Others				4.0	4.0
Daily intake of selenium				45.9	38.7
Corrections by cooking				3.5	2.8
Daily intake of selenium after corrections				42.4	35.9

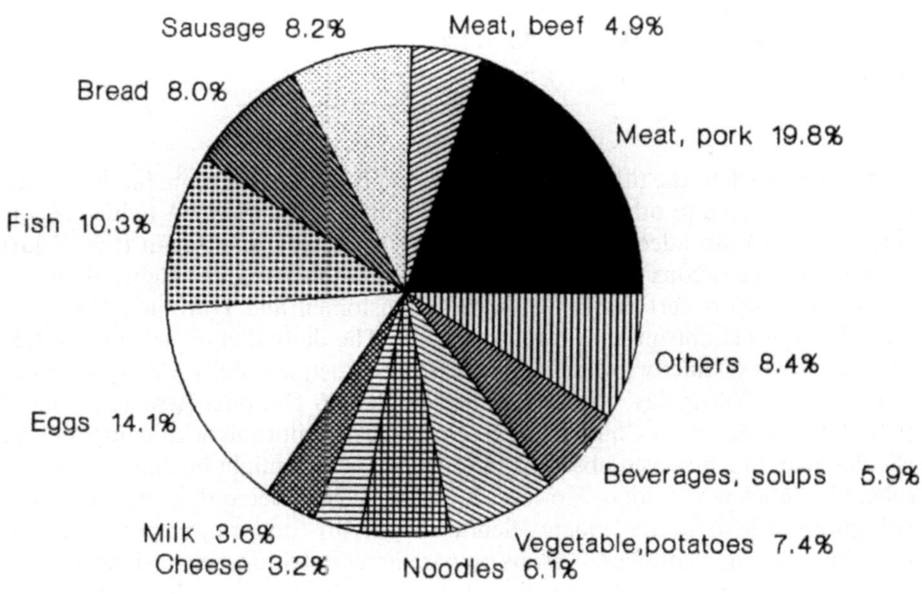

Fig. 1. Daily dietary intake of selenium by single food items in percent of the whole daily selenium intake by food

patients with acute myocardial infarction (Oster et al. 1986a), patients with coronary heart disease (Oster et al. 1987), patients with liver cirrhosis and dialysis patients.

References

Deutsche Gesellschaft für Ernährung (1984) Ernährungsbericht 1984. Deutsche Gesellschaft für Ernährung, Frankfurt

Hartfiel W, Bahners N (1988) Selenium deficiency in the Federal Republic of Germany. Biol Trace Element Res 15:1–12

Koivistoinen P, Huttunen JK (1986) Selenium in food and nutrition in Finland. An overview and action. Ann Clin Res 18:13–17

Oster O, Prellwitz W (1982) A methodological comparison of hydride and carbon furnace atomic absorption spectroscopy for the determination of selenium in serum. Clin Chim Acta 124:277–291

Oster O, Prellwitz W, Kasper W, Meinertz T (1983) Congestive cardiomyopathy and the selenium content of serum. Clin Chim Acta 128:125–132

Oster O, Drexler M, Schenk J, Meinertz T, Kasper W, Schuster CJ, Prellwitz W (1986a) The serum selenium concentration of patients with acute myocardial infarction. Ann Clin Res 18:36–42

Oster O, Schmiedel G, Prellwitz W (1986b) Ein Vergleich des Selengehaltes von Säuglingsnahrung. In: Welz B (ed) Fortschritte in der atomspektrometrischen Spurenanalytik, vol 2. VCH Verlagsgesellschaft, Weinheim, p 409

Oster O, Prellwitz W, Luley C, Meinertz T, Geibel A, Kasper W (1987) The relationship between selenium and coronary heart disease. In: Brätter P, Schramel P (eds) Trace element analytical chemistry in medicine and biology, vol 4. De Gruyter, Berlin, p 579

Oster O, Schmiedel G, Prellwitz W (1988a) Correlations of blood selenium with hematological parameters in West German adults. Biol Trace Element Res 15:47–81

Oster O, Schmiedel G ,Prellwitz W (1988b) The organ distribution of selenium in German adults. Biol Trace Element Res 15:23–45

Schrauzer GN, White DA (1978) Selenium in human nutrition: dietary intakes and effects of supplementation. Bioinorg Chem 8:303–318

Thomson CD, Robinson MF (1980) Selenium in human health and disease with emphasis on those aspects peculiar to New Zealand. Am J Clin Nutr 33:303–323

US National Research Council (1980) Committee on Dietary Allowances "Recommended Dietary Allowances", 9th edn (1980). Natl Acad Sci, Washington

Westermarck T (1977) Selenium content of tissues in Finnish infants and adults with various diseases, and studies on the effects of selenium supplementation in neuronal ceroid lipofuscinosis patients. Acta Pharmacol Toxicol 41:121–133

Blood Selenium Levels of Norwegians Compared to Their Se Intakes

K. Bibow[1,4], H. M. Meltzer[2], H. Holm[2], and G. Norheim[3]

1 Introduction

Norwegian soils are known to be poor in selenium. Normally this would imply low blood-Se levels in the population. Several investigations (Blekastad et al. 1984; Meltzer et al. 1985; Ringstad et al. 1987) have, however, shown that the serum-Se levels of Norwegians are remarkably uniform and relatively high ($110-127\ \mu g\ l^{-1}$) when compared to results from the other Nordic countries (Mutanen et al. 1985; Åkesson and Öckerman 1985). Fifty percent of our bread flour consists of imported American or Canadian wheat with approximately 0.6 ppm Se, which is four to five times higher than our locally grown grain. In addition, our fish consumption is on an average slightly higher than in most other European countries.

To obtain more information on the factors influencing the Se status of Norwegians, a Se supplementation study was conducted by H. M. Meltzer and co-workers in the fall of 1985 (Meltzer 1988). As part of this study 16 women (age 20–53), working or studying in the department, collected duplicate food portions for 1-week. Three collections were extended to a 2-week period.

2 Results and Discussion

The food samples were analyzed by INAA and the blood samples by both AAS and INAA. The results are given in Table 1.

The average Se intake ($60 \pm 17\ \mu g\ l^{-1}$) is comparable to that of Finns before 1985, the serum levels ($118 \pm 10\ \mu g\ l^{-1}$) are comparable to American values, far above the Finnish levels.

As Fig. 1 shows, the correlation between Se intake and whole blood Se is rather weak, $r = 0.37$ (Se intake vs. serum Se: $r = 0.40$). The amount eaten is only weakly reflected in the blood-Se levels.

[1] Department of Chemistry, University of Oslo, 0172 Oslo, Norway
[2] Institute for Nutrition Research, University of Oslo, Oslo, Norway
[3] The Veterinary College of Norway, Oslo, Norway
[4] To whom correspondence should be addressed

Table 1. Se intake, blood-Se and serum-Se for 16 women

	Mean	SD	Range
Se intake (μg day^{-1})	60	17	31 – 93
Se intake (μg 10 MJ^{-1})	80	21	51 – 130
Blood (μg l^{-1})	151	14	130 – 180
Serum (μg l^{-1})	118	10	100 – 130

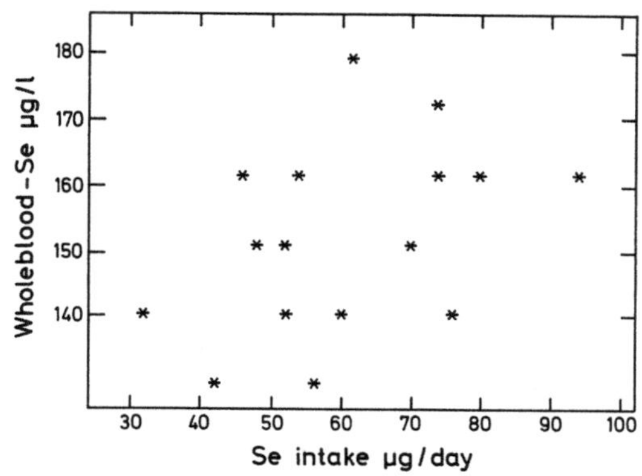

Fig. 1. Initial whole blood Se levels plotted against Se intake as determined by INAA analysis of duplicate portions (r = 0.37, NS)

Some possible explanations are:

1. One week is too short a time to obtain a representative food sample.
2. The blood levels fluctuate.
3. Different chemical forms of Se give different responses to the blood levels.
4. The Se intake is not the only nutritional component which determines the blood-Se levels.

To check the effect of sampling time, three persons collected duplicate portions for 2 weeks. The individual fluctuations from week to week in energy, fat and protein intake were small, and the Se intake fluctuated less than 6%. For the assessment of Se intake, 1-week sampling time seems to be sufficient.

Fluctuating blood levels do not seem to be a problem. Several studies have shown that when blood samples are taken under the same conditions, deviations from week to week are less than 10%.

The third possible explanation is, however, of great importance. It is well known that different chemical forms of Se give different biological responses. This was clearly demonstrated in the Se-supplementation part of the present study (Meltzer 1988). The effects of supplementation on whole blood-Se levels with two different forms of organic Se are shown in Fig. 2.

Supplementation with 200 μg of Se-rich pea flour for 12 weeks produced virtually no effect on the blood levels. Some of the participants were then given another tablet with the same amount of Se, but this time in a yeast-bound form. In 5 weeks, the blood levels increased from 150 to 210 μg l^{-1}. It might seem that

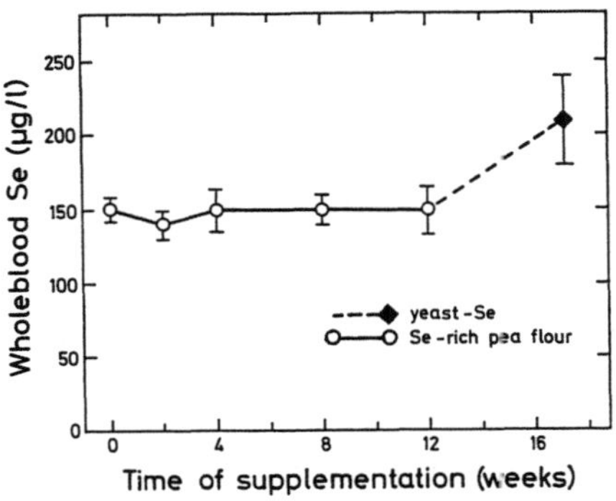

Fig. 2. Effect on whole blood-Se of supplementation with Se-rich pea flour (*solid line*) and yeast Se (*dashed line*) (n = 4). Another yeast Se formulation was also tested, but this gave a response identical to the pea flour

Table 2. Regression statistics of whole blood-Se with nutritional components

	R squared	Significance level
Se intake	0.16	0.12
÷ Protein	0.29	0.10
+ Riboflavin	0.60	0.010
+ Vitamin A	0.71	0.005
+ Vitamin C	0.80	0.003

the pea flour Se was not absorbed. Analysis of urine samples, however, showed that the Se from the pea flour was well absorbed (>60%), but the retention time must have been very short. Thus, we have demonstrated that organically bound Se from different food sources can have a very different bioavailability. Consequently, information on the total Se intake gives very little information on the Se status as long as we do not know the chemical forms of the Se eaten. Differences in the Se bioavailability of foods may therefore explain some of the differences between Norway and the other Nordic countries.

As to the fourth explanation: The Se intake is not the only nutritional factor which determines the blood-Se levels. The absorption of Se is known to be affected by the amount of sulfuric amino acids and the amount of protein in the diet. The supply of other nutrients important for antioxidative defence, such as riboflavin, vitamin A, vitamin C and vitamin E might influence the Se requirement. Multiple regression of the blood-Se with respect to different nutrients is shown in Table 2.

Only 16% of the variations in the blood-Se levels could be explained by the intake of Se. When the intake of protein, riboflavin, vitamin A and vitamin C, was accounted for, 80% of the variations could be explained.

These results indicate that it is important to study the composition of the food intake, as many components of the food influence the blood-Se levels. The single most important factor influencing the Se status of Norwegians may be the high bread consumption, making imported Se-rich wheat a major source of Se. But the effects of a "good" food intake, with moderate amounts of protein and relatively high amounts of naturally occurring vitamins, should also be taken into account.

References

Åkesson B, Öckerman PA (1985) Selenium status in vegans and lactovegetarians. Br J Nutr 53:199–205

Blekastad V, Jonsen E, Steinnes E, Helgeland K (1984) Concentrations of trace elements in human blood serum from different places in Norway determined by neutron activation analysis. Acta Med Scand 216:25–29

Meltzer HM (1988) Selenium – a bioavailability study. Thesis, Inst Nutr, Uni Oslo

Meltzer HM, Botten G, Aaseth JO (1985) Metabolism and biological function of selenium. J Norw Med Assoc 105:979–984

Mutanen M, Alfthan G, Arvilommi H, Koivistoinen P, Varo P (1985) Correlation between dietary selenium, platelet GSH-Px activity and plasma selenium level. Näringsforskning 4:135–138

Ringstad J, Jacobsen BK, Thomassen Y (1987) The Tromsø heart study: relationships between the concentration of selenium in serum and risk factors for coronary heart disease. J Trace Element Electrolyt Health Dis 1:27–31

The Intake of Selenium and Other Nutrients in a Group of Norwegian Rheumatics

H. M. MELTZER[1,5], K. BIBOW[2], R. RØNNEBERG[3], M. HAUGEN[4], and H. HOLM[1]

1 Introduction

Ten years ago, Aaseth et al. (1978) presented the first results showing that selenium metabolism was somehow affected in rheumatoid arthritis. These findings were later confirmed by several others (Jameson 1983; Tarp et al. 1983).

The present study is part of a larger project studying the selenium status and effects of selenium, vitamin E and omega-3-fatty acid supplementation to rheumatics.

There are several reasons why Norwegian rheumatics constitute a suitable study group for such purposes. To mention a few:

1. As reflected in blood values, the Norwegian selenium intake seems by European standards to be more than adequate, average serum levels in Norwegians typically ranging from 110 to 150% of those of other countries.
2. As illustrated in Fig. 1., Norwegians seem to utilize the Se in their foods well. The figure shows the Se intake and serum values in Finns and Norwegians. Whilst the intake in the two countries seems to be comparable, the blood levels of Norwegians are strikingly much higher (Meltzer 1988; Mutanen et al. 1985).
3. The Se-blood levels of Norwegians normally fall within a quite narrow range, thus making it easier to identify groups with aberrating Se intake or metabolism.

For these reasons, it has been assumed that the lowered blood-Se values among Norwegian rheumatics is due to the illness.

Knowing that the inflamed joints of rheumatics are at times subject to a continuous respiratory burst from the granulocytes and macrophages creating large amounts of oxidants (Babior 1984), one may expect the selenium status of rheumatoid arthritis (RA) patients to be somehow compromised. But the illness often also affects the eating habits of these patients, making them eat less than others. Many patients also try out various dietary regimens, so we may not a priori rule out the possibility that the lowered Se status of these patients may at least partly be attributed to below-average intake. Also, the possibility of nutrient interactions, for instance, must be taken into consideration. The intake of toxic

[1] Institute for Nutrition Research, Univ. Oslo, 0172 Oslo, Norway
[2] Department of Chemistry, University Oslo, 0172 Oslo, Norway
[3] Collett-Marwell Hauge A/S, Pb 205, 1371 Asker, Norway
[4] Oslo Sanitetsforenings Reumatismesykehus, Akersbakken, 0172 Oslo 1, Norway
[5] To whom correspondence should be addressed

Fig. 1. Selenium intake versus serum selenium values in Norwegians and Finns

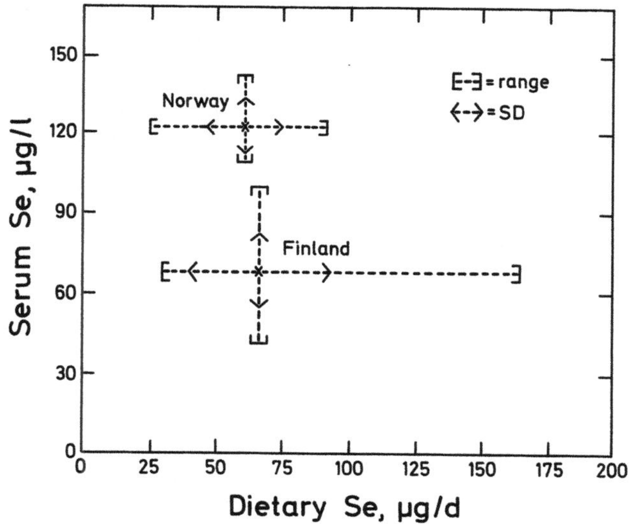

heavy metals must be assessed, as must the intake of trace elements important for other enzymes involved in the antioxidant defense system. From these premises, the twofold aim of the study emerged:

1. To test whether substitution therapy with vitamin E, selenium and EPA (eicosa pentanoic acid, Max EPA) could modulate the antioxidant defense systems in a positive way, thus making it possible to reduce doses of anti-inflammatory and immunosuppressive drugs.
2. To assess the nutrient intake of rheumatics, in order to determine whether there was any correlation between eating habits, disease activity and the effect of the supplements.

2 Patients and Methods

All of the participants had classical RA according to the ARA criteria (minimum three of the following criteria: SR > 28 mm/t, morning stiffness > 45 min, swollenness > three joints, pain > six joints).

Thirteen of the participants in the study, five males and eight females, aged 50–70, for 1 week collected duplicate portions and registered their intake for the dietary survey. After homogenization the samples were freeze-dried. Standard techniques were used for energy, protein and fat analyses. Carbohydrates were calculated by difference, fatty acids analyzed by gas chromatography and trace elements by AAS (day samples) and INAA (week samples).

Table 1. Intake of nutrients and heavy metals in 13 Norwegian rheumatics

	Average	Range
Energy (kcal)	1674 (±369)	1100–2320
Protein (g)	59.5 (±12.9)	34.8–87.1
Fat (g)	65.5 (±19.3)	32.7–104.7
Carbohydrate (g)	211.7 (±52.0)	145.4–323.5
Energy% protein:	14.5 (±11.6)	10.9–16.9
Energy% fat:	35 (±5.4)	26.4–45.2
Energy% KH:	50.5 (±6.6)	37.9–62.7
PS-ratio	0.51	0.18–1.3
Selenium (μg)	52.6 (±20.1)	36–59
Zinc (mg)	6.2 (±1.7)	3.5–10.1
Iron (mg)	9.2 (±5.1)	4.4–20.8
Copper (mg)	1.6 (±0.7)	0.9–3.5
Magnesium (mg)	261 (±90.7)	160–489
Cadmium (μg)	17.2 (±12.1)	7.1–53
Lead (μg)	90.1 (±26.0)	53–141

3 Results

Table 1 summarizes the results. The difference in Se intake between males and females is not significant when the intake is calculated per kilo body weight. The only significant difference between the sexes is in energy percentages from carbohydrate and fat.

4 Discussion

The average Se intake is almost identical to the intake of healthy Norwegians found using the same method (Bibow et al. 1984; Meltzer 1988), thus ruling out the possibility that the lowered blood selenium levels of rheumatics are due to low Se intake.

Furthermore, the trace element intake is not marginal, if compared with the RDA of ESADDI of the US National Research Council. The zinc intake is not exceptionally low for Norwegians, and there are hardly any signs of marginal zinc intake in Norway. The intake of toxic heavy metals are well below the WHO limits. It is not likely that an analysis of the food records will reveal instances of excessive intakes or strong deficiencies. To sum up, the patients eat nutritionally acceptable diets, their main problem being that they eat too *little*.

The illness itself and biochemical side effects thereof, remain the main explanation for the lowered Se status of Norwegian rheumatics.

However, the role of heavy metals in Se metabolism should be further investigated. With our patients, the Se intake was roughly proportional to the Cd+Pb-load (data for Hg are missing). The absorption and effect of the heavy metals in RA-patients may be different from healthy persons. Increased loads of heavy metals may be more critical where selenium status is compromised.

References

Aaseth J, Munthe E, Førre Ø, Steinnes E (1978) Trace elements in serum and urine of patients with rheumatoid arthritis. Scand J Rheumatol 7:237–240

Babior BM (1984) The respiratory burst of phagocytes. J Clin Invest 73:599–601

Bibow K, Riis G, Salbu B (1984) A Study of trace elements in Norwegian diets by the duplicate portion technique. Naringsforskning 4:84–88

Jameson S (1983) Kliniska erfarenheter av GSH-Px-mätningar och selenterapi motsvarande RDA. Symposium Selen i Kosten. Pharmacia Norden AB, Uppsala, pp 76–83

Meltzer HM (1988) Selenium – a bioavailability study. Thesis, University of Oslo, Institute of Nutrition Research

Mutanen M, Alfthan G, Arvilommi H, Koivistoinen P, Varo P (1985) Correlation between dietary selenium, platelet GSH-Px activity and plasma selenium level. Naringsforskning 4:135–138

Tarp U, Overvad K, Hansen JC, Thorlin EB, Graudal H (1983) Selenium deficiency in severe rheumatoid arthritis. Scand J Rheumatol (Suppl) 49:35

Se Supplementation of Fertilizers Has Increased the Se Intake and Serum-Se Concentration of Finnish People

A. ARO[1,3], G. ALFTHAN[1], and P. VARO[2]

1 Introduction

In studies on the mineral composition of Finnish foods in the 1970s low selenium (Se) concentrations of domestic foods were observed (Varo and Koivistoinen 1980) and the mean Se intake of the population was estimated to be around 25 μg day^{-1}. As the results of some epidemiological studies suggested that very low serum-Se concentrations, indicating low Se intakes, were associated with increased risk of cardiovascular death or the risk of cancer, concern was expressed on the quality of domestic foods in Finland. Consequently, a decision was made to supplement multimineral fertilizers with Se starting in 1984 (Koivistoinen and Huttunen 1986). This decision was preceded, among other measures, by a series of agricultural studies showing that in Finland the soil-Se is poorly available for plants. It is assumed that the main factors immobilizing Se are low pH, reducing conditions and high iron content in the soil. Furthermore, it was shown that Se as sodium selenate is incorporated into plants, more readily to the leafy parts than to grains (Yläranta 1985). For this reason, fertilizers which are used for the production of hay and fodder were supplemented with 6 mg kg^{-1} of Se while fertilizers used for the production of cereals were supplemented with a higher dose, 16 mg kg^{-1} fertilizer.

To eliminate the possibility of any unaccounted hazards of this unique experiment, the Ministry of Agriculture appointed an expert group to monitor the effects of Se supplementation on the Se concentration of fertilizers, soil and water, animal feeds, grain crops, basic foods and human subjects. In this study we report the effects of the supplementation on the Se concentration of common foods and human sera.

2 Material and Methods

Samples of 11 basic foods (wheat flour, wheat bread, rye flour, rye bread, milk, Edam cheese, eggs, beef, pork, bovine liver, porcine liver) were collected at 2-month intervals from local retail stores in the Helsinki area. The preparation

[1] Department of Biochemistry, National Public Health Institute, 00280 Helsinki, Finland
[2] Department of General Chemistry, University of Helsinki, 00710 Helsinki, Finland
[3] To whom correspondence should be addressed

and storing of the samples have been described in detail elsewhere (Varo et al. 1988). In addition, some other foods like potatoes, roots, vegetables, and fish were also analyzed, but less frequently.

Serum samples were obtained two to three times annually from 24 healthy employees of the National Public Health Institute representing the urban population of Helsinki. They had participated in a follow-up of various serum parameters for more than 10 years (Alfthan 1988). There were 8 men and 16 women with a mean age of 37 years. Additionally, another reference group, comprising 45 members of farmer families from the rural community of Leppävirta in eastern Finland, were recruited to give two to three annual blood samples for the determination of serum-Se concentration. This group was comprised of 26 men and 19 women, and the mean age was 45 years. None of these subjects consumed vitamin or mineral supplements.

Se was analyzed with electrothermal atomic absorption spectrometry, foods after pretreatment and extraction (Kumpulainen et al. 1983), and serum directly after dilution with nickel nitrate and nitric acid (Alfthan and Kumpulainen 1982). A rigid quality control scheme was adhered to throughout the study (Varo et al. 1988).

3 Results

The serum-Se concentration of both the urban and the rural subjects increased from 65–70 $\mu g\,l^{-1}$ in 1984/85 to about 110 $\mu g\,l^{-1}$ in 1987/88. The values started to increase in late 1985 and reached a new steady state in June 1987 (Fig. 1).

The Se concentration of some basic foods is shown in Fig. 2. The Se concentration of spring wheat increased 20-fold. However, the Se concentration of wheat flour showed a smaller increase, because the Se level of winter wheat remained much lower (Varo et al. 1988). Also, the milling proportions between imported

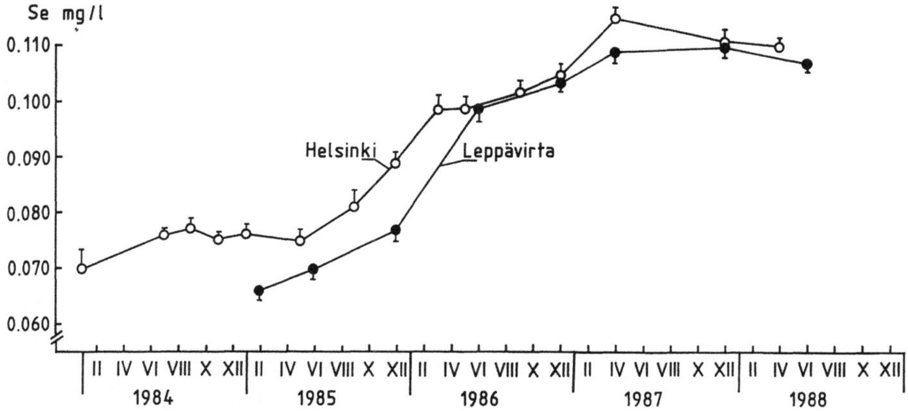

Fig. 1. Serum selenium concentration (mean ± SEM) of 24 urban (Helsinki) and 45 rural (Leppävirta) people

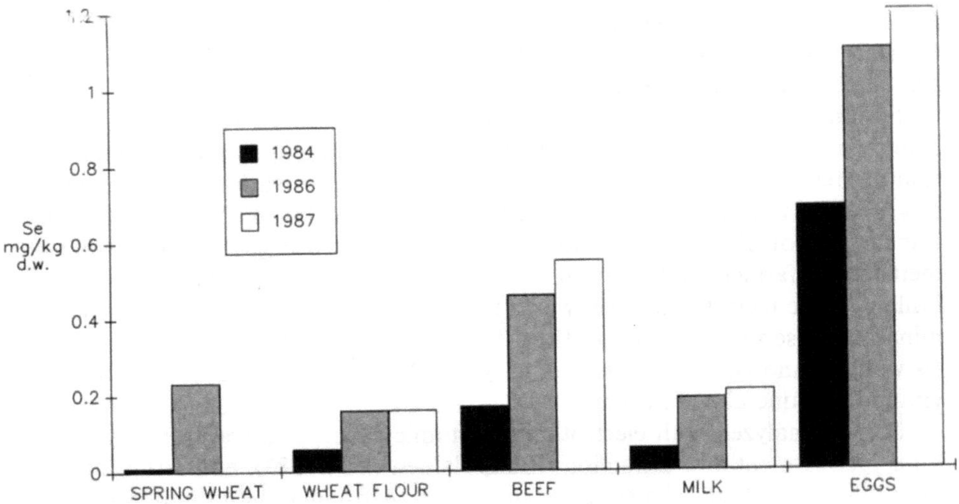

Fig. 2. Selenium concentration of some foods in Finland before (spring 1984) and during (spring 1986 and 1987) the supplementation of fertilizers with sodium selenate

and domestic grains differed slightly during the period of the study. Beef (and similarly pork) and milk products showed a threefold increase of Se concentration. Eggs had a relatively high Se concentration already before the supplementation of fertilizers, but the concentration still doubled. The relative increase of liver Se (both bovine and porcine) was lower than that of corresponding meats.

The estimated Se intake provided by a typical Finnish diet at an energy level of 10 MJ day^{-1} has increased from about 40 µg day^{-1} in 1984 to more than 100 µg day^{-1} in 1987 (Fig. 3). Compared to the intake in the 1970s, the increase is fourfold. In addition to the change in the total intake, the contribution of different foods as sources of Se has changed considerably. In 1975/77 fish accounted for one-third of the total Se intake while in 1987 the contribution of fish, although similar in absolute terms, has been reduced to less than 10%. Milk products, meat and cereals are the most important sources of Se in Finland today. As practically all foods contribute significantly to the total intake, it is currently impossible in Finland to compose a diet which would be deficient in Se.

4 Summary and Conclusions

Supplementation of fertilizers with sodium selenate has increased the mean serum-Se concentration of Finnish people from 70 to 110 µg l^{-1}. The Se intake has increased two- to threefold between 1984 and 1987. Thus, the intervention has been an effective way of improving the Se status of people in Finland. The predicted and observed effects have been in good agreement, and the Se intake is now within the limits generally considered safe and adequate.

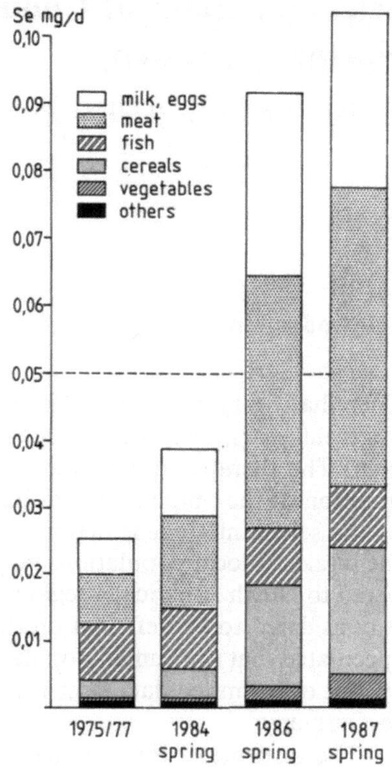

Fig. 3. Selenium intake from different foods in Finland estimated on the basis of a typical diet yielding 10 MJ day^{-1} energy

At the end of each year the expert group of the Ministry of Agriculture gives a recommendation on the level of Se supplementation for the following growing season. So far, the results have been considered excellent, and the expert group has seen no need for adjustments in the original Se levels.

References

Alfthan G (1988) Longitudinal study on the selenium status of healthy adults in Finland during 1975–1984. Nutr Res 8:467–476

Alfthan G, Kumpulainen J (1982) Determination of selenium in small volumes of blood plasma and serum by electrothermal atomic absorption spectrometry. Anal Chim Acta 165:221–227

Koivistoinen P, Huttunen JK (1986) Selenium in food and nutrition in Finland. An overview on research and action. Ann Clin Res 18:13–17

Kumpulainen J, Raittila AM, Lehto J, Koivistoinen P (1983) Electrothermal atomic absorption spectrometric determination of selenium in foods and diets. J Assoc Off Anal Chem 66: 1129–1135

Varo P, Koivistoinen P (1980) Mineral element composition of Finnish foods XII. General discussion and nutritional evaluation. Acta Agric Scand Suppl 22:165–171

Varo P, Alfthan G, Ekholm P, Aro A, Koivistoinen P (1988) Selenium intake and serum selenium in Finland: effects of soil fertilization with selenium. Am J Clin Nutr 48:324–329

Yläranta T (1985) Increasing the selenium content of cereal and grass crops in Finland. Academic diss, Yliopistopaino, Helsinki

Selenium Status of Children and Adults in Northern Germany

B. HEINZOW[1,2], H. JESSEN[1], S. MOHR[1], and D. RIEMER[1]

1 Introduction

There has been an increasing interest in the essential trace element selenium (Se) in recent years, less in its toxicity than in its essential function in animals and man. The therapeutic window between toxicity and deficiency is small and the question of the optimal nutritional intake is still under debate. Due to variations of the selenium content in soil and hence food of different geographical areas, the intake of local populations may differ considerably. Northern Europe is compared to North America, a region of low supply, and the soil content in Germany is considered to be deficient on the average (Hartfiel and Bahners 1987). It was speculated that the human population might be exposed to the risk of Se deficiency, but only limited data exist concerning the Se status of the general population in Germany.

The Institute of Environmental Toxicology has been studying the exposure of the general population in Schleswig-Holstein, the northern federal state of Germany, to trace elements, heavy metals and other environmental pollutants. Selenium levels in blood samples collected between 1986 and 1988 from children (age 2–15) and adults (age 20–70) living in different regions in Schleswig-Holstein, both urban and rural, were determined by means of hydride-generation atomic absorption spectrometry (Perkin Elmer AAS 4000 with HMS 20). Nycomed Seronorm (Norway) was used as reference material for an internal laboratory quality control.

2 Results and Discussion

Results are summarized in Table 1. The obtained mean levels in blood are comparable to the Se status in Scandinavia and other parts of Europe (Thorling and Overvad 1987) but they are lower than reported values from the USA. Slightly lower levels were found in Amrum, the sandy North Sea island, in comparison to the other sites ($p < 0.05$).

[1] Institute of Environmental Toxicology of Schleswig-Holstein, Fleckenstrasse 4, 2300 Kiel, FRG
[2] To whom correspondence should be addressed

Table 1. Selenium status in adults and children from rural and urban regions in Schleswig-Holstein

Location and year	Region	Adults/ children	n	Se concentration ($\mu g\,l^{-1}$)[a] mean	range
Amrum 1986	Rural (island)	Adults	85	80	39 – 145
		Children	25	77	45 – 125
Eutin 1986	Rural (country town)	Children	43	104	75 – 135
Flensburg 1987	Urban (city)	Adults	93	89	34 – 135
Pinneberg 1988	Urban (city)	Adults	27	101	70 – 140
		Children	51	83	62 – 108

[a] whole blood

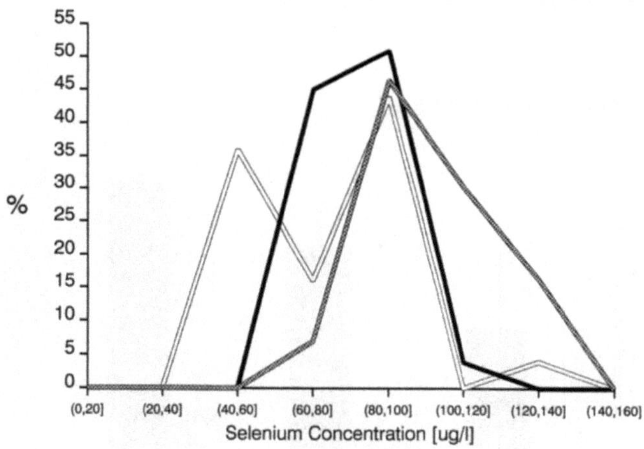

Fig. 1. Frequency distribution of selenium concentrations in blood of children. ▧ Amrum 1986; ■ Pinneberg 1988; ⊠ Eutin 1986

The concentration of selenium in blood is affected by several factors including age and dietary habits. The values are nearly normally distributed with a range of 30 – 145 $\mu g\,l^{-1}$ (Figs. 1 and 2). Children had a slightly lower selenium status compared to adults, no age dependency was found within the age group 20 – 70 years; this is in accordance with other reports. In this study there was no difference between men and women, smokers and nonsmokers. Fish consumption had no significant influence on the Se status, but was associated with higher blood mercury levels. Sufficient evidence exists on beneficial and adverse effects of interactions between essential and toxic elements. It was therefore of interest to correlate the Se status with the blood levels of other trace elements, heavy metals and other environmental pollutants. Statistical analysis revealed a significant correlation ($p < 0.05$) between selenium and mercury in blood (Fig. 3). Interestingly, this correlation was apparent in populations from the North Sea

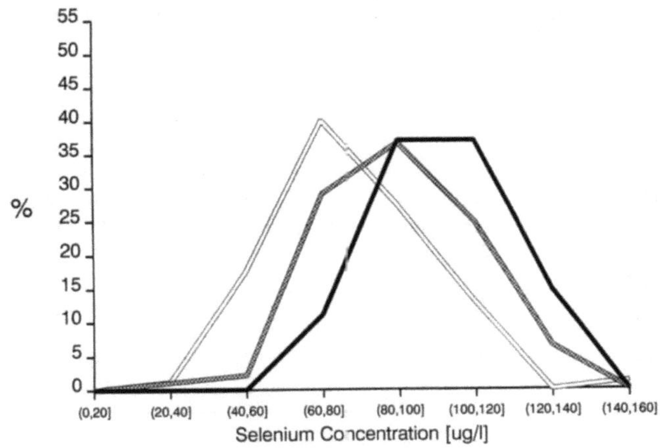

Fig. 2. Frequency distribution of selenium concentrations in blood of adults. ▩ Amrum 1986; ▩ Flensburg 1987; ■ Pinneberg 1988

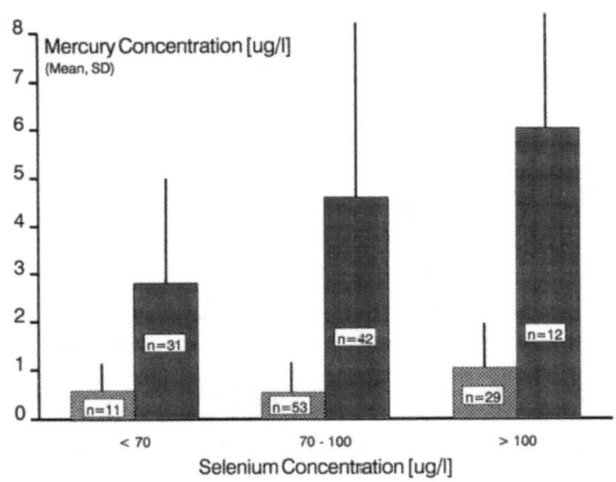

Fig. 3. Correlation between mercury and selenium in blood of adults. ▩ Flensburg; ▩ Amrum

island Amrum with high and from the urban region of Flensburg with low mercury blood levels as well. It is speculated that kinetic interactions between these elements are the underlying cause of this finding, possibly an enhancement of the binding characteristics of mercury in blood in relation to selenium status (Fang et al. 1976).

No significant correlation was found between Se status and arsenic in urine, lead, cadmium, zinc and copper, persistent chlorinated compounds (HCB, DDT, HCH, PCBs) in blood and cholesterol, triglycerides and retinol in serum.

Considering levels between 40 and 400 ng ml^{-1} as being "safe", it is concluded that, in general, the Se supply meets the requirements and no overt signs of deficiency were found. Optimal intake can be defined as freedom from signs of deficiency or toxicity; another view, however, aims at chemopreventive effects of antioxidants. This concept has received widespread attention in western industrialized communities and supplements to the normal diet have been introduced in some countries. At levels found in most of the participants of this study the activity of selenium-dependent enzyme systems might be below maximum and in 5% of the population the Se status might be suboptimal. However, more epidemiological information is needed before recommendations on supplementation can be made.

References

Fang SC, Chen RW, Fallin E (1976) Influence of dietary selenite on binding characteristics of rat serum protein to mercurial compounds. Chem Biol Interact 15:51–57

Hartfiel W, Bahners N (1987) Selenmangel in der Bundesrepublik Deutschland. VitaMinSpur 2:125–131

Thorling EB, Overvad K (1987) Selenium status in Denmark. A comparison with 16 sites in Europe. Environ Health Ser 20:37–38 (WHO, Copenhagen)

Urinary Excretion of Selenium (Se) and Trimethyl-selenonium (TMSe) by NZ Women During Long-Term Supplementation with Selenate or Selenomethionine (Semet)

M. F. ROBINSON[1,4], C. P. JENKINSON[1], G. LUZHEN[2], C. D. THOMSON[1], and P. D. WHANGER[3]

The unusual situation in New Zealand adds one more variable to our understanding of how selenium is metabolized, though the problem is complicated by the scarcity of information about dietary forms, functions, metabolism and excretion. We had noted that New Zealanders, and also the Chinese in Keshan disease areas, excreted a smaller proportion of the selenium presented to their kidneys, i.e., they had lower renal plasma clearances, than North American residents (Robinson et al. 1985). Evidence of urinary metabolites is scanty, although it is almost 20 years since Byard (1969) and Palmer et al. (1969) identified trimethyl selenonium (TMSe) in rat urine; a major metabolite when the diet was relatively high in Se (>0.5 ppm Se), but minor when the urine was low in Se. Although TMSe has been identified in human urine (Burk 1976), its importance is not clear, varying from less than 1 to 50% of total Se in the urine, partly because of the difficulty in separating TMSe by the HPLC, dual column, or reineckate methods or a combination of these, and ultimately estimating it precisely.

We decided to exploit the NZ scene once again in a collaborative study with the Oregon group (P. Whanger and colleagues). We followed the urinary excretion of Se by 36 women in a double blind supplementation trial for 32 weeks with 200 μg Se day^{-1} as selenate mixed with plain yeast (selenate), Se-enriched yeast (Semet) or plain yeast (placebo). The subjects were requested not to eat the Se-rich foods, fish, liver, kidney and brazil nuts, for the 2 days before the time blood samples and 24-h urines were collected at 0, 2 and 4 weeks, then monthly for 28 weeks. The completeness of urine collections was checked from creatinine excretions, and three subjects were excluded because their creatinine excretion was less than 7 mmol day^{-1} on more than two occasions.

Total Se was determined in individual urines and also in pooled group urines, and TMSe in pooled group urines for most stages by a dual-column ion-exchange method, a modification of the method of Nahapetian et al. (1984). Our choice was limited by the very low concentration of TMSe in normal NZ urines, with the consequent large size of sample required for the reineckate method, and the problem of desalting adequately for the HPLC technique (Kraus et al. 1985). Good agreement was reached between HPLC and the chosen dual column techniques for TMSe-rich urine collected from a subject after ingesting TMSe

[1] Department of Human Nutrition, University of Otago, Dunedin, New Zealand
[2] Institute of Nutrition and Food Hygiene, Beijing, China
[3] Oregon State University, Corvallis, OR 97331, USA
[4] To whom correspondence should be addressed

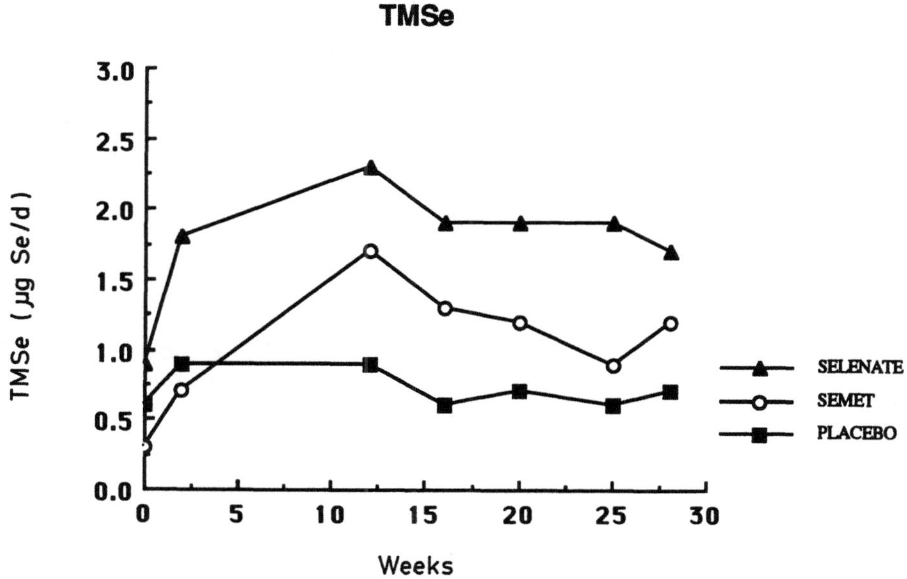

Fig. 1. Trimethylselenonium (TMSe) excretion (µg day^{-1}) in pooled urines during supplementation with 200 µg day^{-1} as selenate mixed with plain yeast (▲), plain yeast placebo (■), and Se-enriched yeast, Semet (○)

(144 ng Se ml^{-1}; 139 ng Se ml^{-1} urine respectively). Blanks, standards and recoveries were set up with each set of analyses for each stage of supplementation. Blank values (mean 0.12 ± 0.08 SD ng Se ml^{-1}, $n = 29$), which were always subtracted, were comparable with TMSe found in some urines of placebo and presupplementation groups (0.2–0.6 ng Se ml^{-1} urine), and were all close to the lower limit of sensitivity of the method (Thomson and Robinson 1986). Replicate analyses of urines from supplemented subjects gave CV of 10 to 20%; mean recoveries of standards and TMSe added to urines varied between 81 and 94%. Urinary TMSe appeared to increase with supplementation (Fig. 1), was greater at each stage for the selenate than for the Semet groups, and remained so throughout the study. However, the increases were small and showed little difference when expressed as a percentage of the total Se in the urine at about 1%. This was a slightly lower proportion of TMSe/Se than that found for the placebo group (3 to 6%), and also for urines from subjects in Keshan areas of China. Whether there is a real difference between placebo and supplemented groups is not clear, particularly as TMSe was barely detectable for the placebo groups. Our study supports the findings of Sun et al. (1987), who had found similarly low values of about 1% of TMSe/Se for US residents, confirming TMSe as a minor urinary component.

Returning to the response in total urinary Se, plasma Se soon reached a plateau at about 100 ng ml^{-1} with selenate in 4 weeks dosing, whereas plasma-Se continued to increase for 20 weeks with selenomethionine to about 200 ng ml^{-1}. The much greater urinary excretion with selenate than selenomethionine reflected

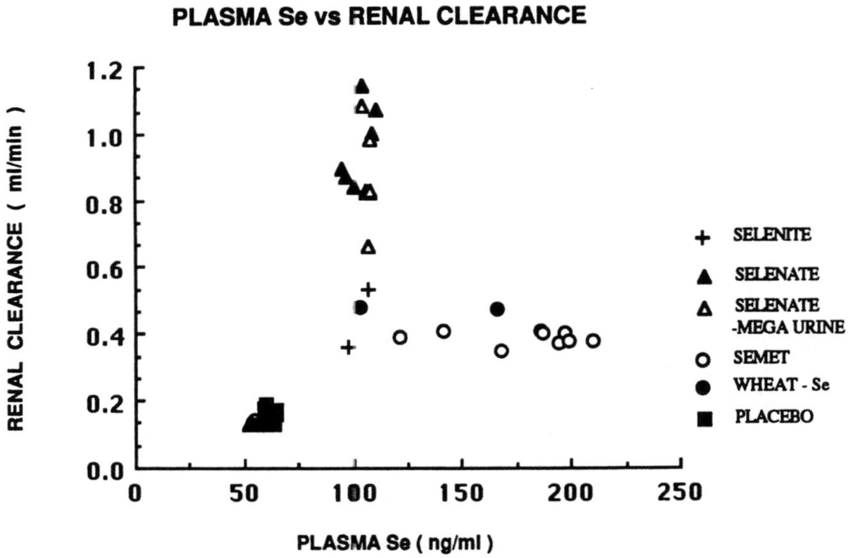

Fig. 2. Plasma Se (ng ml^{-1}) and renal plasma clearance (ml min^{-1}) for New Zealand women during supplementation with 200 µg Se day^{-1} as selenate mixed with plain yeast (▲), Se-enriched yeast, Semet (○), selenite (+), high Se wheat bread (●)

the much higher proportion of the dose recovered on the first day (71 vs. 22%) observed in our single, 1-mg dose studies, when peak excretions were also reached in 3 h (Thomson and Robinson 1986). This possibly accounts for the wide variations in mean selenate excretions, confirmed by results for pooled urines in this study. Examination of individual results showed that several subjects had excretions well over 200 µg Se day^{-1} (mega-urines), reflecting some irregularities in taking the supplements, e.g., late in the evening or possibly two instead of the prescribed one capsule each day in the morning. Such irregularities are not apparent with selenomethionine where the peak excretion is much less, and hence insensitive to irregular dosing by the subjects.

A close relationship was observed for selenomethionine between 24-h urine and plasma-Se concentrations ($r = 0.96$; $p < 0.001$), confirming our earlier observations with daily dosing with 100 µg Se as selenomethionine (Robinson et al. 1978). All these changes in plasma and urinary Se with supplementation are brought together in the renal plasma clearances, which increased during the first 2 weeks of supplementation to reach a plateau at about 0.4 ml min^{-1} for selenomethionine, but continued up to over twice that value for selenate at 0.8 to 1.10 ml min^{-1} depending on the number of irregular dosers at each stage. The difference in response to the two supplements is shown in Fig. 2. Comparison of these findings with those of our earlier studies (Robinson 1988) with daily supplements of 200 µg Se day^{-1} as selenite showed similar responses in plasma-Se for selenite and selenate, but only a moderate increase in clearance (0.5 ml min^{-1}) compared to selenate. Further daily supplements of high Se wheat bread (also 200 µg Se day^{-1}) showed clearances of 0.5 ml min^{-1}, with plasma Se at 166±11 ng ml^{-1}, thereby showing a response between those for selenate and selenomethionine, the latter accounting for 50% or more of wheat Se.

Thus, Fig. 2 summarizes the responses of Dunedin women to supplements of 200 µg Se supplied in the four prime forms of selenate, selenite, selenomethionine and food Se (as in bread). Further work is required on the nature of the metabolites and their distribution in the plasma and whether TMSe may become a major urinary component at intakes approaching toxic levels.

Acknowledgements. We are most grateful for the assistance of Dr. H.E. Ganther, Dr. M. Janghorbani, Dr. V.C. Morris and their colleagues for establishing the techniques for TMSe, to Dr. O.E. Olsen for a generous donation of TMSe, and to Mr. P.K. Huemmer for technical assistance. The work was supported by the Medical Research Council of New Zealand and by USDA Grant (86-CRCR-1-2084).

References

Burk RF (1976) Selenium in man. In: Prasad A (ed) Trace elements in human health and disease, vol 2. Academic Press, London New York, pp 105–133

Byard JL (1969) Trimethyl selenide, a urinary metabolite of selenite. Arch Biochem Biophys 130:556–560

Kraus RJ, Foster SJ, Ganther HE (1985) Analysis of trimethylselenonium in urine by high-performance liquid chromatography. Anal Biochem 147:432–436

Nahapetian AT, Young VR, Janghorbani M (1984) Measurement of trimethyl-selenonium ion in human urine. Anal Biochem 140:56–62

Palmer IS, Fischer DD, Halverson AW, Olson OE (1969) Identification of a major selenium excretory product in rat urine. Biochim Biophys Acta 177:336–342

Robinson MF (1988) The New Zealand selenium experience. Am J Clin Nutr 48:521–534

Robinson MF, Rea HM, Friend GM, Stewart RDH, Snow PC, Thomson CD (1978) On supplementing the selenium intake of New Zealanders. Br J Nutr 39:589–600

Robinson JR, Robinson MF, Levander OA, Thomson CD (1985) Urinary excretion of selenium by New Zealand and North American human subjects on differing intakes. Am J Clin Nutr 41:1023–1031

Sun XF, Ting BTG, Janghorbani M (1987) Excretion of trimethylselenonium ion in human urine. Anal Biochem 167:304–311

Thomson CD, Robinson MF (1986) Urinary and fecal excretion and absorption of a large supplement of selenium: superiority of selenate over selenite. Am J Clin Nutr 44:659–663

Selenium (Se) and Glutathione Peroxidase (GSH-Px) in Blood Components of NZ Women During Long-Term Supplementation with Selenate or Selenomethionine (Semet)

C. D. THOMSON[1,3], M. F. ROBINSON[1], and P. D. WHANGER[2]

1 Introduction

The low selenium (Se) soil results in low intakes for New Zealand residents, and although Se-responsive conditions have not yet been identified, some groups may be at risk of deficiency under special circumstances. Intakes of Se may be increased by dietary means, self-medication or controlled supplementation. However, knowledge of absorption, availability and metabolism of Se is necessary to make recommendations on dietary intakes, maximum safe levels and supplementation should it become desirable.

Selenite, selenate, selenomethionine (Semet) and Se-yeast have been suggested as suitable forms for prolonged Se supplementation. We have had experience with dosing of small groups of subjects taking selenite and Semet daily at 100 µg for up to 44 weeks (Thomson et al. 1978, 1982). Blood-Se levels were greater with Semet but both were equally effective in raising erythrocyte glutathione peroxidase (GSH-Px) activities. Selenite was adopted for human use by New Zealand self-dosers for relief of muscular pains because of its successful use in treating animal deficiencies of Se, but there is a growing concern about the long-term effects of prolonged selenite supplementation.

Selenate is chemically more stable then selenite and was used instead of selenite as a control standard in the Finnish Se-supplementation studies (Levander et al. 1983). There is also increasing emphasis on the use of selenate for agricultural purposes. However, there is little information on the metabolism of selenate or the long-term effects of its supplementation. The aim of this study was to compare the long-term effects of selenate and selenomethionine on Se concentrations and GSH-Px activities in blood components of a low Se population.

2 Methods

Thirty-six New Zealand women aged 18 – 23 years participated in a double-blind Se trial for 32 weeks. The subjects were randomly assigned to three groups and received a daily supplement of 200 µg Se in tablets containing selenate mixed with

[1] Department of Human Nutrition, University of Otago, Dunedin, New Zealand
[2] Oregon State University, Corvallis, OR 9733, USA
[3] To whom correspondence should be addressed

plain yeast (selenate), Se-enriched yeast (major form as Semet) or plain yeast (placebo). The subjects were requested not to eat the Se-rich foods, fish, liver kidney and brazil nuts, for 2 days before the time blood samples were taken.

Blood was drawn on the day before supplementation (week 0) at 2 and 4 weeks, and monthly for a further 28 weeks. Heparinized blood was separated into plasma and erythrocytes and whole blood, plasma and erythrocytes assayed for Se and GSH-Px activities. Blood for platelet preparation was collected for GSH-Px assays as described previously (Thomson 1985).

Se was analyzed by the automated fluorimetric method (Brown and Watkinson 1977) and GSH-Px assayed by a modification (Thomson et al. 1982) of the method of Paglia and Valentine (1967). Protein in plasma was measured by the standard Biuret method (Gornall et al. 1949) and in platelets by the method of Lowry et al. (1951)

3 Results and Discussion

3.1 Selenium Concentrations

Table 1 gives Se concentrations in whole blood and plasma and GSH-Px activities in whole blood, plasma and platelets after 32 weeks of supplementation. Plasma-Se rose to reach a plateau at a concentration of about 100 ng Se ml^{-1} in the selenate group at around 2 weeks, but rose to about 200 ng Se ml^{-1} with Semet plateauing at around 12 weeks. The trends in Se concentrations in plasma were similar to those observed after supplementation with selenate and Se-rich yeast

Table 1. Se concentration and GSH-Px activities after supplementation with selenate and selenomethionine[a]

Week	Selenate		Semet		Placebo	
	0	32	0	32	0	32
Se						
Whole blood (ng ml^{-1})	66 ± 10	109 ± 14	71 ± 9	274 ± 42	76 ± 13	75 ± 10
Plasma (ng g^{-1} protein)	54 ± 7	107 ± 15	55 ± 9	193 ± 28	59 ± 10	63 ± 8
GSH-Px						
Whole blood (U g^{-1} Hb)	23 ± 5	35 ± 7	24 ± 4	34 ± 7	27 ± 5	27 ± 4
Plasma (U g^{-1} protein)	3.0 ± 0.5	4.5 ± 0.6	3.0 ± 0.2	4.0 ± 0.7	2.9 ± 0.5	3.1 ± 0.5
Platelets (U g^{-1} protein)	332 ± 77	648 ± 116	304 ± 108	478 ± 146	338 ± 63	325 ± 76

[a] Values are means ± SD

by Finnish men who also received 200 μg daily but for a shorter period of 11 weeks (Levander et al. 1983).

Whole blood Se in the New Zealand women also plateaued at a concentration similar to that of plasma Se at 4 to 7 weeks in the selenate group. It is interesting that this is the level of Se concentration reached in studies of selenite supplementation at a level of 100 μg Se day^{-1} (Thomson et al. 1982). On the other hand, whole blood Se continued to rise throughout the study after Semet supplementation, reaching 270 ng ml^{-1} at 32 weeks.

3.2 Glutathione Peroxidase Activities

GSH-Px activities in whole blood rose to about 150% of the presupplementation value with both selenate and Semet plateauing at week 12. The similar response of GSH-Px activities to different forms of Se was also observed in studies with selenite and Semet (Thomson et al. 1982). In the present study a similar trend was seen in plasma GSH-Px activities, but with an earlier plateau at 2 weeks. This increase in plasma GSH-Px was not seen in Finnish men (Levander et al. 1983).

In contrast, platelet GSH-Px activities plateaued at 200% of the presupplementation value at 2 to 4 weeks in the selenate group, but only to 160% after Semet supplementation ($P < 0.001$). This difference persisted throughout the supplementation period, in contrast to the Finnish study in which platelet GSH-Px in the Se-yeast group rose more slowly than that of the selenate group but approached a similar level by the end of the 11-week supplementation period (Levander et al. 1983).

One striking observation was the large individual variation in response of GSH-Px activities to Se supplementation, and this was not always consistent for one subject with respect to platelet and whole-blood GSH-Px activities.

When mean GSH-Px in whole bood for each week was plotted against mean whole-blood Se concentration, a good correlation was found for both selenate and Semet ($r = 0.87$; $r = 0.73$).

In an earlier supplementation study we found an extremely good relationship between platelet GSH-Px activities and both Se and GSH-Px in liver (Thomson et al. 1988). This was with selenite dosing but the correlation was still very good when the Se-supplemented group was not included in the analysis, indicating that the relationship still applied with food Se. We believed that study validated the use of platelet GSH-Px for assessing Se status. Because the response of platelet GSH-Px depends on the form of Se given, the present study supports the suggestion of Levander et al. (1983) that this measure may also be useful for assessing the bioavailabilities of Se in different foods.

Acknowledgements. This work was supported by the Medical Research Council of New Zealand and by USDA Grant (86-CRCR-1-2084). We are very grateful to the subjects, and to Mr. Peter Huemmer and Ms. Trudy Milne for their technical assistance.

References

Brown MW, Watkinson JH (1977) An automated fluorimetric method for the determination of nanogram amounts of selenium. Anal Chem Acta 89:92–97

Gornall AG, Bardawill CJ, David MM (1949) Determination of serum proteins by means of the Biuret reaction. J Biol Chem 177:751–766

Levander OA, Alfthan G, Arvilommi H, Gref CG, Huttunen JK, Kataja M, Koivistoinen P, Pikkarainen J (1983) Bioavailability of selenium to Finnish men as assessed by platelet glutathione peroxidase activity and other blood parameters. Am J Clin Nutr 37:887–897

Lowry OH, Rosebrough NJ, Farr AL, Randall RJ (1951) Protein measurement with the folic phenol reagent. J Biol Chem 193:265–275

Paglia DE, Valentine WN (1967) J Lab Clin Med 70:158–169

Thomson CD (1985) Selenium-dependent and non-selenium-dependent glutathione peroxidase in human tissues of New Zealand residents. Biochem Int 10:673–679

Thomson CD, Burton CE, Robinson MF (1978) On supplementing the selenium intake of New Zealanders. 1. Short experiments with large doses of selenite or selenomethionine. Br J Nutr 39:579–587

Thomson CD, Robinson MF, Campbell DR, Rea HM (1982) Effect of prolonged supplementation with daily supplements of selenomethionine and sodium selenite on glutathione peroxidase (EC 1.11.1.9) activity in blood of New Zealand residents. Am J Clin Nutr 36:24–31

Thomson CD, Stevens SM, Rij AM van, Wade CR, Robinson MF (1988) Selenium and vitamin E supplementation: activities of glutathione peroxidase in human tissues. Am J Clin Nutr 48:316–323

Comparison of Human Response to Low Doses of Inorganic and Organic Selenium

E. JOHANSSON[1,5], L.-O. PLANTIN[2], V. GALGAN[3], T. WESTERMARCK[4], and U.-R. NORDBERG[4]

1 Introduction

Man's response to low doses of selenium and the significance of its chemical form for physiological requirements needs further clarification. As the concentration may differ in the cells and body fluids following administration of inorganic and organic selenium (Robinson et al. 1978; Johansson et al. 1983), we investigated some effects of known forms of inorganic and organic selenium in blood plasma, erythrocytes and urine.

2 Material and Methods

Selenium (100 µg) was given orally, every day to two groups, A and B; each had seven volunteers on a plain diet.

Group A. Six women and one man (mean age 29.7 years) were given Formulation A consisting of sodium selenite only.

Group B. Six men and one woman (mean age 26.3 years) were given Formulation B consisting of 50 µg selenium as sodium selenite and 50 µg selenium as l-selenomethionine, plus vitamins, E 10 mg, B_2 1.5 mg, B_6 2 mg and B_{12} 2 µg to assist the metabolism of glutathione peroxidase (E-GSH-Px).

Fourteen days before the trial, the day of onset, 4 months of trial and 1 month after cessation of the trial, the selenium concentration was estimated in blood plasma and erythrocytes by neutron activation analysis (Plantin et al. 1981) and the activity of GSH-Px in erythrocytes after Paglia and Valentine (1967) and Beutler et al. (1977) and in the urine by atomic absorption spectrophotometry by continuous hydride formation with FIA (Galgan and Frank 1988).

[1] Institute of Radiation Research, Department of Physical Biology, Box 535, 75121 Uppsala, Sweden
[2] Karolinska Institute, Research Center, Huddinge Hospital, 14186 Huddinge, Sweden
[3] Chemical Department, SVA, 75007 Uppsala, Sweden
[4] Gillobacken, 02400 Kyrkslätt, Finland
[5] To whom correspondence should be addressed

3 Results

The results are summarized in Table 1. The activity of E-GSH-Px at 14 days of trial is shown in Fig. 3.

Table 1. The concentration of selenium in the blood plasma (P), erythrocytes (E) and 24 hourly output of urine (U) and the activity of the selenium-dependent enzyme, E-GHS-Px

		P-Se (ng g^{-1})	E-Se (ng g^{-1})	U-Se (μg Se/24 h)	E-GSH-Px (U g^{-1} Hb)
Trial					
14 days before	A[a]	61.9	98.7	–	27.8
	B	66.6	98.2	–	22.7
Onset	A	73.2	–	21	28.8
	B	73.1	–	34	21.6
1 m	A	–	–	–	31.0
	B	–	–	–	22.7
4 m	A	65.1	101.2	64	25.4
	B	101.6	124.3	60	28.9
1 m after	A	59.2	98.8	29	30.7
	B	75.1	141.7	32	22.7

[a] Group A on Formulation A; group B on Formulation B

4 Discussion

In blood plasma selenium carriers such as albumin, hemoglobin and GSH show early changes in selenium concentration but the functional information of selenium status as indicated by GSH-Px may be limited in low dose experiments.

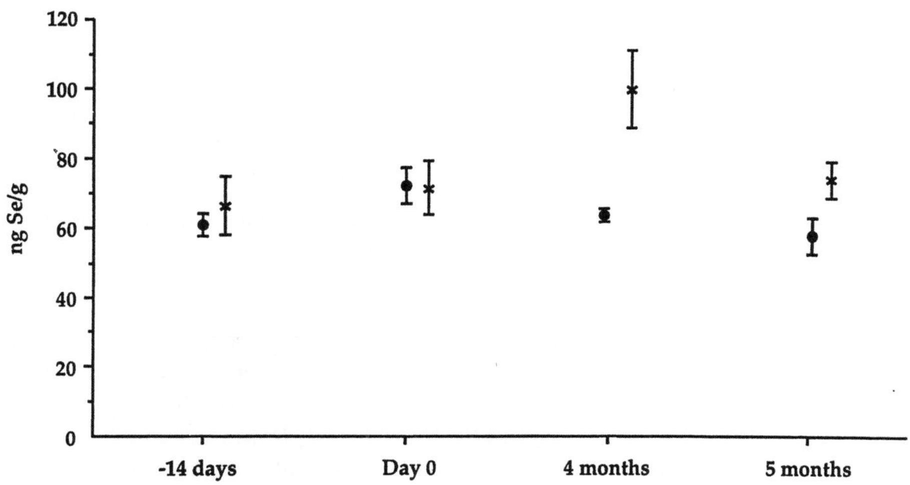

Fig. 1. Selenium concentration in plasma after 100 μg Se as sodium selenite or a mixture of sodium selenite and l-selenomethionine and vitamins. (●) Sodium selenite group A; (×) the mixture group B

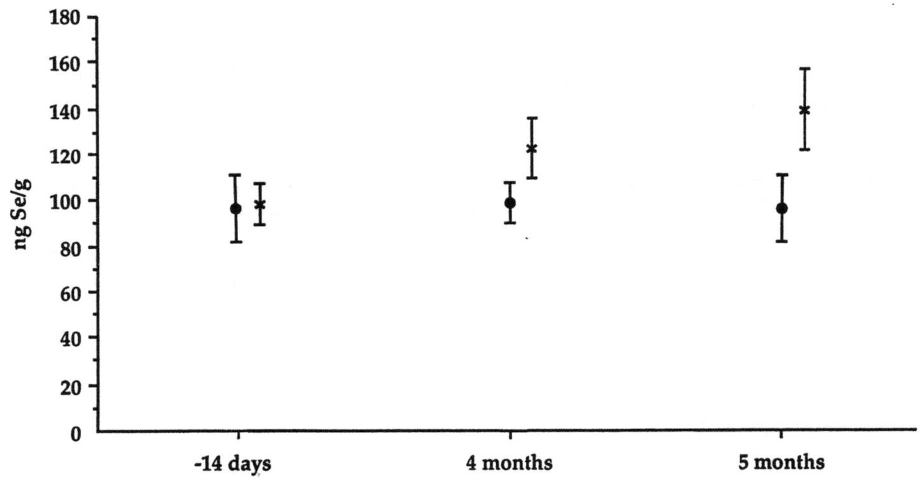

Fig. 2. Erythrocyte selenium concentration after 100 µg selenium as sodium selenite or sodium selenite and l-selenomethionine and vitamins. (●) Sodium selenite group A; (×) the mixture group B

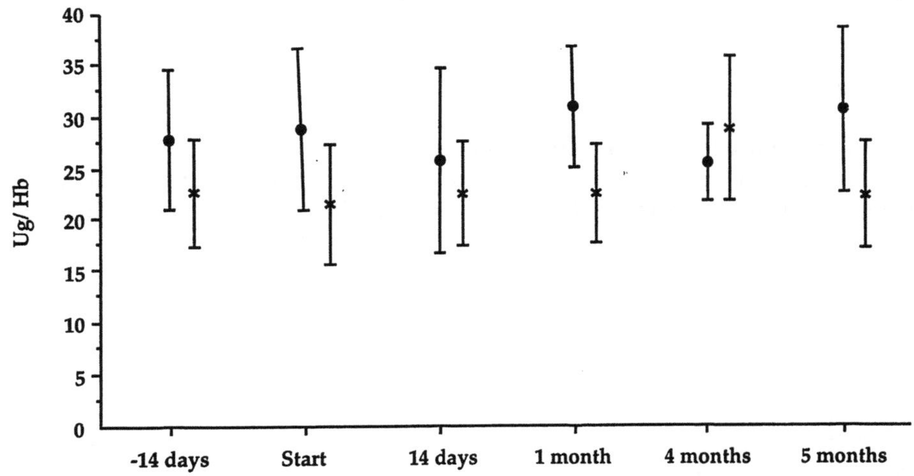

Fig. 3. E-GSH-Px activity after oral sodium selenite or sodium selenite and l-selenomethionine and vitamins. (●) Sodium selenite group A; (×) the mixture group B

We took blood samples, 14 days before onset, at onset, 4 months after trial and 1 month after cessation of the trial. The response of sodium selenite as shown by the selenium concentration in plasma in group A was low (Fig. 1). Group B given Formulation B showed a significantly increased plasma selenium concentration ($P<0.001$) in contrast to group A, indicating that inorganic selenium may be stored in different compartments than organic selenium.

The uptake of selenium in erythrocytes represents longer exposure time, the life span of a red cell being about 4 months. The assimilation from sodium

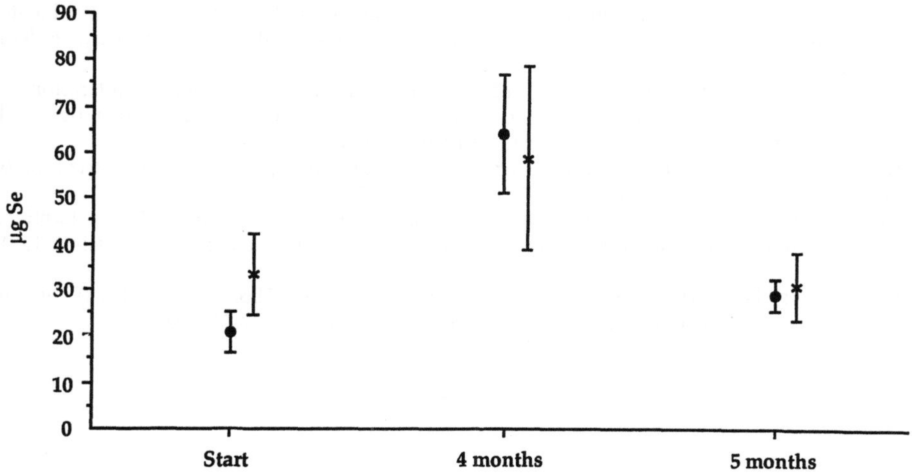

Fig. 4. Selenium excretion in 24 hourly urine after Formulation A or B

selenite in group A was low (Fig. 2), but the uptake in group B given Formulation B was significantly higher ($P<0.001$) and stayed at a higher level, 1 month after cessation of supplementation than in group A. The greater selenium uptake by erythrocytes in group B indicates different handling of inorganic and organic selenium compounds and supports the observation of Beilstein and Whanger (1986) that l-selenomethionine is associated with the heme. Figure 3 shows that group A showed a low response to Formulation A but group B given Formulation B showed a significant increase ($P<0.05$). It is suggested that vitamins B_2, B_6 and B_{12} assist in the synthesis of the active part, l-selenocysteine of GSH-Px and that vitamin E spares the selenium.

Significant differences in the plasma GSH-Px were not shown during the trial, the dose of selenium being probably too low. The retention of selenium in urine in group B was greater than in group A (Fig. 4). Inorganic selenium compounds may form more water-soluble intermediates than organic selenium, which is incorporated mainly in proteins.

A combination of low doses of inorganic and organic selenium plus vitamins when administered orally gave a greater response in plasma, erythrocytes and E-GSH-Px than sodium selenite alone.

Acknowledgement. Financial support from Astra Läkemedel AB is gratefully acknowledged.

References

Beilstein MA, Whanger PD (1986) Deposition of dietary organic and inorganic selenium in rat erythrocyte proteins. J Nutr 116:1701–1711

Beutler E, Blume KG, Kaplan JC, Löhr GW, Valentine WN (1977) International Committee for Standardization in Haematology: Recommended methods for red cell enzyme analysis. Br J Haematol 35:331–340

Galgan V, Frank A (1988) Automated system for determination of selenium in biological materials. Abstr 5th Int Workshop Trace Element Anal Chem Med Biol, Neuherberg, Federal Republic of Germany, April 15–18

Johansson E, Lindh U, Landström E (1983) The incorporation of selenium and alterations of macro- and trace element levels in individual blood cells following supplementation with sodium selenite and vitamin E. Biol Trace Element Res 5:433–447

Paglia DE, Valentine WN (1967) Studies on the quantitative and qualitative characterization of erythrocyte glutathione peroxidase. J Lab Clin Med 70:158–169

Plantin L-O, Meurling S, Strandvik B (1981) Selenium levels in Swedish children determined with a simple neutron activation analysis method. In: Mineral elements '80, Helsinki 1981, p 453

Robinson MF, Rea HMF, Friend GM, Stewart RDH, Snow PC, Thomson CD (1978) On supplementing the selenium intake of New Zealanders. Br J Nutr 39:589–600

The Relativity Between Some Epidemiological Characteristics of Kaschin-Beck Disease and Selenium Deficiency

YI-FANG JIANG[1] and GUANG-LU XU[1,2]

1 Introduction

In 1972–1973 it was found that rats fed with cereals and drinking water from Kaschin-Beck disease (KBD) areas died of acute massive liver necrosis, and that the selenium contents in grains were lower in the endemic areas than in the non-endemic areas, suggesting that there might be a selenium deficiency in the environment and *milieu interne* of humans in KBD areas. Since the late 1970s a comprehensive geo-eco-epidemiological survey has been carried out in 26 provinces of China. The results indicate that selenium is the most related element among the 20 elements analyzed, indicating that there is a close relationship between selenium deficiency and KBD (Kaschin-Beck disease research group of Xi'an Medical University 1975; Mo 1984). In 1974, Li et al. in Gansu province began to use sodium selenite as a means of treatment in early-stage KBD patients and obtained encouraging results. In this chapter results of investigations on the relativity of some epidemiological characteristics between selenium deficiency and KBD are briefly discussed.

2 The Comparison of Geographical Characteristics of Selenium-Deficient and KBD Areas

KBD is an endemic disease that occurs mainly in agricultural and rural areas in China. It has been reported that KBD areas are distributed over more than 300 counties in 14 provinces in China, including Heilongjian, Jiling, Liaoning, Hebei, Shandong, Henan, Shanxi, Shaanxi, Gansu, Qinghai, Sichuan provinces, as well as in Neimong and Xizang autonomous regions. Linking up the provinces and regions on a map, the KBD areas can be seen as a geographical belt, due to its special pattern of distribution. This is one of the most prominent epidemiological features of KBD. Actually, the belt mentioned above is a zone made up of many KBD and KBD non-affected areas. If one KBD non-affected area is surrounded by several KBD areas, it could be called a "health island". Conversely, if one KBD

[1] Research Laboratory of Keshan Disease, Xi'an Medical University, Xi'an 710061, Shaanxi, The People's Republic of China
[2] To whom correspondence should be addressed

Table 1. Se status in the environment and in the children

Sample	Selenium level (ppb)				
	Non-low Se zone		Low Se zone	KBD areas	KBD non-affected area
	South-East zone	North-West zone			
Soil	230	190	130	89	190
Water	–	–	–	0.18	0.97
Corn	53	49	16	3.6	>3.6
Wheat	52	106	22	4.1	>4.1
Rice	64	87	20	20	20
Blood ($ng\ ml^{-1}$)	–	–	–	17	76
Hair	378	377	129	110	200
Urine ($\mu g\ day^{-1}$)	>6	>6	–	<6	>6
Intake ($\mu g\ day^{-1}$)	>30	>30	–	20	30

area is surrounded by several KBD non-affected areas, it could be called a "disease island".

Three different eco-geographical zones with different selenium levels have been designated: the South-East zone with a high selenium level, the Middle zone with a low selenium level and the North-West zone with a high selenium level (Tan et al. 1987). The results from epidemiological investigations indicate that the Middle zone and KBD areas generally overlap each other in China. All KBD are distributed over low selenium environments in the low selenium zone. The selenium status in the environments of the three different zones as well as the selenium status in children are listed in Table 1 (Tan et al. 1987; Jiang and Bai 1985).

The situation of the KBD geographical belt has been relatively stable for more than 50 years. The KBD and KBD non-affected areas are recognized without much doubt, although the health condition of the population in KBD areas could change from one year, or season, to the next. There is little evidence that KBD can spread from a KBD area to a KBD non-affected area, differing from infectious diseases. Some cities in China, for example Xi'an and Harbin, which are surrounded by KBD areas, have always belonged to KBD non-affected areas. The situation of the low selenium zone has also been stable for at least 30 years, based on information from studies of white muscle disease of livestock in the KBD zone of China. The stability of the KBD zone shows a close relationship to that of the low selenium zone. The low selenium zone could be considered as an essential condition for forming the KBD geographical belt according to the recent bio-epidemiological surveys.

3 The Comparison of Time Variation of Selenium-Deficient and KBD Areas

It has been found that in KBD areas both the condition of the individual patient and the incidence of KBD measured by X-ray detection among the childhood

Table 2. The relationship between patients' condition and Se status of the childhood population

Locality sample	Selenium level			
	Spring	Summer	Autumn	Winter
Shaanxi hair (ppb)	175 ±14 (17)[a]	257 ±20 (20)	211 ±19 (20)	156 ±18 (17)
Shaanxi urine (μg day^{-1})	4.7± 0.5 (16)	4.9± 0.4 (24)	5.3± 0.6 (18)	4.1± 0.3 (17)
Henan blood (ng ml^{-1})	19 ± 2.7 (5)	25 ± 1.7 (12)	26 ± 1.8 (5)	22 ± 1.3 (21)
Patients' condition	Worse	Improved	Improved	Worse
Heilongjiang hair (ppb)	76 ± 6 (15)	–	67 ±10 (15)	59 ± 7 (15)
Incidence of KBD	79.1% (139)	–	70.5% (149)	72.8% (151)
Patients' condition	Worse		Worse	Worse

[a] $\bar{X} \pm SE$ (N)

population can vary, depending on seasons or years. Time variation of KBD is characterized by irregular intervals ranging from about 3 months to more than 1 year. Furthermore, the patients' condition may worsen in some KBD areas, and at the same time it may improve in other KBD areas. It is interesting to note that both situations mentioned above can be reversed within a short period of time without any medical treatment. Some epidemiological information has indicated that the selenium status of both the individual patient and the childhood population fluctuates with various seasons or years, and that there is a close relationship between the changes of the patients' conditions and the selenium status. Generally, when selenium levels of both the individual patient and the population as a whole were high, then the patients' condition improved and the incidence of KBD in the population decreased. The contrary, however, was the case when selenium levels were low. In addition, when the selenium status was stable, no change in the patients' condition was found (Table 2; Bai et al. 1983; Bao et al. 1984; Jiang 1984).

4 The Correlation of Population Distributions in Selenium-Deficient and KBD-Affected Areas

The prevalence of KBD is mainly among the agricultural populations in the KBD zone, especially among the rural children ranging from 6–13 years of age. It is known that children with KBD have a low nutritional selenium status. The relationship between low selenium children and children with KBD have been reported as follows:

1. Low selenium environment, low selenium population and low selenium rate (Jiang et al. 1981).

The low selenium zone in China is actually composed of many areas with variant selenium levels. These areas have been generally divided into three different categories, depending on the selenium levels in samples of water, soils and cereals: low selenium, normal selenium and high selenium areas. Only in the low selenium environment were there a considerable number of patients with KBD. Consuming drinking water and locally produced food in the low selenium environment can affect the selenium status of the population living in this environment. The selenium content of drinking water is very low in this area. The selenium status of the population in low selenium areas is predominately affected by the selenium levels in food. In fact, sources of food and diet were so variant that three groupings became necessary: (1) most of the population was in a low selenium state; (2) part of the population was in a low selenium state and (3) little of the population was in a low selenium state. Jiang and Bai (1985) devised a new term, i.e. a low selenium rate, which is an epidemiological index for the effect of a low selenium environment on the low selenium status of the population living in this environment, and formulated a preliminary standard based on selenium levels in blood, hair and urine as well as in the daily dietary intake. More than ten functional parameters in children with different levels of urinary selenium excretion were measured. Many of the metabolic parameters measured (Creatinine, hydroxyproline, hydroxylysine, CPK, GOT, GPT, r-GT, LDH, HBDH, cAMP, cCMP, and three components of urinary glycosaminoglycans) were similar to those of children with normal selenium levels, but only in children with an excretion of more than 6 μg Se per 24 h urine (Jiang et al. 1981, 1984, 1987; Jiang and Bai 1985; Zhang and Jiang 1984). Daily urinary selenium excretion could, therefore, be considered as an index for judging low selenium or non-low selenium children.

The food of children in China with a low selenium status is mainly composed of low selenium cereals except rice (corn, wheat and highland barley), vegetables, very little meat and eggs and few, if any, marine products. Eating more meat and eggs or non-low selenium cereals (e.g. rice or marketed grain) can raise the selenium concentration of children in a low selenium environment to comparable levels of a non-low selenium environment. The elevated low selenium rate is the most prominent feature of selenium distribution in low selenium populations.

2. The relationship between urinary selenium levels and X-ray detection rate of KBD.

Although children with KBD always have a low selenium nutritional status, not all show KBD clinically. Jiang et al. investigated the relationship between the urinary selenium excretion rate per 24 h and X-ray detection rate of KBD, and found a dose-response curve. The regression equations were obtained from KBD and KBD non-affected areas in Shaanxi province: $Y_1 = 6.10 - 3.86\,X_1$ and $Y_2 = 5.93 - 3.18\,X_2$. Y and X are probability units translated from the X-ray detection rate of KBD and the logarithmic urinary selenium content respectively. The ED_{50} were identical, approximating to 2.0 μg per 24 h urine (Jiang et al. 1984).

The results mentioned above indicate that low selenium children are vulnerable to KBD, and that selenium deficiency may play an important role in the etiology or pathogenesis of KBD.

Table 3. The effect of corn of different sources on the incidence of KBD

Locality	Source	Clinical detection rate of KBD, % (N)	X-ray detection rate of KBD, % (N)
Heilongjiang	Locally grown	44.0 (125)	75.8 (149)
Heilongjiang	Marketable	0.8 (118)	4.6 (87)

Table 4. The effect of exchanging grain on the incidence of KBD

Locality	Date	Number	X-ray negative detection rate of KBD (%)	X-ray detection rate of epiphysis on right hand (%)
Heilongjiang	Before[a]			
	Oct. 1971	140	28.6	61.4
	After[b]			
	March 1974	148	49.3	1.4
	Dec. 1974	178	49.4	0
	June 1975	200	60.5	0
	July 1976	95	68.4	3.2
	Oct. 1977	145	78.6	0.7

[a] Before exchanging, the grain was locally grown in a KBD area.
[b] After exchanging, the grain was marketable grain from KBD non-affected areas.

Table 5. The effect of different kinds of cereals on the incidence of KBD

Locality	Date	Cereal	Number	Clinical detection rate of KBD (%)	X-ray detection rate of KBD (%)
Heilongjiang	Dec. 1979	Corn	55	58.2	76.4
Heilongjiang	Dec. 1979	Rice	51	0	0

3. The relation between the selenium level of food and the incidence of KBD.

The main pathogenic factor in KBD incidence might be the consumption of locally produced cereals except rice (corn and wheat) in KBD areas. Some investigations revealed that the selenium status of children in KBD areas is predominately affected by the selenium levels of cereals in the daily diet. Generally, low selenium children can obtain 3–4 μg Se day^{-1} from corn or wheat. There is a close relation between the selenium level of cereals and the incidence of KBD as illustrated in Tables 3 and 4 (Kaschin-Beck Disease Research Group of Harbin Medical College 1973, 1978). Jiang (1984) analyzed selenium contents in two different varieties of corn, one being a locally grown corn in a KBD area and the other a market grain. The selenium contents were 3.8 ± 2.8 and 14.2 ± 4.1 ppb respectively ($p < 0.01$). The hair selenium contents in the two different groups of children eating these corns were 67.3 ± 40.0 and 135.8 ± 40.0 ppb respectively ($p < 0.01$).

Table 6. The effect of wheat of different origins on the incidence of KBD

Locality	Year	Origin	Clinical detection rate of KBD, % (N)	X-ray detection rate of KBD, % (N)
Heilongjiang	1954	KBD area	0 (48)	–
	1964	Locally grown	1.4 (210)	–
	1974	Locally grown	9.6 (276)	–
	1981	Locally grown	26.1 (463)	49.0 (49)

Table 7. Se availability in the feeds with variant formulae

Year	Group	Rice (%)	Corn (%)	Wheat (%)	Soybean (%)	Vitamin compound (%)	Salt compound (%)	Se (ppb)	Incidence of liver necrosis in rats (%)
1984[a]	1	39	25	25	6.5	1.0	3.5	14.1	13.2
1984	2	39	25	25	6.5	1.0	3.5	26.5	5.0
1983[b]	3	–	50	39	6.5	1.0	3.5	31.4	45.0
1983	4	–	50	39	6.5	1.0	3.5	33.0	31.6
1983	5	–	50	39	6.5	1.0	3.5	46.1	10.0

[a] In 1984, the grain came from south-west China.
[b] In 1983, the grain came from north-west China.

The effect of various kinds of cereals on the incidence of KBD is shown in Table 5 (Wang 1982). Selenium contents in both corn and rice produced in the same KBD area were analyzed. The selenium contents of corn and rice were 4.3 and 20.4 ppb respectively ($p < 0.01$). Hair selenium was 68 and 178 ppb respectively ($p < 0.01$) in the two different groups of children (Chu 1982). Another example is illustrated in Table 6. The results showed that wheat selenium contents from KBD non-affected and KBD areas (locally grown) were 12.8 ± 4.9 and 5.1 ± 0.5 ppb respectively ($p < 0.01$). Hair selenium contents in the two different groups of children were 175.9 ± 12.1 and 43.2 ± 17.4 ppb respectively ($p < 0.01$).

The results mentioned above reveal that selenium levels of cereals of different origins varied, which could affect the incidence of KBD.

5 The Selenium Availability in Various Kinds of Cereals

The bioavailability of selenium from different kinds of food may be a complex function of the criterion studied, as well as the level of dietary selenium needed and other dietary factors (Mutanen et al. 1986). Bioavailability of selenium differed in food with variant formulae, based on the experiments in the incidence of acute massive liver necrosis in rats (Table 7; Kaschin-Beck Disease Research Laboratory of Xi'an Medical University 1987). As shown in Table 7, the selenium bioavailability in groups 1 and 2 was higher than those in groups 3, 4 and 5, which suggested that the selenium bioavailability may be higher from rice than from wheat or corn. That both the selenium content and the selenium bioavailability from rice were higher than those of corn and wheat in KBD areas probably explain why KBD has not spread over the low selenium regions in south-west China: the staple food of the residents in the region is mainly rice.

References

Bai C (1983) Studies on seasonal changes of selenium contents in environments and *milieu interne* of Kaschin-Beck disease. Acta Acad Med Xi'an (Suppl):1–4

Bao BJ, Huang WH, Hou SF, Zhu ZY (1984) A study on season dynamic of the selenium nutrition state and vitamin E content of residents in Kaschin-Beck affected and non-affected regions. Chin J Endemiol 3:137

Chu QL (1982) Studies on differences between the rice and maize grown in Kaschin-Beck endemic area. Acta Harbin Med College 16:56–64

Jiang YF, Yang JB (1981) Studies on hydroxyproline contents in urine of children with different patients' condition from Kaschin-Beck affected area. Acta Harbin Med College 15:70–75

Jiang YF (1984) A study on differences in fungi and chemical elements between local maize from Kaschin-Beck disease endemic areas and marketable maize from non-endemic areas. Acta Harbin Med College 18:137

Jiang YF, Bai C (1985) Muscular metabolism in the children with low selenium state. Acta Acad Med Xi'an (Suppl):17–23

Jiang YF, Bai C, Zhang SY, Li YL, Yu ZD, Cao JL (1984) Studies on urinary creatinine contents in children with different selenium levels in Kaschin-Beck disease affected and non-affected areas. Acta Acad Med Xi'an 5:43–48

Jiang YF, Bai C, Yu ZD, Cao M, Jiang Z, Zhang ZQ (1987) Studies on hydroxylysine contents in daily urine of children with different selenium levels in Kaschin-Beck disease endemic and non-endemic areas. J Xi'an Med Univ 8:148–151

Kaschin-Beck Disease Research Group of Harbin Medical College (1973) Investigations of Kaschin-Beck disease in Heilongjiang province. Monography of Harbin Medical College on scientific researches, pp 41–44

Kaschin-Beck Disease Research Group of Xi'an Medical College (1975) Advances in studying Kaschin-Beck disease in the past 5 years. Acta Acad Med Xi'an (4):33–41

Kaschin-Beck Disease Research Laboratory of Harbin Medical College (1978) 5-year field survey of the effect of exchanging grains on Kaschin-Beck disease control in Heilongjiang. Monogr Harbin Med College Sci Res, pp 11–20

Kaschin-Beck Disease Research Laboratory of Xi'an Medical University (1987) The effect of low selenium diets from three different regions on biochemical metabolism and pathomorphological changes in rats. J Xi'an Med Univ (Suppl):35–49

Li CZ (1979) X-ray studies on the effect of selenium and vitamin E in treating the early stage Kaschin-Beck disease patients. Zhong hua Yixue Zazhi 59:169–171

Mo DX (1984) Advances of studies on the relationship between Kaschin-Beck disease and chemical elements in etiology. J Commun Endemic Dis (2):8–13

Mutanen M, Koivistoinen P, Morris VC, Levander OA (1986) Nutritional availability to rats of selenium in four seafoods: crab, oyster, shrimp and Baltic herring. Br J Nutr 55:219–225

Tan JA, Zheng DX, Hou SF, Zhu WY, Li RB, Zhu ZY, Wang WY (1987) Selenium ecological chemico-geography and endemic Keshan and Kaschin-Beck disease in China. In: Combs GF Jr, Spallholz JE, Levander OA, Oldfield JE (eds) Proc 3rd Int Symp Selenium Biol Med, part B. AVI Press, New York, pp 859–876

Wang JD (1982) Epidemiological investigation of cause of high incidence of Kaschin-Beck disease in Hei River region. Acta Harbin Med Univ 16:56–64

Zhang SY, Jiang YF (1984) Studies on contents of three components of glycosaminoglycans extracted from urine of children with different selenium levels in Kaschin-Beck disease and non-endemic areas. J Commun Endemic Dis (2):42–44

The Selenium and Phospholipid (Phosphatidylcholine) Deficiency and Low Temperature of Environment May be the Pathogenetic Basis of Kaschin-Beck Disease

FANG SHENG LI[1,2], JIN-YANG GUAN[1], YOU-JIN DUAN[1], LI-MING ZOU[1], XU-JUAN YAN[1], PING MA, QIAN SUN, LI LI[1], and SHU-YING LI[1]

1 Introduction

Kaschin-Beck disease (KBD), an endemic, deforming osteoarthropathy, which occurs in remote and secluded places among the mountainous areas of China and in the eastern part of the Soviet Union, seriously impairs the development of children. The main pathological changes of KBD is the necrosis of chondrocytes and early calcification. Furthermore, with the occurrence of thickened joints, deformed fingers, and atrophied muscles, the development of children is stopped.

The endemic areas are distributed from north-east to south-west China covering over 25 latitudinal degrees, and reaching ca. 500 m above sea level the in north-east and increasing gradually to the south-west to nearly 2500 m in the Sichuan plateaus. There the average temperature is $0-16\,^{\circ}C$; even in summer, it reaches no more than $20-24\,^{\circ}C$ (Li et al. 1984a). It has been clarified that the endemic area of KBD was deficient in Se as in the case of Keshan disease. Because of its climatic conditions, the variety of crops and vegetables is restricted, therefore, most inhabitants have an unbalanced diet year-round. Interestingly, some inhabitants in endemic areas would not suffer from this disease, if they would always eat fish or meat, such as the fishermen or hunters living in the same endemic area. Through our investigation of the Se content, the membrane lipid composition, erythrocyte form and the fragility of KBD patients and animal model test (Li et al. 1984b; Duan et al. 1984; Li 1985; Li et al. 1985) have been elucidated, especially, the decrease in Se and membrane phospholipid (PL), i.e., phosphatidylcholine (PC), are noteworthly. The latter is not only an important component in the membrane but also an essential factor for maintaining the activity of alkaline phosphatase (ALP), a membrane-binding enzyme concerned in the calcification process of bone tissue (Tiruppathi et al. 1987). As mentioned above, we have investigated KBD in the Inner Mongolia endemic area and have performed a dynamic study of an animal model fed a diet collected from a KBD endemic area.

[1] Institute of Basic Medicine, Liaoning Academy of Medical Sciences, 6-18-6 Hoping Dajie, Shenyang 110005, The Peoples' Republic of China
[2] To whom correspondence should be addressed

2 Material and Methods

Blood specimens were collected from children 10–15 years of aged with KBD living in Huluenbeier Meng of Inner Mongolia and from a control group. The animals used in this study were 110–170 g Wistar rats, divided randomly into two groups: the test group, fed a diet prepared with cereals of the endemic area, and the other as a control, fed diets from non-endemic areas. The composition of the diet (%) was: corn 60; wheat 16; bean-cake flour 15; salt 2; lard 5; cod-liver oil 2. Food and water (deionized) were supplied ad libitum for 5 months, and analyzed each month for Se content (Wilkie and Young 1970), glutathione peroxidase (GSH-Px) activity (Hafeman et al. 1974) and plasma enzymogram. Membrane total lipids of erythrocytes were extracted (Folch et al. 1957), and the phospholipid (Stewart 1980), cholesterol (Zak 1957) and fluorescent chromolipid (Jain and Hochstein 1980) contents were determined. Fractionation of phospholipids was performed by thin layer chromatography (Li 1981) and detected with a Chromatoscanner CS-910 (Shimadzu).

3 Results and Discussion

3.1 The Selenium Level and Membrane Lipid Composition of Kaschin-Beck Diseased Children

As detected in KBD endemic areas of Shaanxi province, the Se level in hair and blood of children living in Inner Mongolia was also lower than that in non-endemic areas (Table 1), and the GSH-Px activity was decreased. However, there are differences in the membrane lipid composition between KBD and controls (Table 2). It has been shown that the phospholipid (PL) content, particularly the phosphatidylcholine (PC), decreased significantly (Table 3). The molar ratio of Ch (cholesterol)/PL and SM (sphingomyelin)/PC increased (Tables 2 and 3), which showed that age changes occurred in the biomembrane (Barenholz and Thompson 1980), e.g., increased acanthocytes (Duan et al. 1984), reduced membrane fluidity (Yang et al. 1984), and increased fragility (Li et al. 1984b). Moreover, an increase in the activity of some enzymes in the plasma enzymogram

Table 1. The selenium level in hair and blood and glutathione peroxidase (GSH-Px) activity in erythrocytes of Kaschin-Beck diseased children (Mean ± SD)

Groups	n	Selenium content (ppm)		GSH-Px activity ($U\,ml^{-1}$)
		Hair	Erythrocytes	
Kaschin-Beck disease	20	0.132 ± 0.042	0.021 ± 0.008	159.5 ± 20.6
Control in endemic area	14	0.144 ± 0.034	0.053 ± 0.031**	187.6 ± 22.4*
Control in non-endemic area	17	0.396 ± 0.055***	0.104 ± 0.018***	231.5 ± 22.8***

* $p<0.05$; ** $p<0.01$; *** $p<0.001$

Table 2. The lipid component in the erythrocyte membrane of Kaschin-Beck disease and control children (Mean ± SD)

Groups	n	Phospholipid (PL) (μg/10^7 cell)	Cholesterol (Ch) (μg/10^7 cell)	Ch/PL Molar ratio
Kaschin-Beck disease	10	1.44 ± 0.14[a]	1.35 ± 0.06	2.0
Control in endemic area	6	2.48 ± 0.13	1.54 ± 0.04	1.1
Control in non-endemic area	7	2.72 ± 0.16	1.49 ± 0.03	1.0

[a] Compared with control in endemic area, $p<0.01$.

Table 3. Composition of the main phospholipid in the erythrocyte membrane of Kaschin-Beck disease and control children (Mean ± SD)

Groups	n	Sphingomyelin (SM) (μg/10^7 cell)	Phosphatidyl-choline (PC) (μg/10^7 cell)	Phosphatidyl-ethanolamine (PE) (μg/10^7 cell)	SM/PC Molar ratio
Kaschin-Beck disease	15	0.44 ± 0.07	0.28 ± 0.04[a]	0.70 ± 0.05	1.41
Control in endemic area	15	0.40 ± 0.04	0.47 ± 0.07	0.50 ± 0.09	0.76
Control in non-endemic area	15	0.67 ± 0.07	0.53 ± 0.07	0.30 ± 0.07	1.12
Normal adults	7	0.79 ± 0.14	0.57 ± 0.09	1.17 ± 0.17	1.23

[a] Compared with control in endemic area, $p<0.05$.

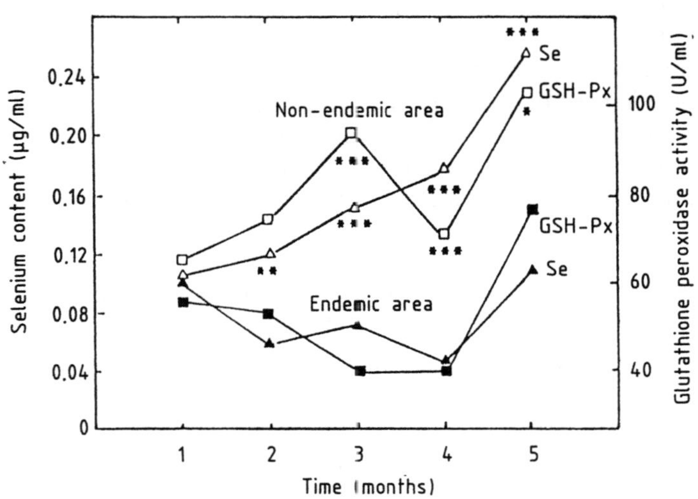

Fig. 1. Time course of selenium content and glutathione peroxidase activity in erythrocytes of rats fed a diet made from cereals originating from endemic (▲ ■) and non-endemic (△ □) areas. $*p < 0.05$; $**p < 0.01$; $***p < 0.001$

occurred (Shi et al. 1984; Pan 1961; Sun et al. 1985). Similar changes have been discovered also in cartilage tissue of KBD patients (Li et al. 1984c).

3.2 The Dynamic Observations on an Animal Model Fed a Diet from a Kaschin-Beck Disease Endemic Area

We had carried out a KBD animal investigation with a semipurified diet low in Se and PL for young dogs and guinea pigs, and a series of changes appeared similar to that of KBD patients andwhich could be controlled by supplementation of Se and soybean phospholipids (Li et al. 1984d). In the present study rats were fed a diet prepared from cereals produced in endemic areas. The changes in Se level and membrane lipid composition of erythrocytes in the rats were detected

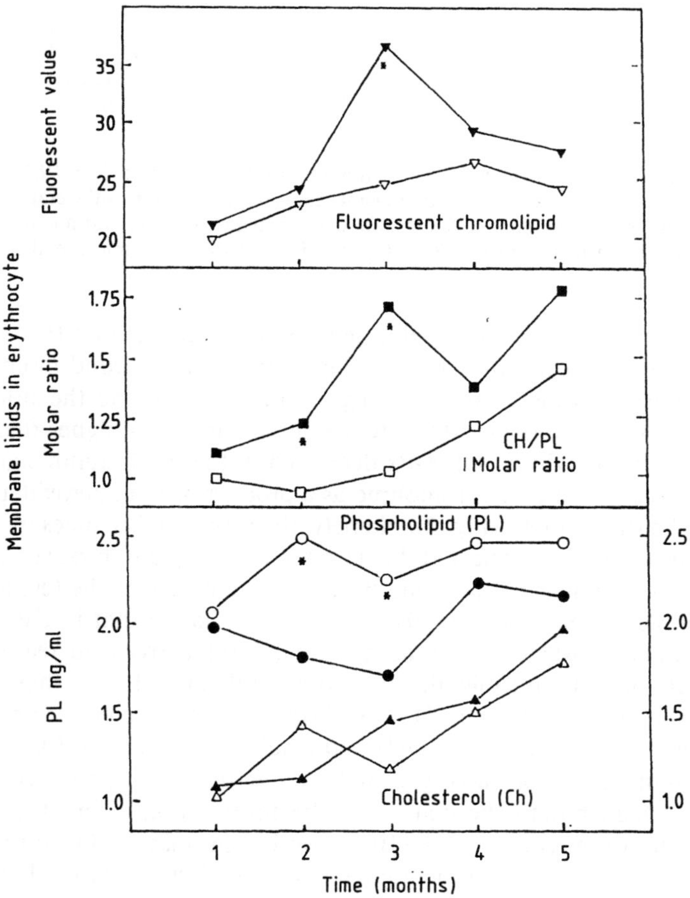

Fig. 2. Changes of phospholipid (PL ● ○), cholesterol (Ch ▲ △) content with time, molar ratio of Ch/PL (■ □) and fluorescent chromolipid (▼ ▽) in erythrocytes of rats fed a diet prepared from Kaschin-Beck disease endemic (*filled symbols*) and non-endemic (*open symbols*) cereals. $*p < 0.05$

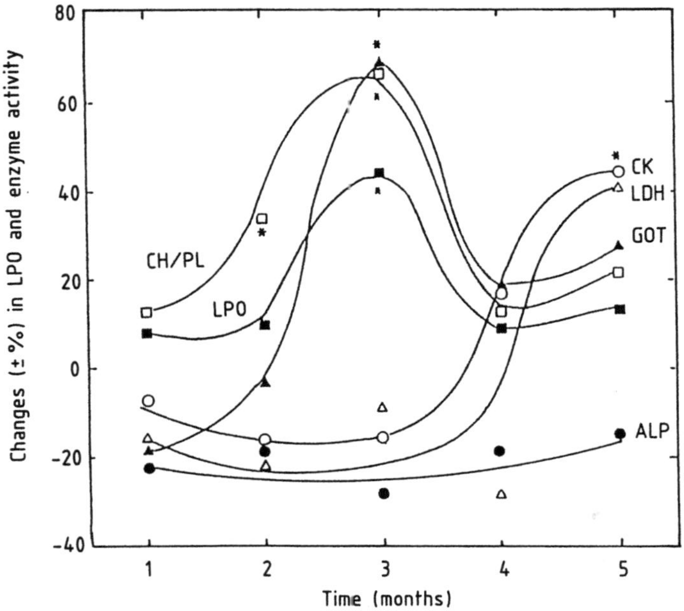

Fig. 3. Dynamic changes of the membrane lipid composition (*Ch/PL*), lipid peroxide (*LPO*) in erythrocytes and activity of plasma enzymes glutamate oxalacetate transaminase (*GOT*), creatine phosphokinase (*CK*), lactate dehydrogenase (*LDH*) and alkaline phosphatase (*ALP*) of rats fed a diet prepared from Kaschin-Beck disease endemic cereals. *$p < 0.05$

in the first month after being fed a KBD endemic diet; the contents of Se and PL were all lower than that in the control in the second month (Figs. 1 and 2). At the same time, as an age change in the biomembrane, the molar ratio of Ch/PL and SM/PC in the erythrocyte membrane lipid was higher than that of the control. The activity of GSH-Px decreased in the third month significantly (Fig. 1), and the fluorescent chromolipid as a product of lipid peroxidation also increased in the third month (Fig. 2). Namely, the process of changes in the lipid composition of the membrane and the decrease in Se content in rats of the test group were almost parallel. At the same time, in the animals of the test group, a change in the plasma enzymogram also appeared similar to that in KBD patients, i.e., the glutamate oxalacetate transaminase (GOT) activity increased gradually after beginning the endemic diet, and rose significantly higher than that in the control in the third month (Fig. 3). The plasma activity of creatine phosphokinase (CK), lactate dehydrogenase (LDH) and alkaline phosphatase (ALP) decreased at the beginning and increased gradually during the feeding process. The LDH and CK were higher than the control in the fourth month, but the ALP activity was similar to the control in the fifth month. This may be due to the fact that CK and ALP exist as integral membrane enzymes (Tiruppathi et al. 1987; Müller et al. 1986; Ogata et al. 1987). Therefore, it is possible that in the early stage of pathogeny some changes in the structure and function of the membrane occurred, thus increasing the permeability and allowing the GOT (as an enzyme in the cytoplasma) to escape through the membrane and to penetrate into the blood.

However, membrane-bound enzymes may be released after deterioration of the membrane, than they fall away and finally are released from the cells into the blood. The appearance of the above mentioned changes, in turn, not only supports the essence of this disease, which may be a pathogenetic membrane, but also establishes the important action of phosphatidylcholine, which has a close relationship with tissue ALP and the low temperature of the environment of the pathogenetic process of KBD. As shown in studies in Tiruppathi et al. (1987), the activity of ALP in several mammalian tissues is severely affected by phosphatidylcholine or choline content of the membrane and by temperature. Therefore, the pathogeny of KBD may originate from selenium deficiency, phosphatidylcholine and the low temperature of the environment.

References

Barenholz Y, Thompson TE (1980) Sphingomyelin in bilayers and biological membranes. Biochim Biophys Acta 604:129–158

Duan YJ, Li L, Li FS (1984) Change in erythrocyte shape of Kaschin-Beck disease children. A collection of investigations on Kaschin-Beck disease in Yongshou. PHP, Beijing, pp 353–357

Folch J, Lees M, Stanley H (1957) A simple method for the isolation and purification of total lipids from animal tissues. J Biol Chem 226:497–509

Hafeman DJ, Sunde RA, Hoekstra WG (1974) Effect of dietary selenium on erythrocytes and liver glutathione peroxidase in the rat. J Nutr 104:580–587

Jain SK, Hochstein P (1980) Membrane alteration in phenylhydrazine induced reticulocytes. Arch Biochem Biophys 201:683–687

Li FS (1981) The TLC and density scanning of erythrocyte membrane phospholipids. Annu Rep Liaoning Basic Med Inst Shenyang, pp 148–150

Li FS (1985) Selenium nutrition and Kaschin-Beck disease and Keshan disease in China. Proc 2nd Symp Trace Nutr Research, Tokyo, pp 9–19

Li FS, Bai ZC, Liu GL, Cheng BR (1984a) The pathogeny of Kaschin-Beck disease – a comprehensive ecologic effect of endemic area. J Ecol 6:10–14

Li FS, Bai QF, Guan LY, Duan YJ, Li L (1984b) Changes of membrane composition in erythrocytes of Kaschin-Beck disease children and their pathogenetic significance. A collection of investigations on Kaschin-Beck disease in Yongshou. PHP, Beijing, pp 338–344

Li FS, Zou LM, Duan YJ (1984c) The molecular biological feature of changes in cartilage of Kaschin-Beck disease and aetiological discussion. A collection of investigations on Kaschin-Beck disease in Yongshou. PHP, Beijing, pp 372–375

Li FS, Wei FQ, Guan JY (1984d) Metabolic changes in lipid in animals fed with low selenium. Acta Pharm Sin 5:269–273

Li FS, Guan JY, Zou LM, Duan YJ, Li L (1985) Membrane damage in Kaschin-Beck disease and its aetiological significance. In: IPCS/WPRO/PROC Meeting on Kaschin-Beck Disease, Beijing

Müller M, Cheneval D, Carafoli E (1986) The mitochondrial creatine phosphokinase is associated with inner membrane cardiolipin. Adv Exp Med Biol 194:151–155

Ogata S, Hayashi Y, Yasutake K, Ikehara Y (1987) Chemical identification of lipid components in the membranous fraction from of rat liver alkaline phosphatase. J Biochem 102:1609–1615

Pan FT (1961) Biochemical analysis on blood and urine of Kaschin-Beck disease patients. The collection of materials of researches, vol 18. Medical College, Shenyang, pp 15–29

Shi GZ, Li ZQ, Wang QL (1984) Plasma enzyme activity and some components of Kaschin-Beck disease children. A collection of investigations on Kaschin-Beck disease in Yongshou. PHP, Beijing, pp 287–315

Stewart JCM (1980) Colorimetric determination of phospholipids with ammonium ferrothiocyanate. Anal Biochem 104:10–14
Sun K, Li LQ, Wang ZW (1985) The serum enzymogram of Kaschin-Beck disease in Heilongjiang Province. Chin J Endemiol (Spec ed) 4:156–159
Tiruppathi C, Alpers DH, Seetharam B (1987) The role of choline on the activity-temperature relationship of brush-border alkaline phosphatase. Biochim Biophys Acta 898:283–292
Wilkie JB, Young M (1970) Improvement in the determination of selenium in biological materials. Agric Food Chem 18:94–948
Yan FY, Huang F, Lin ZH (1984) Changes in erythrocyte membrane of Kaschin-Beck disease children. A collection of investigations on Kaschin-Beck disease in Yongshou. PHP, Beijing, pp 366–371
Zak B (1957) A simple rapid microtechnic for serum total cholesterol. J Lab Clin Med 27:583–588

Low Selenium State and Increased Erucic Acid in Children from Keshan Endemic Areas – A Pilot Study

M. D. LARYEA[1,3], Y. F. JIANG[2], G. L. XU[2], D. FROSCH[1], and I. LOMBECK[1]

1 Introduction

For over 50 years a cardiomyopathy is known to occur endemically in distinct regions of China. In the 1970s the low selenium content of soil, food and tissues was discovered and since then the prevention of Keshan disease by selenium supplementation is known. Still the pathogenesis of Keshan disease is not fully explained by selenium deficiency alone, as in other parts of the world healthy or diseased people with low selenium intake seldom suffer from cardiomyopathy. It remains questionable whether an additional environmental, infectious or nutritional factor may contribute to the pathogenesis of the disease. Mostly children of peasants who only live on locally grown food are affected. Knowing some of the nutritional habits we tried to find an additional factor which accentuates selenium deficiency and estimated in a pilot study the fatty acids in plasma of healthy children from Keshan areas.

2 Material and Methods

In nine school children (6–10 years) from Huang-Ling county, a selenium-deficient area 200 km north of Xi'an, 5 ml of EDTA blood were taken in the fasting state. The children were not suffering from infectious, metabolic or cardiac diseases. A short questionnaire about their nutrition revealed that their food consisted mainly of wheat and potato products. Sometimes bean curd, seldom meat and never fish or milk were ingested. The main source of fat consisted of rape seed oil. The estimation of the selenium content in plasma, whole blood and hair was performed in duplicates by atomic absorption spectrophotometry plus hydride generation after stepwise digestion of the sample with nitric and perchloric acid.

One ml of EDTA plasma was immediately stored in brown vials with 5 ml chloroform/methanol 1 : 1. Plasma lipids were extracted, separated into fractions by thin layer chromatography and methylated using methanolic hydrochloride.

[1] Universitäts-Kinderklinik Düsseldorf, Moorenstr. 5, 4000 Düsseldorf, FRG
[2] Xi'an Medical University, The People's Republic of China
[3] To whom correspondence should be addressed

The fatty acid methylesters formed were analyzed by capillary column gas chromatography. Peaks were identified by comparison of retention times with standard substances and if necessary by mass spectrometry. The values were compared to those of 11 healthy, age-matched German children by U-test.

3 Results

1. Selenium in plasma and whole blood of Chinese children was very low, amounting to about 11%, and 14% respectively of those in healthy children (Table 1). Hair selenium was also markedly reduced ranging from 54 to 97 ng g^{-1}.

2. The analysis of plasma fatty acids of the phospholipid and cholesterylester revealed that linoleic acid, the essential fatty acid of the n6-series, was lower in the Chinese children than in German children and the desaturation product, alpha-linolenic acid, 18:3n6, was higher. Arachidonic acid, 20:4n6, and the sum of the polyunsaturated fatty acids of both fractions did not differ between the children (Table 2).

Marked differences were observed with respect to the long-chain n9-series. In Düsseldorf children the oleic acid was higher probably because of the intake of margarines containing seed oils. In all samples of the Chinese children erucic acid, 20:1n9, was present in the phospholipid and cholesterylester fractions. This fatty acid was not detectable in any of the German sampels (detection limit 0.01%). The metabolite of erucic acid, eicosenoic acid, 20:1n9, was also elevated in the Chinese samples, whereas it was not detectable or found only in traces in the German samples, again in both fractions. The long-chain polyunsaturated fatty acids of the n3-series showed also a significant increase in the Chinese children which can be explained by the food habits, i.e., the intake of vegetables plus erucic acid-containing rape seed oil.

4 Discussion

The pilot study revealed besides very low selenium contents of plasma, whole blood and hair, erucic and eisocenoic acid in plasma phospholipids and cholesterylesters from healthy school children in a Keshan endemic area. According to a short questionnaire, all children ingested rape seed oil. Rape seed oil containing erucic acid is known to cause cardiac lipodosis, focal necrosis and fibrosis in the heart of most animal species including monkeys. Greater amounts of erucic acid given in the diet overload limited capacities for erucic acid shortening and oxidation, a temporary lipidosis of the heart develops. Usually after some weeks an adaptive decrease of the lipid accumulation in the heart occurs (Beare-Rogers et al. 1971).

Although no direct proof has been presented that erucic acid is cardiotoxic also for humans, in many countries the erucic content of oils destined for human

Table 1. Selenium content

	Chinese children		German children	
	x	SD	x	SD
Plasma Se (ng ml^{-1})	8.0	2.3	72	11
Whole blood Se (ng ml^{-1})	11.4	2.7	84	7
Hair Se (ng g^{-1})	70.8	14.4	429	117

Table 2. Fatty acid pattern of plasma phospholipids and cholesterylesters

	German children		Chinese children	
	Mean	Range	Mean	Range
Phospholipids				
16:0	28.03	(24.46 – 31.38)	29.26	(27.96 – 30.45)
18:0	15.32	(13.77 – 18.27)	14.81	(11.88 – 18.04)
18:2n6	21.35	(17.47 – 24.32)	19.49	(17.72 – 21.32) $p<0.05$
18:3n6	0.06[a]	(n.d. – 0.13)	0.19[a]	(0.07 – 0.33)
20:4n6	8.80	(6.72 – 11.24)	7.92	(6.67 – 9.48)
18:1n9	14.13	(11.63 – 17.60)	11.93	(9.71 – 14.57) $p<0.025$
20:1n9	0.17[a]	(n.d. – 0.32)	0.72[a]	(0.38 – 1.31) $p<0.001$
22:1n9	n.d.	–	0.36	(0.09 – 0.85) $p<0.001$
18:3n3	0.13[a]	(n.d. – 0.20)	0.50[a]	(0.28 – 0.98) $p<0.001$
20:5n3	0.09[a]	(n.d. – 0.33)	0.86[a]	(0.58 – 1.33) $p<0.001$
22:6n3	2.71	(1.31 – 4.52)	2.39	(1.28 – 3.43)
Sum PUFA	37.76	(32.06 – 41.48)	38.10	(36.97 – 41.27)
Cholesterylesters				
16:0	12.11	(9.86 – 17.17)	12.27	(10.72 – 13.58)
18:0	1.65	(1.17 – 2.84)	2.11	(1.46 – 2.56)
18:2n6	50.60	(46.12 – 56.76)	44.46	(41.96 – 50.26) $p<0.001$
18:3n6	0.62[a]	(0.43 – 0.82)	1.51[a]	(1.05 – 1.77) $p<0.001$
20:4n6	5.77	(3.98 – 7.58)	5.86	(4.15 – 6.86)
18:1n9	20.48	(17.73 – 23.89)	19.53	(16.99 – 25.48) $p<0.05$
20:1n9	0.05[a]	(n.d. – 0.25)	0.23[a]	(0.09 – 0.57) $p<0.001$
22:1n9	n.d.[b]	–	0.26	(0.14 – 0.51) $p<0.001$
18:3n3	0.44	(0.23 – 0.61)	1.21	(0.76 – 1.77) $p<0.001$
20:5n3	0.04[a]	(n.d. – 0.24)	0.99[a]	(0.65 – 1.33) $p<0.001$
22:6n3	0.50	(0.19 – 0.74)	0.89	(0.38 – 1.20)
Sum PUFA	59.14	(54.11 – 64.91)	56.40	(52.54 – 62.76)

[a] Median values.
[b] n.d. = Not detectable.

consumption has been limited to less than 5% since 1975. Therefore, only a few data about erucic acid in humans were published in the early 1970s (Cocchi et al. 1974; Gatti and Michalek 1975; Maronesi et al. 1972; Trémelières et al. 1972). Levels of 0.3 – 3.8% (Gatti and Michalek 1975) erucic acid in blood lipids were reported. These results were obtained by analyzing total plasma lipids. In addition, Cocchi et al. 1974 presented data on the erucic acid of plasma phospholipids and total erythrocytes. These cited data are not comparable to ours. The technique has changed in the last 15 years. The published data were obtained using packed column chromatography. We used capillary gas chromatography which resulted in a high resolution of fatty acids also including those present in small amounts.

The Chinese children from an area where Keshan disease occurs have both a low selenium status and increased levels of erucic and eicosenoic acid in plasma phospholipids and cholesterylesters. It remains to be elucidated whether the amount of erucic acid ingested is high enough to directly cause cardiac lipidosis or whether it leads in combination with reduced selenium glutathione peroxidase activity to irreversible damage of the myocardium. It was observed that in Huang-Ling county the incidence rate of cardiomyopathy increases after festivals, e.g., the spring festival, when the ingestion of rape seed oil is high. The endemic character of the disease could be explained by the variation of the two factors, the increase of decrease or the selenium content of grains during different seasons and the variation of the intake of erucic acid.

References

Beare-Rogers JL, Nera EA, Heggtveit HA (1971) Cardiac lipid changes in rats fed oils containing long-chain fatty acids. Can Inst Food Techn J 4:120–124

Cocchi M, Biagi PL, Maranesi M, Cipolla IM, Tavernese G, Lorusso S, Turchetto E (1974) Sulla presenza di acido erucico nei lipidi e nei fosfolipidi plasmatici umani. Boll Soc It Biol Sper 50:290–294

Gatti GL, Michalek H (1975) Ingestions on rape seed oil toxicology. Drug Res 25:1639–1642

Maronesi, M, Barzanti V, Cocchi M, Tavernese G, Cipolla IM, Turchetto E (1972) Sulla presenza di acido erucico nel sangue umano. Boll Soc It Biol Sper 48:1205–1207

Trémolières A, Collomb MH, Trémolières J (1972) Effets métaboliques de l'huile de colza chez l'homme. Cah Nut Diét VII:67–71

Selenium in Children with Acute Leukaemia During Chemotherapy

D. FROSCH[1,3], S. KNAB[1], H. JÜRGENS[1], H. MENZEL[2], and I. LOMBECK[1]

1 Introduction

In animals selenium was reported to show an inhibitory effect on carcinogenesis and to reduce the growth rate of tumors (Shamberger 1979). Epidemiological studies led to the hypothesis of an inverse correlation between average daily selenium intake and human cancer mortality in some countries (Clark 1985; Burk 1986).

The question whether selenium plays a role in human cancer patients before the development of the disease, during the clinical course or during chemotherapy is still open. Therefore the selenium state was investigated in children with acute leukaemia, the prevalent malignancy during childhood.

Table 1. Patients, diagnosis, and therapeutic scheme[a]

Name	Age (year)	Sex	Diagnosis	Treatment according to
K.C.	3.0	F	ALL	CoALL 82 Ind. I/1
H.M.	3.3	M	ALL	CoALL 82 Ind. I/1
D.S.	3.3	F	ALL, ZNS-Rez.	ALL BFM 83 f. ZNS-Rez.
S.S.	3.9	F	AML	ALL BFM 83 Ind.
R.N.	7.5	F	Biph. L.	CoALL 82 Ind. II/1 day 1 – 7
				AML BFM 83 Ind. from day 8
J.C	7.9	M	ALL	CoALL 82 Ind. II/1
F.A.	10.9	F	ALL	CoALL 82 Ind. II/1
G.T.	11.1	M	Biph. L.	CoALL 82 Ind. II/1 day 1 – 33
				AML BFM 83 Ind. from day 34
S.H.	15.7	M	ALL	CoALL 82 Ind. II/1

[a] Abbrevations: ALL = acute lymphatic leukaemia; AML = acute myeloic leukaemia; biph. L. = acute biphaenotypical leukaemia; ALL, ZNS-Rez. = relapse of ALL in central nervous system; CoALL 82 Ind. I/1 = cooperative ALL therapy study, induction therapy for low risk group; CoALL 82 Ind. II/1 = cooperative ALL therapy study, induction therapy for high risk group; AML BFM 83 f. ZNS-Rez. = multicenter therapy study for treatment of ALL in children with relapse in central nervous system; AML BFM 83 Ind. = multicenter therapy study for treatment of AML in children, induction therapy.

[1] Universitätsklinik Düsseldorf, Kinderklinik I, Moorenstr. 5, 4000 Düsseldorf, FRG
[2] Institut für Toxikologie, Universität Düsseldorf, 4000 Düsseldorf, FRG
[3] To whom correspondence should be addressed

2 Methods

Blood samples were taken from nine children with acute leukaemia (Table 1) on admission and during chemotherapy for 47 to 125 days. Selenium contents of plasma and whole blood were measured by hydride generation atomic absorption spectrophotometry after stepwise wet digestion with nitric and perchloric acid.

Glutathione peroxidase activity in plasma and red blood cells was assayed by a coupled test system using t-butyl-hydroperoxide as acceptor substrate.

3 Results

The selenium state of leukaemic children on admission did not differ from age-matched, healthy control children (Fig. 1).

The selenium content of whole blood rose during 2 months of therapy to 122% (range 103–149%) of the values on admission (Fig. 2).

The values of the erythrocyte glutathione peroxidase activity rose to 133% (range 118–152%) of the initial values during the same time (Fig. 2).

In eight patients plasma selenium values increased during the first days of induction therapy. This was followed by a plateau or a slight decrease during the second to fourth week and a somewhat higher peak in the fifth to seventh week. The total increase of plasma selenium ranged from 166 to 300% of the initial values.

After the seventh week plasma selenium contents slightly decreased but finally these eight children showed no lower values than on admission (92–200%; Fig. 3).

Only the patient D. S. showed a different curve: there was a decrease to about 60% during the first weeks followed by an increase to 97% in the fifth week with a slow decrease afterwards to 68% of the values on admission.

Glutathione peroxidase activity in plasma revealed a similar pattern to plasma selenium, here we also found a steep increase during the first weeks. Peak values of 130 to 280% of the initial activities were also found in the fifth to seventh week. A slow decrease afterwards resulted in plasma peroxidase activities of 60 to 145% of the values on admission.

The patient D. S. showed only a single small peak in the first week (120% of the initial plasma glutathione peroxidase activity), then all values were lower than on admission.

The increase of selenium contents and glutathione peroxidase activities was similar in all patients with different types of leukaemia and even under different therapeutic regimens with the exception of D. S. This girl was treated because of the first relapse of acute lymphatic leukaemia with a solitary manifestation in the central nervous system.

There was no correlation between the administration of different therapeutic drugs and the changes of selenium contents and glutathione peroxidase activities.

Leucopenia, anaemia and thrombocytopenia occurring naturally or induced by chemotherapy were observed in all children, but on different days of the obser-

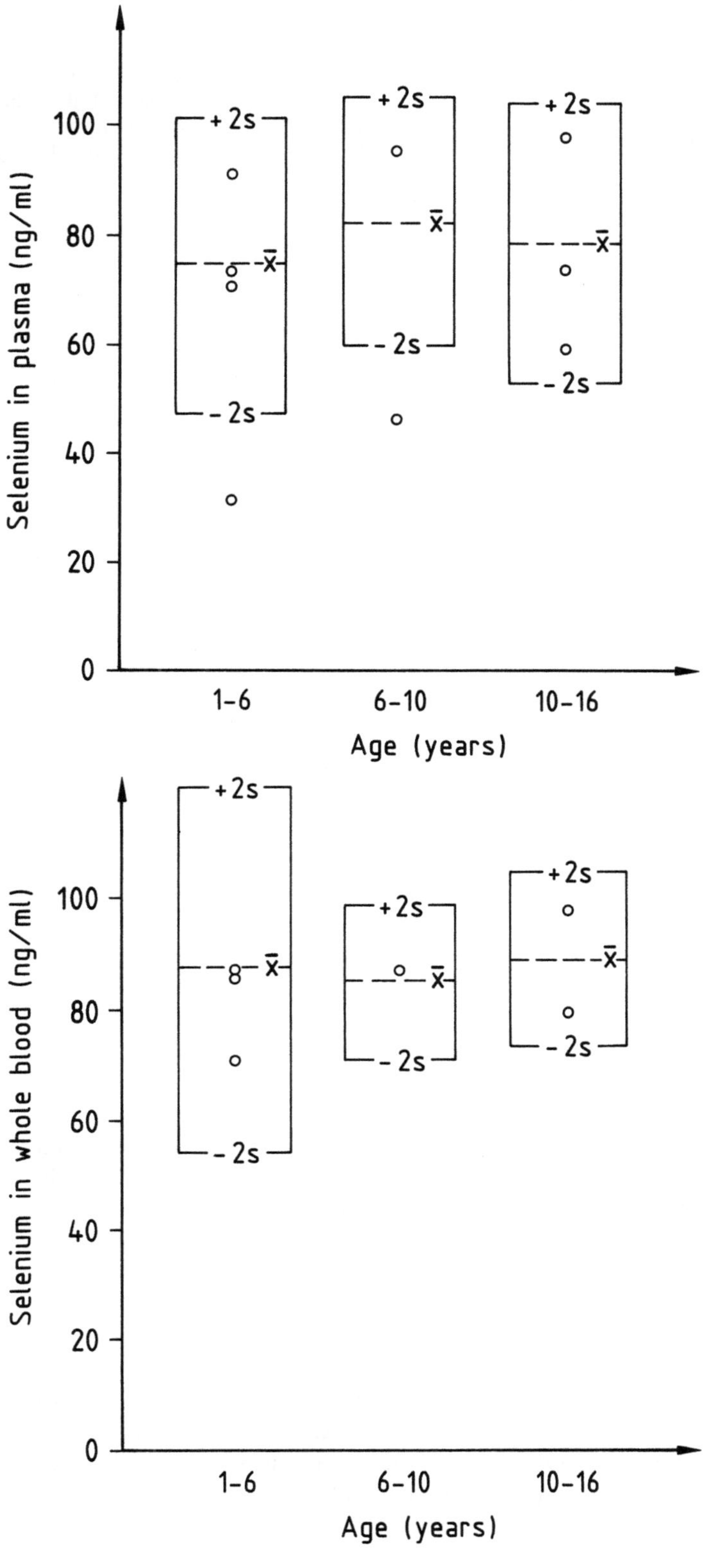

Fig. 1. Selenium contents of plasma and whole blood in leukaemic children on admission (○) and of age-matched, healthy children (┌─┐)

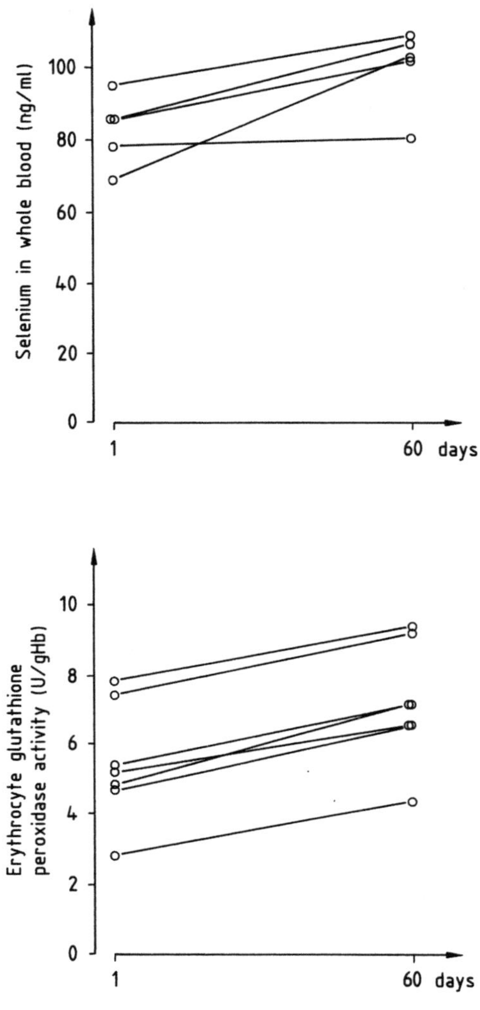

Fig. 2. Selenium contents of whole blood and glutathione peroxidase activity in erythrocytes of leukaemic children during chemotherapy (induction period)

vation period. There was also no correlation to an increase or decrease of selenium or glutathione peroxidase activity. Blood or blood fractions administered did not alter selenium contents.

4 Discussion

With respect to selenium status and glutathione peroxidase activity the children with systemic leukaemia responded quite uniformly to induction therapy. The increase of blood selenium values cannot be explained by an increased selenium intake. Due to poor appetite and vomiting, the children usually ate less under chemotherapy. In addition, most of them were on parenteral nutrition during the

Fig. 3. Increase of selenium contents in plasma of eight leukaemic children during chemotherapy (induction period)

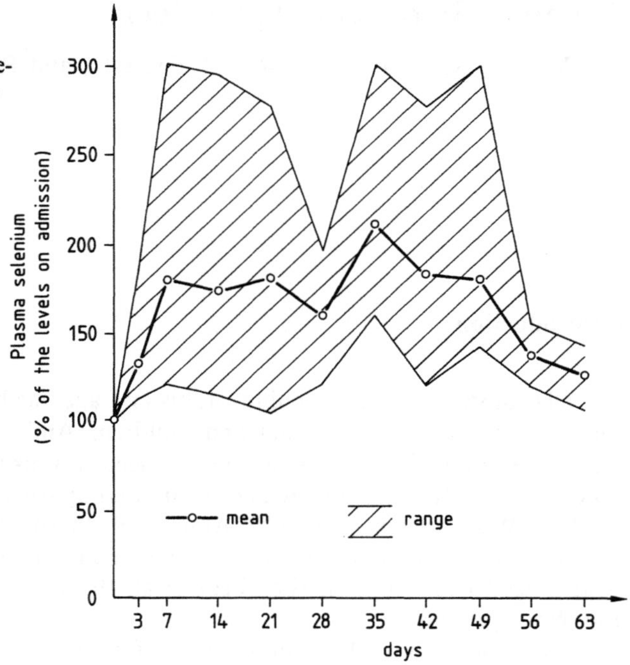

first weeks. Solutions for i.v. feeding are known to contain only negligible amounts of selenium.

Our hypothesis is that the increase of selenium reflects a redistribution of the element in the body due to cell destruction induced by chemotherapy. Concomitantly there was an increase of glutathione peroxidase activities in plasma and red blood cells.

Our investigation did not show low blood selenium contents during any phase of leukaemic treatment. Therefore, there is no need for selenium supplementation in pediatric patients with leukaemia.

References

Burk RF (1986) Selenium and cancer; meaning of serum selenium levels. J Nutr 116:1584–1586

Clark LC (1985) The epidemiology of selenium and cancer. Fed Proc 44:2584–2589

Shamberger RJ (1979) Relationship of selenium to cancer. I. Inhibitory effect of selenium on carcinogenesis. J Am Cancer Inst 44:931–936

Selenium Status and Age Effects

J. L. VALENTINE[1,3], H. K. KANG[2], B. FARAJI[1], and P. A. LACHENBRUCH[1]

1 Introduction

The relationship between selenium status and age has been studied mostly in the younger age groups of infants and children. Amin et al. (1980) found that premature infants have lower selenium concentrations than full-term infants and children. No difference was evident from a comparison of full-term infants and children (Westermarck et al. 1977; Amin et al. 1980). Others have shown that the lowest blood selenium and glutathione peroxidase activity (GSH-Px) is found in infants aged 1 to 4 months (Lombeck et al. 1977; Verlinden et al. 1983b; Butler and Whanger 1987).

Some evaluations of selenium status for very young children compared to adults have been done. Lombeck et al. (1977) observed a clear age dependency for serum selenium with an increase in age from infancy to adulthood. Their work did not indicate significant differences for a comparison of GSH-Px activity across the same age groups. In an earlier study, Levine and Olson (1970) found that values of selenium in blood and plasma of nine well-nourished children, aged 1 to 4 years, were lower than those of three adults. These well-nourished children had been clinically diagnosed as having pneumonia, diphtheria, post-kwashiorkor, and congenital heart disease. One 12 year-old child with unknown diagnosis had values for blood and plasma selenium that were in the same range as the adults'. However, all sample sizes were small. Lowered selenium concentration of whole blood of children compared to adult values was also reported by Westermarck et al. (1977).

A recent study by Verlinden et al. (1983a) could find no relationship to age for blood selenium or glutathione peroxidase activity in healthy Belgian adults, aged 19 to 58 years. In a subsequent study (Verlinden et al. 1983b) comparing whole blood selenium concentrations in elderly subjects, age greater than 60 years, to those reported in their prior study of healthy Belgian adults, no significant difference was found.

As only a few evaluations of selenium status in older children and adults have been performed, we report here our evaluations of selenium status in two large

[1] School of Public Health, Center for Health Sciences, University of California, Los Angeles, CA 90024, USA
[2] VA Office of Environmental Epidemiology, Armed Forces Institute of Pathology, Washington, DC 20306, USA
[3] To whom correspondence should be addressed

American populations which were stratified according to age. Blood, urine, and hair selenium concentrations and glutathione peroxidase activity are used as selenium status indicators.

2 Methods

2.1 Participant Characterization

Participants were selected from Casper, Wyoming and Sun Valley, Nevada. Prior analysis of the Casper and Sun Valley drinking water showed no undue exposure to selenium. Levels in the drinking water were less than the EPA standard of $10\ \mu g\ l^{-1}$ (Valentine et al. 1987).

A random selection of households receiving the water supply was done. No participant was included unless their current drinking water source had been used for at least 1 year.

Questionnaire administration to each participant was performed to determine possible selenium exposure due to occupation and age at the time of sampling. Each participant was sampled for whole blood by venipuncture, 24-h urine collection, and a scalp-to-length hair sample. No hair samples were obtained from Sun Valley participants for selenium measurement.

2.2 Laboratory Methods

Selenium concentrations in whole blood, 24-h urine, and hair samples were analyzed according to the procedure published previously (Valentine et al. 1988). Samples were prepared for instrumental analysis by wet ashing 10 to 20 ml of blood, 6 to 20 ml of urine, and 0.5 to 2.0 g of hair in concentrated nitric acid and perchloric acid. Digests of approximately 1 to 2 ml were diluted in 6 N HCl to an approximate 30 ml volume. From this final solution hydrogen selenide gas was generated using sodium borohydride. The hydrogen selenide formed was analyzed on a Perkin-Elmer Model 603 atomic absorption spectrophotometer. Standards including a reagent blank were analyzed using the same procedure. Accuracy for this element was within $\pm 5\%$ of an EPA-prepared true value sample. Mean recovery percentages of selenium added to distilled, deionized water and carried through the entire procedure (acid digestion, hydride generation) were within $\pm 10\%$ of added amounts.

Glutathione peroxidase activity of whole blood was analyzed by the procedure of Thomson et al. (1977). Glutathione peroxidase hemolysate preparation was as detailed by Valentine et al. (1988). Hemolysate preparation involved taking 0.5 ml of whole blood and washing it three times with 1-ml portions of isotonic saline. An incubation mixture was prepared that contained 2 μmol reduced glutathione (GSH); 0.5 units glutathione reductase, Type III, purified from yeast, 150 units mg^{-1} protein (Sigma Chemical Co.); 1 μmol NaN_3; and 0.1 μmol reduced nicotinamide adenine dinucleotide phosphate (NADPH) in 0.02 M sodium phosphate buffer (pH 7.0) with 6 mM Na_2EDTA. Exactly 2.7 ml of the incubation mixture was added to 0.1 ml of hemolysate and allowed to equilibrate at 25 °C for 10–15 min. The mixture was transferred to the quartz cuvette of a Coleman Hitachi Model 124 double beam spectrophotometer. The reaction was initiated with the addition of 0.1 ml of 2.2 mM hydrogen peroxide and the conversion of NADPH to NADP recorded at an absorbance of 340 nm between 2 and 4 min. Nonenzymatic oxidation of GSH was determined by the substitution of an equal volume of water for hemolysate. Enzyme activity is reported as micromoles of NADPH oxidized per minute per milliliter blood (EU ml^{-1}).

2.3 *Statistical Method*

Student's t-test was employed for statistical evaluation of differences between means for the various age groupings for each of the selenium status indicators. Linear regression analysis of each indicator with age was performed for the combined groups of all participants from a given community. The corresponding level of significance for the t-value and the regression coefficients were determined using statistical tables from Dixon and Massey (1969).

3 Results

A characterization of participants as determined from the questionnaire was previously published (Valentine et al. 1988) and is given in Table 1. All participants were of the Caucasian group with the exception that 4 of the 86 participants from Sun Valley were Spanish or not identified. None of the participants had selenium-exposing occupations [e.g., refinery production of selenium, glass or ceramics manufacture, semiconductor manufacture or research, or the production of topically applied selenium ointment and veterinary usage (National Research Council 1976)] or any undue exposure that would lead to high body burden levels of selenium.

The normal levels of selenium in the various tissues used as selenium-status indicators are given in Table 2. Urinary selenium ranged from 14.3 to 30.2 $\mu g\, l^{-1}$ for Sun Valley and 26.0 to 61.8 $\mu g\, l^{-1}$ for Casper participants. These values are

Table 1. Characterization of participants

Demographic variable	Sun Valley	Casper
Age (years)	35 ± 17[a]	37 ± 18
Sex		
M	35 (41%)	33 (48%)
F	51 (59%)	36 (52%)
Years lived in community	4.8 ± 4.0	11.0 ± 8.8

[a] Mean ± SD.

Table 2. Selenium concentrations in blood, hair, urine, and GSH-Px by age groups

	Under 20		20 to 29		30 to 39		40 to 49		50 to 59		Over 60	
	n	Mean ± SD	n	Mean ± SD	n	Mean ± SD	n	Mean ± SD	n	Mean ± SD	n	Mean ± SD
Sun valley participants												
Urine Se (μg l^{-1})	20	30.2 ± 14.1	8	29.6 ± 14.1	15	16.2 ± 13.3	11	17.8 ± 12.7	11	23.0 ± 16.6	6	14.3 ± 6.9
Blood Se (μg/100 ml)	20	12.2 ± 2.9	8	13.0 ± 1.7	17	11.6 ± 3.6	13	12.6 ± 4.0	14	11.0 ± 4.0	7	13.8 ± 3.4
GSH-Px (EU ml^{-1})	18	2.05 ± 0.50	8	2.22 ± 0.43	15	2.11 ± 0.57	14	2.19 ± 0.62	14	2.25 ± 0.33	7	2.21 ± 0.37
Casper participants												
Urine Se (μg l^{-1})	15	47.1 ± 23.2	10	61.8 ± 43.7	14	30.6 ± 17.6	10	28.6 ± 17.9	12	24.6 ± 14.0	7	26.0 ± 11.8
Blood Se (μg/100 ml)	13	6.3 ± 1.8	10	8.2 ± 3.3	13	6.7 ± 1.6	11	7.0 ± 1.0	12	8.3 ± 2.6	7	8.9 ± 2.4
GSH-Px (EU ml^{-1})	14	2.11 ± 0.52	10	2.18 ± 0.79	13	2.13 ± 0.73	11	2.47 ± 1.0	12	2.42 ± 0.74	7	2.06 ± 0.50
Hair Se (μg g^{-1})	13	0.28 ± 0.19	9	0.38 ± 0.26	9	0.33 ± 0.21	10	0.30 ± 0.14	11	0.47 ± 0.64	5	0.30 ± 0.05

similar to those reported in a review by Robberecht and Deelstra (1984). Urinary selenium concentrations were higher in the under-20 and 20-to-29 age groups of both communities when compared to the older age groups. These differences were found to be statistically significant ($p < 0.005$).

Blood selenium concentrations for the Sun Valley participants ranged from 11.0 to 13.8 μg/100 ml across all age groups. A statistically significant difference between the groups was found for the under-20 group with the elderly over-60 group at $p < 0.05$ using Student's t-test. The concentration of selenium in blood compared well to previous levels determined for normal, healthy individuals (Tulley and Lehmann 1982; Verlinden et al. 1983b). Blood selenium concentrations of the Casper participants were lower than those determined for participants from Sun Valley. An accurate assessment of their dietary intakes (24-h dietary recall or food frequency) was not obtained for participants, so the difference cannot be explained. The blood selenium concentrations of the Casper participants do fall within the range of normal as reported by Welz and Melcher (1987). Relationships to age that were statistically significant were determined for the under-20 age group with the 20-to-29 , 50-to-59, and over-60 groups using t-test comparisons. Again blood values tended to be lower in the youngest group compared to the elderly.

Hair selenium concentrations were found to range from 0.28 to 0.47 μg g^{-1} across all age groups and participants. These values are in agreement with mean values reported by others (Lombeck et al. 1978; Thimaya and Ganapathy 1982; Chen et al. 1982; Lombeck et al. 1987). They are also similar to values reported by Valentine et al. (1988) for mothers and children in California. No variation with age could be determined.

Glutathione peroxidase activity values showed little variation across age groups. No statistically significant difference ($p < 0.05$) between communities or between age groups was noted. Mean values of GSH-Px ranged from 2.05 to 2.42 EU ml^{-1} for both community participants.

Linear regression analysis of each selenium-status indicator with age was performed. The coefficients obtained are presented in Table 3. Only the regression analysis of urinary selenium concentration with age gave statistically significant coefficients of ($r = -0.304$, $p < 0.01$ for Sun Valley participants and of $r = -0.40$, $p < 0.001$ for Casper participants).

Table 3. Linear regression analysis of selenium-status indicators with age

	Sun Valley			Casper		
Regression	n	r	*p*-value	n	r	*p*-value
Urine vs age	71	−0.304	<0.01	68	−0.400	<0.001
Blood vs age	79	+0.004	>0.05	66	+0.216	>0.05
GSH-Px vs age	76	+0.13	>0.05	67	+0.095	>0.05
Hair vs age	–	–	–	56	−0.081	>0.05

4 Discussion

The characterization of selenium distribution with age for several selenium-status indicators is needed for a good representation of the selenium state of healthy individuals. Evaluations of selenium distribution in blood, hair, and GSH-Px activity has been reported by other researchers. Most, however, have never included all three status indicators in the same study for all individuals selected. No published reports for urinary selenium status distribution with age were found.

We have evaluated the selenium concentrations in blood, hair, and urine by age groups. A similar evaluation of GSH-Px activity with age was presented. It seems warranted that these four selenium-status indicators are used in such assessments as any one by itself may not truly represent the selenium status of the individual in all cases (Valentine et al. 1978).

We were able to find age relationships with the selenium-status indicators of urinary and blood selenium. An age effect for urinary selenium was also reported by Robberecht and Deelstra (1984) when they cited the dissertation work of Verlinden (1981), where the concentration of selenium in urine was found to be lower in older people when compared to young adults. Robberecht and Deelstra attributed this difference to a lowered selenium intake by the elderly. Lane and co-workers (1983) have commented that in their study of free-living and institutionalized elderly subjects, adequate selenium status existed; a positive correlation between erythrocyte selenium and dietary selenium was determined.

The determination of selenium status utilizing hair samples may represent as useful an indicator as urine since both are noninvasive techniques. Hair has an advantage in its ability to be easily sampled and stored. We were unable to find a correlation of hair selenium concentration with age. In contrast, studies by Thimaya and Ganapathy (1982) found significantly higher levels of selenium in the hair of adolescents (11 – 15 years), a decline in selenium hair concentrations with ages 16 – 40 years, and the lowest selenium hair concentrations in the elderly group (61 – 70 years).

Whole blood selenium concentrations were related to age in our study by t-test comparisons. These findings differ from those reported by Verlinden et al. (1983 a, b). They found no relationship between age and whole blood selenium concentrations in healthy adults and in the elderly. They did, however, find a relationship with plasma and red cell selenium, with plasma selenium lower in the elderly than in other adults and red cell selenium showing the converse. Our population was composed mostly of healthy adults and elderly with the youngest person in each community group of participants being 8 years to age. Only four in each community had ages less than 12 years. Our work supports the age affect as noted by others (Levine and Olson 1970; Westermarck et al. 1977; Lombeck et al. 1977) for a comparison of children to adult levels of blood selenium.

In summary, the relationship of selenium status of age for the most utilized body burden indicators (blood, hair, urine, and GSH-Px) remains questionable when compared across all age groups. The majority of the literature tends to support an increase in blood selenium concentrations with age, especially when very young children are included. Our study evaluated mostly adolescents and healthy adults. We were able to relate age only to the body-burden indicators of urinary and blood selenium concentrations. As urinary selenium excretion has been little utilized in age relationship studies, additional research with this body fluid is warranted.

References

Amin S, Chen SY, Collipp PJ, Castro-Magana M, Maddaiah VT, Klein SW (1980) Selenium in premature infants. Ann Nutr Metab 24:331–340

Butler JA, Whanger PD (1987) Dietary selenium requirements of pregnant women and their infants. In: Combs GF Jr, Levander OA, Spallholz JE, Oldfield JE (eds) Selenium in biology and medicine. Van Nostrand Rheinhold, New York, pp 688–700

Chen SY, Collipp PJ, Boasi LH, Isenschmid DS, Verolla RJ, San Roman GA, Yeh JK (1982) Fluorometry of selenium in human hair, urine and blood. Ann Nutr Metab 26:186–190

Dixon WJ, Massey FJ (1969) Introduction to statistical analysis, 3rd edn. McGraw-Hill, New York

Lane HW, Warren DC, Taylor BJ, Stool E (1983) Blood selenium and glutathione peroxidase levels and dietary selenium of free-living and institutionalized elderly subjects. Proc Soc Exp Biol Med 173:87–95

Levine RJ, Olson RE (1970) Blood selenium in Thai children with protein-calorie malnutrition. Proc Soc Exp Biol Med 134:1030–1034

Lombeck I, Kasperek K, Harbisch HD, Feinendegen LE, Bremer HJ (1977) The selenium state of healthy children. I. Serum selenium concentrations at different ages; activity of glutathione peroxidase of erythrocytes at different ages; selenium content of food of infants. Eur J Pediatr 125:81–88

Lombeck I, Kasperek K, Harbisch HD, Becker K, Schumann E, Schröter W, Feinendegen LE, Bremer HJ (1978) The selenium state of children. II. Selenium content of serum, whole blood, hair and the activity of erythrocyte glutathione peroxidase in dietetically treated patients with phenylketonuria and maple-syrup-urine disease. Eur J Pediatr 128:213–223

Lombeck I, Kasperek K, Mengel H (1987) Selenium status of pediatric patients. In: Combs GF Jr, Levander OA, Spallholz JE, Oldfield JE (eds) Selenium in biology and medicine. Van Nostrand Rheinhold, New York, pp 619–630

National Research Council (1976) Selenium. Natl Acad Sci, Washington DC

Robberecht HJ, Deelstra HA (1984) Selenium in human urine: concentration levels and medical implications. Clin Chim Acta 136:107–120

Thimaya S, Ganapathy SN (1982) Selenium in human hair in relation to age, diet, pathological condition and serum levels. Sci Total Environ 24:41–49

Thomson CD, Rea HM, Doesbury VM, Robinson MF (1977) Selenium concentration and glutathione peroxidase activities in whole blood of New Zealand residents. Br J Nutr 37:457–460

Tulley RT, Lehmann HP (1982) Flameless atomic absorption spectrophotometry of selenium in whole blood. Clin Chem 28:1448–1450

Valentine JL, Kang HK, Spivey GH (1978) Selenium levels in human blood, urine, and hair in response to exposure via drinking water. Environ Res 17:347–355

Valentine JL, Reisbord LS, Kang HK, Schluchter MD (1987) Effect on human health of exposure to selenium in drinking water. In: Combs GF Jr, Levander OA, Spallholz JE, Oldfield JE (eds) Selenium in biology and medicine. Van Nostrand Rheinhold, New York, pp 675–687

Valentine JL, Faraji B, Kang HK (1988) Human glutathione peroxidase activity in cases of high selenium exposures. Environ Res 45:16–27

Verlinden M (1981) The determination of selenium by atomic absorption spectroscopy. Methodology and biomedical implications. Ph D thesis, Univ Antwerp, Wilryk, Belgium

Verlinden M, Sprundel M van, Auwera JC Van der, Eylenbosch WJ (1983 a) The selenium status of Belgian population groups. I. Healthy adults. Biol Trace Element Res 5:91–102

Verlinden M, Sprundel M van, Auwera JC Van der, Eylenbosch WJ (1983b) The selenium status of Belgian population groups. II. Newborns, children, and the aged. Biol Trace Element Res 5:103–113

Welz B, Melcher M (1987) Accuracy of selenium determination in human body fluids using hydride-generation atomic absorption spectrophotometry. In: Combs GF Jr, Levander OA, Spallholz JE, Oldfield JE (eds) Selenium in biology and medicine. Van Nostrand Rheinhold, New York, pp576–585

Westermarck T, Raunu P, Kijarinta M, Lappalainen L (1977) Selenium content of whole blood and serum in adults and children of different ages from different parts of Finland. Acta Pharmacol Toxicol 40:465–475

Part V Selenium and Malignant Growth

The Influence of Dietary Selenium on Colon, Pancreas, and Skin Tumorigenesis

D.F. BIRT[1,2], P.M. POUR[1], and J.C. PELLING[1]

1 Introduction

The influence of dietary selenium on experimental carcinogenesis has been extensively investigated in the mammary glands of mice and rats (reviewed by Medina 1986). Most studies on cancer at this site have consistently reported inhibition of mammary carcinogenesis by a number of agents. Recent investigations of dietary selenium's effects on mammary tumorigenesis were described in the preceding chapter by C. Ip.

Selenium's effects on other models of skin, liver, trachea, lung, colon, and pancreas cancer have also been investigated and, in general, the results have not been uniformly encouraging, but some inhibition has been observed (reviewed by Birt 1986). For example, colon cancer induced by azoxymethane (Soullier et al. 1981), N-nitrosobis(2-oxopropyl)amine (BOP) (Birt et al. 1982), 1,2-dimethylhydrazine (Jacobs et al. 1977), or methylazoxymethanol (Jacobs et al. 1977) was inhibited in rats fed dietary selenium supplements. While liver cancer induced by dimethylaminoazobenzene (Clayton and Baumann 1949) or acetoaminofluorene (Harr et al. 1972) was inhibited by selenium; selenium was not effective against aflatoxin-B_1-induced liver cancer (LeBoeuf et al. 1985).

Our purpose has been to determine the influence or selenium on additional established models of experimental carcinogenesis and thus determine if selenium inhibition of cancer is specific for particular types of sites or cancer or if the inhibitory effects of selenium are general. Our first studies used BOP as a colon carcinogen for rats and as a pancreatic carcinogen for Syrian hamsters. Our most recent work employed the two-stage model of skin carcinogenesis in SENCAR mice using 7,12-dimethylbenz(a)anthracene (DMBA) as the initiator and 12-0-tetradecanoylphorbol-13-acetate (TPA) as the promoter of skin tumorigenesis.

2 BOP-Induced Carcinogenesis in the Colon and Pancreas

Subcutaneous treatment of rats with BOP resulted in colon, liver, and lung cancer (Pour et al. 1979). In Syrian hamsters, however, this carcinogen induced primarily

[1] Eppley Institute for Research in Cancer, University of Nebraska Medical Center, 42nd and Dewey Ave., Omaha NE 68105, USA
[2] To whom correspondence should be addressed

pancreatic cancer (Pour et al. 1978). This is a particularly relevant model for pancreatic cancer because the carcinomas are of ductal/ductular type, and thus similar in morphology to the most common form of the human disease (Pour et al. 1982). The other commonly used models of pancreatic carcinogenesis result in acinar cell carcinomas in rats. Our objective was to compare the effects of selenium in these two BOP-induced models to determine if the carcinogen or the site seemed to be a determinant factor for cancer inhibition.

To investigate the effects of selenium in colorectal carcinogenesis, we treated MRC-Wistar rats weekly with 5 mg BOP kg^{-1} body weight (Birt et al. 1982). Casein-based diets were supplemented with 0.1 or 2.0 ppm selenium as sodium selenite. These levels of selenium were selected because young Wistar rats could not tolerate higher levels without decreases in the growth rate. These levels of selenium resulted in a 33% elevation in erythrocyte glutathione peroxidase activity at week 50 of the experiment (Birt et al. 1982). The rats were killed after 50 injections of BOP and the intestine was examined histologically. Female rats did not develop colon carcinomas in either dietary group with this protocol, but colon carcinoma yield was reduced by about 50% in the male rats fed the high selenium dosage (Table 1). The high selenium diet was better tolerated by the males and survival was actually better among these rats than among the males fed the low selenium diet. Survival of female rats was not influenced by dietary selenium. Interestingly, male rats treated with BOP and fed diet supplemented with 0.1 ppm selenium developed a 14% incidence of primary lung adenocarcinoma while those fed the 2.0 ppm supplemented diet did not develop lung tumors (Birt et al. 1982).

Determination of single-strand breaks in colonic mucosa from rats fed the two selenium dosages and treated with BOP indicated a more rapid repair of alkali-labile sites induced by BOP in the animals fed the high selenium diet than in those fed the control diet (Birt et al. 1982). Although single-strand breaks are a useful indicator of DNA damage, they do not appear to result in the colon tumors induced by BOP.

In studying selenium effects on pancreatic cancer, we treated Syrian hamsters with a single dosage of BOP (20 mg kg^{-1} body weight) (Birt et al. 1986b). Diets containing 0.1 or 5.0 ppm selenium from sodium selenite were fed in three protocols: (1) before BOP treatment; (2) after BOP treatment; or (3) before and after BOP treatment. This feeding protocol was employed to examine the effects of

Table 1. Colon adenocarcinoma in male rats fed two selenium levels and treated weekly with 5 mg BOP kg^{-1}

Weekly injections	Dietary selenium (ppm)	No. effective animals	No. TBA[a]	Incidence (%)	No effective animals[b]
BOP	0.1	29	12	41	1.0^{B}
BOP	2.0	30	8	27	0.5^{A}
0.9% Saline	0.1	30	0	0	0.0
0.9% Saline	2.0	30	0	0	0.0

[a] TBA = tumor-bearing animal
[b] X^2 test, A<B ($P<0.05$)

Table 2. Pancreatic carcinoma in hamsters fed three levels of selenium before and after treatment with 20 mg BOP kg^{-1} at 8 weeks of age[a]

Dietary Se (ppm)	No. effective animals	No. TBA[b]	Incidence (%)	No. effective animals
0.0	133	9	7	0.09
0.1	66	4	6	0.06
5.0	131	12	9	0.15

[a] Data were pooled across males and females and across the selenium level fed before or after BOP.
[b] TBA = tumor-bearing animal.

Table 3. Pancreatic carcinoma in hamsters fed selenium containing low and high fat diets and treated with four weekly injections of 5 mg BOP kg^{-1}

Sex	Dietary level ppm Se	Se form	Dietary fat	No. effective animals	Incidence (%)	No. effective animals[a]
Male	0.1	Na Selenite	Low	18	61	1.39[A]
	0.1	Na Selenite	High	19	53	1.42[A]
	2.5	Na Selenite	Low	23	78	1.83
	2.5	Na Selenite	High	18	61	2.44[B]
	5.0	Na Selenite	Low	21	62	1.52
	2.5	Seleno methionine	Low	16	50	1.38
Female	0.1	Na Selenite	Low	18	33	0.50[C]
	0.1	Na Selenite	High	16	69	1.00
	2.5	Na Selenite	Low	21	33	0.71
	2.5	Na Selenite	High	19	42	0.80
	5.0	Na Selenite	Low	23	22	0.30
	2.5	Seleno methionine	Low	18	39	1.11[D]

[a] Data presented on groups treated with four weekly injections of 5 mg BOP kg^{-1} only, X^2 test, A<B, C<D ($P<0.05$).

selenium on initiation or postinitiation of carcinogenesis. In this experiment, the incidence of carcinomas was not affected by selenium (Table 2). However, there were effects of selenium levels on adenoma yield (Birt et al. 1986b), but the relevance of these observations to malignant tumorigenesis is unclear since these adenomas are not precursors of adenocarcinomas.

Further studies were designed using hamsters treated with four BOP doses rather than one dosage of BOP. In addition, the hamsters were fed a high fat diet following BOP treatment to increase the carcinoma yield (Birt et al. 1988). In this experiment, diets containing selenium supplements of 0.1, 2.5, or 5.0 ppm selenium from sodium selenite were fed from before BOP treatment until the end of the experiment 70 weeks after the first BOP treatment. In addition, a 2.5 ppm selenium supplement from selenomethionine was fed in the low fat diet to test the efficacy of an organic form of selenium in modifying the tumor yield. We included this treatment because we had previously shown that feeding selenium supplements as selenomethionine resulted in a greater increase in the level of pan-

Table 4. Acinar cell nodules

Sex	Dietary level ppm Se	Se form	Dietary fat	No. effective animals	Incidence[a] (%)	No. effective animals[b]
Male	0.1	Na Selenite	Low	18	11	0.2
	0.1	Na Selenite	High	19	26	1.4
	2.5	Na Selenite	Low	23	0	0.0
	2.5	Na Selenite	High	18	11	0.9
	5.0	Na Selenite	Low	21	5	0.1
	2.5	Seleno methionine	Low	16	0	0.0
Female	0.1	Na Selenite	Low	18	11^{A}	$0.6^{B,C}$
	0.1	Na Selenite	High	16	62^{B}	6.2^{D}
	2.5	Na Selenite	Low	21	14	0.2^{A}
	2.5	Na Selenite	High	19	36	1.7^{C}
	5.0	Na Selenite	Low	23	23	$0.4^{A,B}$
	2.5	Seleno methionine	Low	18	22	$0.8^{B,C}$

[a] X^2 test, A<B ($P<0.05$).
[b] X^2 test, A<B<C<D ($P<0.05$). Values with two letters are not significantly different from values with either of the letters.

creatic selenium than feeding the same level of selenium as sodium selenite (Julius et al. 1983).

There was no evidence of cancer inhibition with any of the treatments (Table 3). In contrast, feeding the high selenium-high fat diet to male hamsters resulted in a twofold increase of pancreatic cancer yield (number of carcinomas per effective animal) compared with male hamsters fed the low selenium-high fat diet or the low selenium-low fat diet. Furthermore, feeding 2.5 ppm selenium as selenomethionine to female hamsters resulted in a doubling of carcinoma yield, in comparison with females fed a diet containing 0.1 ppm selenium as sodium selenite. It is noteworthy that a level of 5.0 ppm selenium, which is not toxic to hamsters (Birt et al. 1983, 1986a), did not influence pancreatic carcinogenesis when fed in a low fat diet. Neither body weight nor survival was influenced by the selenium treatment protocols.

Another important observation in this experiment was an inhibition of acinar cell nodules in hamsters fed the selenium-supplemented diets (Table 4). These lesions are similar to the atypical acinar cell nodules observed in the rat models of pancreatic cancer but they do not progress to carcinomas in the BOP-hamster model. Work by O'Connor et al. (1985) indicated that acinar cell nodules induced in the rat by azaserine were also inhibited by dietary selenium supplements.

3 Skin Carcinogenesis Studies

The two-stage skin tumorigenesis model in the SENCAR mouse is particularly useful for separating carcinogenesis into the initiation and promotion stages. Thus, the influence of dietary modifications can be studied as factors in the initia-

Table 5. Skin papilloma yield after 12 weeks of promotion on mice fed selenium during DMBA-induced initiation or TPA-induced promotion, or throughout tumorigenesis

Dietary selenium fed during treatments with:		No. effective animals	No. papillomas/ effective animal[a]
DMBA (ppm Se)	TPA (ppm Se)		
0.1	0.1	30	17[A]
4.0	0.1	29	17[A]
0.1	4.0	29	23[B]
2.0	2.0	28	24[B]
4.0	4.0	30	26[C]

[a] Data presented on DMBA plus TPA-treated groups, ANOVA only. Results: A<B<C ($P<0.05$). Analysis conducted on data from weeks 6 through 12. The 0.1 ppm Selenium diet represents a control level.

tion or the promotion phases of carcinogenesis. Treatment with only the initiator or only the promoter will result in very few, if any, papillomas, whereas treatment with both initiator and promoter results in a high yield of papillomas and, eventually, carcinomas. Since the effect of dietary selenium on two-stage tumorigenesis had not been examined with this model, we conducted such experiments. All of our experiments will this model have used selenium from sodium selenite.

In our first experiment, we treated the shaved backs of 8- to 9-week-old female SENCAR mice with 100 nmol of DMBA in 0.2 ml acetone. Beginning 1 week later, we treated twice weekly with 3.2 nmol TPA in 0.2 ml acetone for a total of 20 weeks.

Selenium was fed in a semipurified diet at a level of 2.0 or 4.0 ppm. In one group, 4.0 ppm selenium was fed to mice 1 week preceding and 1 week following initiation with DMBA. In another group, this level of selenium was fed from the time of the first treatment with TPA until the end of the experiment, and in two additional groups selenium was fed at a level of 2.0 or 4.0 ppm for 1 week prior to DMBA initiation until the end of the experiment. None of these dietary protocols influenced body weight gain. The tumor data at week 12 of promotion are shown in Table 5. Feeding the selenium supplement preceding and at the time of DMBA had no influence on tumor rates. However, feeding the supplement during TPA treatments resulted in an elevation of tumor yield. A dose-response elevation in tumor rates was observed in the groups given 2.0 or 4.0 ppm selenium throughout the experiment.

The purpose of our next experiment was to determine the influence of selenium, alpha-tocopherol, 4-hydroxyphenyl retinamide (4-HPR), and combinations of these agents on two-stage skin tumorigenesis in the SENCAR mouse model. Tumors developed early and tumor yield was high in the experiment described above, and for this reason we conducted this study with a lower dosage of DMBA (10 nmol) and the same TPA dosage as our earlier experiment. Experimental diets were formulated with selenium alone at 2.0 ppm, selenium plus alpha-tocopherol (1 g kg^{-1}), and selenium plus 4-HPR (0.75 mmol kg^{-1}). These diets were fed starting 1 week prior to DMBA initiation and throughout the entire carcinogenesis protocol. The results of this experiment, shown in Table 6, indicate

Table 6. Skin papilloma yield after initiation with DMBA and 20 weeks of promotion with TPA on mice fed selenium, selenium combined with alpha-tocopherol, or selenium, alpha-tocopherol, and 4-HPR throughout tumorigenesis

Dietary supplement	No. effective animals	No. papillomas/effective animal[a]
Control	120	6.6
Selenium (2 ppm) (Se)	28	6.2
Se + α-tocopherol (1 g kg^{-1}) (E)	29	6.0
Se + E + 4-HPR (0.75 mmol kg^{-1})	30	6.1

[a] Data presented on DMBA plus TPA-treated groups only. No significant differences were observed.

that with this carcinogenesis protocol, selenium alone, selenium in combination with alpha-tocopherol, or selenium in combination with alpha-tocopherol and 4-HPR had no influence on skin tumor yield.

These results suggest that selenium may enhance skin tumor promotion when the dosage of initiator is high, but not when low dosages of DMBA are given. We are currently conducting a study with DMBA dosages of 0, 10, or 50 nmol and selenium dosages of 2.0 and 4.0 ppm. Preliminary results of this experiment are confirming our hypothesis.

4 Summary and Conclusions

Our experiments on the influence of dietary selenium on pancreatic and skin tumorigenesis suggest that selenium is not an effective inhibitor of these forms of cancer. Furthermore, under some conditions, dietary selenium supplements can enhance tumorigenesis at these sites.

Pancreatic carcinogenesis was enhanced in male hamsters only when a carcinogen treatment was administered which resulted in an elevation in cancer yield and when a high fat diet was fed. In female hamsters, pancreatic carcinogenesis was enhanced only with this carcinogenesis protocol and when selenium was fed in the form of selenomethionine.

It appears that the influence of selenium on BOP-induced carcinogenesis differs according to the target site or the animal used, since colon carcinogenesis in male rats was inhibited by supplements of 2.0 ppm selenium, but pancreatic cancer was not inhibited under any of the conditions studied in our laboratory. It further appears that the form of pancreatic cancer induced (ductular or acinar cell) may be a factor in the efficacy of selenium in preventing this cancer. Acinar cell nodules, the precursors to acinar carcinomas in rats, were inhibited in both rats (O'Connor et al. 1985) and hamsters; ductular carcinomas, which have a morphology similar to the most common forms of human cancer, were enhanced under specific conditions, but not inhibited under any of the conditions studied.

The results of our studies on skin tumorigenesis suggest that the dosage of initiating carcinogen may be an important factor in determining the effect of selenium on tumor promotion in this model. Using another skin tumor model, Overvad et al. (1985) reported that selenium could inhibit UV light-induced mouse skin tumorigenesis when the carcinogenesis protocol involved a high exposure rate to UV light, but inhibition was not observed with lower dosages (as reported at the Tübingen meeting, 1988). Further studies employing the two-stage mouse skin model of initiation and promotion are underway in our laboratory to clarify this issue. These results will hopefully lead to a cautious approach to the use of selenium compounds for the prevention or treatment of human disease.

Acknowledgments. Studies conducted in the authors' laboratories were supported by U.S. Public Health Service grants R01 CA24549 and P-30 CA36727, contracts N01 CN 55451-01 and N01 CN 55451-03, from the National Cancer Institute and grant SIG-16 from the American Cancer Society.

References

Birt DF (1986) Update on the effects of vitamins A, C, and E, and selenium on carcinogenesis. Proc Soc Exp Biol Med 183:311–320

Birt DF, Lawson TA, Julius AD, Runice CE, Salmasi S (1982) Inhibition by dietary selenium of colon cancer induced in the rat by bis(2-oxopropyl)nitrosamine. Cancer Res 42:4455–4459

Birt DF, Julius AD, Runice CE, Sayed S (1983) Tolerance of diets deficient or excessive in selenium by Syrian hamsters. Ann Nutr Metab 27:81–93

Birt DF, Julius AD, Runice CE (1986a) Tolerance of low and high dietary selenium throughout the life-span of Syrian hamsters. Ann Nutr Metab 30:233–240

Birt DF, Julius AD, Runice CE, Salmasi E (1986b) Effects of dietary selenium on bis(2-oxypropyl)nitrosamine-induced carcinogenesis in Syrian golden hamsters. J Natl Cancer Inst 77:1281–1286

Birt DF, Julius AD, Runice CE, White LT, Lawson T, Pour PM (1988) Enhancement of BOP-induced pancreatic carcinogenesis in selenium-fed Syrian golden hamsters under specific dietary conditions. Nutr Cancer 11:21–33

Clayton CC, Baumann CA (1949) Diet and azo dye tumors: effect of diet during a period when the dye is not fed. Cancer Res 9:575–582

Harr JR, Exon JH, Whanger PD, Weswig PM (1972) Effect of dietary selenium on N-fluorenyl-acetamide (FAA) induced cancer in vitamin E supplemented rats. Clin Toxicol 5:187–194

Jacobs MM, Jansson B, Griffin AC (1977) Inhibitory effects of selenium on 1,2-dimethylhydrazine and methylazoxymethanol acetate induction of colon tumors. Cancer Lett 2:133–138

Julius AD, Davies MH, Birt DF (1983) Toxic effects of dietary selenium in the Syrian hamster. Ann Nutr Metab 27:296–305

LeBoeuf RA, Laishes BA, Hoekstra WG (1985) Effects of dietary selenium concentration on the development of enzyme-altered liver foci and hepatocellular carcinoma induced by diethylnitrosamine or N-acetylaminofluorene in rats. Cancer Res 45:5489–5495

Medina D (1986) Selenium and murine mammary tumorigenesis. In: Reddy BS, Cohen LA (eds) Diet, nutrition and cancer: a critical evaluation, vol II. CRC Press, Boca Raton, pp 23–41

O'Connor TP, Roebuck BD, Campbell TC (1985) Dietary intervention during the post dosing phase of L-azaserine-induced preneoplastic lesions. J Natl Cancer Inst 75:955–957

Overvad K, Thorling EB, Bjerring P, Ebbesen P (1985) Selenium inhibits UV light induced skin carcinogenesis in hairless mice. Cancer Lett 27:163–170

Pour PM, Salmasi SZ, Runge RG (1978) Selective induction of pancreatic ductular tumors by single doses of N-nitrosobis-(2-oxopropyl)amine in Syrian golden hamsters. Cancer Lett 4:317–323

Pour P, Salmasi S, Runge R, Gingell R, Wallcave L, Nagel D, Stepan K (1979) Carcinogenicity of N-nitrosobis(2-hydroxypropyl)amine and N-nitrosobis(2-oxopropyl)amine in MRC rats. J Natl Cancer Inst 63:181–190

Pour P, Sayed G, Sayed S (1982) Hyperplastic preneoplastic and neoplastic lesions found in 83 human pancreases. Am J Clin Pathol 77:137–152

Soullier BK, Wilson PS, Nigro ND (1981) Effect of selenium on azoxymethane-induced intestinal cancer in rats fed high fat diet. Cancer Lett 12:343–348

Is Selenium Metabolism Necessary for Its Anticarcinogenic Activity?

C. Ip[1]

1 Introduction and Update

At the Third International Selenium Symposium held in Beijing in 1984, I presented a talk entitled, "Susceptibility of Mammary Carcinogenesis in Response to Dietary Selenium Levels" (Ip 1987). Since the current lecture is going to discuss new developments from our laboratory in the past few years, including both published and unpublished data, a brief review and update of the Beijing presentation would be relevant to serve as background materials.

We have studied extensively the relationship between dietary selenium (Se) intake and susceptibility to mammary carcinogenesis induced by dimethylbenz(a)anthracene (DMBA) in rats. The levels of Se under investigation spanned from nutritional to pharmacological ranges. Results of our experiments indicated that selenium deficiency enhanced mammary tumorigenesis, but only in animals that were maintained on a high polyunsaturated fat diet (Ip and Sinha 1981). At the pharmacological range, our laboratory was the first to report a dose-response relationship between dietary selenite supplementation and the inhibition of mammary carcinogenesis (Ip 1981a). To date, considerable evidence exists in the literature that excess Se can protect laboratory animals against preneoplastic lesion development or tumorigenesis induced by chemical carcinogens or viruses (Ip 1986). Numerous studies with a broad spectrum of tumor models have led to the conclusion that the degree of neoplastic suppression is dependent on the strain of animals, the composition of the diet, the form of selenium used (to be discussed later), as well as the specificity and dose of the carcinogen (Ip and Medina 1987).

We initially reported that Se can inhibit both the initiation and promotion/progression phases of DMBA-induced mammary carcinogenesis (Ip 1981b). In a later study, we subsequently demonstrated that the chemopreventive effect during the initiation phase was observed only at near-toxic levels of Se (e.g., 5 ppm). At lower, nontoxic levels (e.g., 2.5 ppm), Se was ineffective if it was supplemented just before and up to 7 days after DMBA administration, although this dose produced significant suppression of tumorigenesis when it was maintained continuously in the postinitiation phase for the duration of the experiment (Ip and Daniel 1985). The protective effect of Se was reversible and an uninterrupted

[1] Department of Breast Surgery, Roswell Park Memorial Institute, Buffalo, NY 14263, USA

intake of Se was necessary to achieve maximal prophylaxis (Ip 1981 b). These findings suggest that the active species of Se with anticarcinogenic potential is present or is generated only when the supply of Se is maintained at a high level. Consistent with the lack of an effect of nontoxic levels of Se on the initiation phase of chemical carcinogenesis, we also failed to find significant changes in DMBA binding to DNA or adduct formation following Se treatment (Ip and Daniel 1985).

At the Beijing Symposium, I also reported that supranutritional levels of dietary vitamin E (1000 ppm or above) were able to enhance the anticarcinogenic potency of selenite (Horvath and Ip 1983), although the same dose of vitamin E by itself was inactive. Conversely, a low or suboptimal intake of vitamin E was found to attenuate the efficacy of Se (Ip 1985). One of the major concerns of the practical aspect of chemopreventive agents has been the fine line separating what is generally regarded as the effective dose and toxic dose (Bertram et al. 1988). Our aim therefore was to exploit the synergism between vitamin E and Se in finding the lowest doses of these agents which could still provide an adequate protection against neoplasia. With the use of a matrix design, we recently showed that a combination of 1 ppm Se (as selenite) and 500 ppm vitamin E was able to produce a 50% reduction in mammary tumorigenesis (Ip 1988 a). This level of Se is five times below the marginal toxicity level of 5 ppm observed in our previous experience (Ip 1981 a). Results of this study suggest that the minimum effective dose has to be systematically established for each micronutrient, and that combination regimen chemoprevention represents a feasible strategy in cancer control if the threshold level of each agent can be determined under a prescribed set of conditions.

2 Inorganic Versus Organic Selenium Compounds in Cancer Chemoprevention

The most common form of Se used in experimental carcinogenesis studies is selenite. Other inorganic Se compounds, such as selenate and selenium dioxide, have also been found to have comparable cancer protective activities (Ip and White 1987). Information on the use of organic Se compounds is not nearly as extensive. Our laboratory has recently reported the results of mammary tumor chemoprevention by selenomethionine and selenocystine (Ip and White 1987). We have compiled the data of a large number of experiments that had been carried out in our laboratory in the last 10 years involving supplementation with either selenite or selenomethionine. On a Se weight basis and over a graded dose range (1 – 5 ppm Se), our analysis showed that selenomethionine is not as active as selenite in cancer chemoprevention. When Se concentrations in blood, liver, and skeletal muscle were measured in rats given either 1, 3, or 5 ppm Se as either selenite or selenomethionine, it was found that tissue Se retention was always higher in animals given selenomethionine compared to those given selenite at each of the three levels tested (unpublished results). The difference was only minimal in blood, but was quite dramatic in the skeletal muscle. For example, at 3 ppm dietary Se, the following results were obtained when the data were expressed as fold increase over the controls (0.1 ppm Se) in tissue Se levels: blood, selenite – 2.2×; selenomethionine – 2.5×; liver, selenite – 2.7×; selenomethionine – 3.6×; skeletal muscle, selenite – 1.7×; selenomethionine – 8×. Thus, the higher burden of tissue Se in selenomethionine-treated rats does not appear to confer a better protection against tumorigenesis. In other words, our data imply that there may be an uncoupling between total tissue Se levels and cancer chemoprevention. Since selenite and selenomethionine are metabolized via different pathways, the question that comes to mind is centered on whether Se metabolism is necessary for its anticarcinogenic activity.

3 Differential Effect of Dietary Methionine on the Biopotency of Selenomethionine and Selenite

Selenomethionine can enter several metabolic routes in the body. First, it can be converted via the transsulfuration pathway to selenocysteine which in turn is degraded to H_2Se by the enzyme selenocysteine lyase. Alternatively, it can be catabolized to methylselenol directly via the transamination pathway. Selenomethionine also readily acylates met-tRNA and can be incorporated into proteins in place of methionine. Selenite is metabolized to H_2Se via selenodiglutathione. Selenite is not converted to selenomethionine in nonruminants and so cannot be compartmentalized into general body proteins, with the exception of Se incorporation into specific selenoproteins (e.g., Se-dependent glutathione peroxidase) via H_2Se as the precursor. In contrast to the assimilatory pathway of Se utilization in the synthesis of selenoproteins (in the form of selenocysteine) with known stoichiometry, the detoxification of H_2Se involves the sequential methylation by S-adenosylmethionine to the mono-, di-, and trimethylated derivatives as excretory products.

Thus, it can be appreciated that the availability of methionine can conceivably regulate the metabolism of both selenomethionine and selenite. We therefore decided to examine the influence of a low methionine intake on the chemopreventive efficacy of selenomethionine versus selenite in the DMBA-induced mammary tumor model. Animals were fed from weaning a purified 20% casein diet with or without 0.3% methionine supplementation. Selenomethionine or selenite, at a final concentration of 3 ppm Se, was added to the diet starting 5 days after DMBA administration.

Results of the carcinogenesis experiment, reported in a recent article in 1988 (Ip 1988b), indicated that suboptimal dietary methionine significantly reduced the protective effect of selenomethionine. In contrast, the efficacy of selenite was affected only marginally by the feeding of a casein-based diet without additional methionine supplementation. In rats given 3 ppm selenomethionine, tissue Se was actually higher in those fed a diet with a suboptimal amount of methionine than in those with an adequate intake of methionine. It is most likely that when methionine is limiting, a greater percentage of selenomethionine is incorporated into body proteins in place of methionine. On the other hand, dietary methionine did not influence the level of tissue Se in animals given selenite. An increase of dietary methionine to 0.6% did not enhance the efficacy of selenomethionine in cancer protection but would allow the use of a higher level of selenite without the accompanying adverse effects.

In summary, our study confirmed the previous conclusion that a high tissue concentration of Se is not necessarily an indicator of reduced susceptibility to carcinogenesis. Rats that were given a low methionine diet and supplemented with selenomethionine actually had the highest Se concentrations in the blood and in the liver, but they also had the highest tumor incidence and yield (Ip 1988b). Thus, indiscriminate compartmentalization of selenomethionine into the general protein pool will lead to an attenuation of its anticarcinogenic activity, despite the maintenance of an elevated tussue titer.

4 Efficacy of Trimethylselenonium Versus Selenite and Their Modulation by Arsenite

As pointed out in the last section, reduction and methylation reactions are important steps in the metabolism of Se. The two major excretory products are dimethylselenide, which is exhaled in the breath under conditions of high Se administration, and trimethylselenonium ion ($TMSe^+$), which is found in the urine. The methylation pathway generally has been regarded as a detoxification mechanism. Arsenite is known to decrease the toxicity of selenite (Hill 1975), but it increases the toxicity of $TMSe^+$ (Obermeyer et al. 1971). Clearly, the interaction of arsenite with Se is dependent on the chemical form of Se. Our study, which was carried out in collaboration with H. Ganther of the University of Wisconsin at Madison, was designed to study (1) whether $TMSe^+$ has any anticarcinogenic action by itself; (2) whether arsenite affects the activity of $TMSe^+$ in chemoprevention; and (3) whether the interaction between arsenite and $TMSe^+$ differs from that between arsenite and selenite.

The results of this experiment (Ip and Ganther 1988) indicated that supplementation of selenite (3 ppm Se) alone produced about a 50% reduction in tumor yield, and arsenite (5 ppm As) reduced the response to selenite. In contrast, arsenite greatly enhanced the protective effect of $TMSe^+$ (40 ppm Se); this combination was nearly as effective as selenite, although either $TMSe^+$ or arsenite alone was inactive. Thus, arsenite has the capacity to influence the anticarcinogenic action of Se, and can either potentiate or attenuate the protective effect depending on the methylation state of the Se compound.

Arsenic is known to interfere with the formation of dimethylselenide from selenite, probably by trapping the intermediary metabolite H_2Se as well as by inhibiting the microsomal enzyme that methylates H_2Se (Hsieh and Ganther 1977). The fact that the anticarcinogenic efficacy of selenite is attenuated by arsenite suggests that the metabolism of selenite via the reduction and methylation pathway is critical for its cancer prevention effect.

We believe that the anticarcinogenic effect obtained with the combination of arsenite and $TMSe^+$ favors the idea that some metabolite prior to $TMSe^+$ is responsible. This argument encompasses two main points: (1) $TMSe^+$ is not metabolically inert, but undergoes demethylation in animals; and (2) remethylation to $TMSe^+$ is inhibited by arsenite. Evidence for demethylation and remethylation of $TMSe^+$ was recently obtained by Foster et al. (1986) using compounds doubly-labeld with ^{75}Se and ^{14}C and measuring the ratio of ^{14}C to ^{75}Se in dimethylselenide and $TMSe^+$ that was excreted.

The enhancement of anticarcinogenic activity of $TMSe^+$ by arsenite would be explained by arsenite-mediated inhibition of methyltransferase activity, preventing remethylation of the demethylated $TMSe^+$ and shifting the balance in favor of dimethylselenide (and possibly methylselenol) accumulation. It follows that monomethylated or dimethylated forms of Se could have greater anticarcinogenic activity than $TMSe^+$, and inhibition of their methylation would enhance this biological activity. Although the active forms of Se responsible for its cancer preventing effects have not been identified, the present study clearly shows that Se metabolism is an important area of research in elucidating the action of Se in chemoprevention.

5 Dietary Protein and Selenite Chemoprevention

Ganther et al. (1966) reported over 20 years ago that selenite volatalization could be increased two- to threefold by increasing the protein content of the basal diet. If the accumulation of certain Se metabolite(s) is responsible for the anticarcinogenic effect, then increasing the protein content of the diet would be expected to potentiate the efficacy of selenite. In this experiment, DMBA-treated rats were maintained on a diet containing either 20, 30, 40, or 50% casein and supplemented with 3 ppm Se as selenite. Preliminary results showed there was a trend toward reducing cancer risk with increasing dietary protein in these animals (unpublished data). Moreover, there was a significant increase in the tumor-free period especially in the 50% casein group. When the difference in methionine content was adjusted between the 20% casein and 50% casein groups by adding extra methionine to the former, the lower tumor yield and longer tumor-free period as a result of selenite treatment were still observed in the 50% casein diet. These experiments indicate that dietary protein is a more effective modulator of selenite chemopreventive efficacy than additional methionine. The protein enhancing effect is consistent with the hypothesis that the generation of some critical Se metabolite(s) via overloading of the normal detoxification pathway is an important mechanism for the manifestation of Se protective responses.

6 Other Mechanistic Studies

An earlier report from Hoekstra's laboratory showed that excess selenite feeding caused a dose-dependent "adaptive" increase in the concentration of reduced glutathione (GSH) in the liver (LeBoeuf and Hoekstra 1985). The increase in GSH was preceded by increases in oxidized glutathione (GSSG) as well as the GSSG to GSH ratio. LeBoeuf and Hoekstra (1985) further speculated that the effects of Se on cell proliferation may be mediated by increased cellular GSSG, thereby inactivating the eukaryotic initiation factor 2 in protein synthesis. We were able to confirm the slight increases in hepatic GSH and GSSG following supplementation with selenite. These increases, however, were not observed with selenomethionine (unpublished data). Thus, the differential effects of selenite and selenomethionine in glutathione metabolism pertubation argue against the significance of this mechanism in Se-mediated cancer prevention.

In an attempt to explore the synergistic interaction of Se and vitamin E, blastogenesis and natural killer cell assays were carried out in our laboratory using spleen cells from rats that were supplemented with either agent alone or in combination (Ip and White 1987). In general, there was no consistent or meaningful changes of blastogenesis in the presence of concanavalin A, phytohemagglutinin (both T-cell mitogens), or lipopolysaccharide (a B-cell mitogen). Stimulatory responses in cytotoxicity against YAC-1 cells were likewise not detected at various effector to target cell ratios. Although the concept of enhanced immunosurveillance as a mechanism of resistance to tumorigenesis is an attractive idea, our in vitro data have failed to implicate either spleen cell proliferative activity or natural killer cytotoxic responses as playing a positive role in the inhibition of neoplastic development by selenite and/or vitamin E supplementation.

7 Conclusion

The search for indirect mechanisms with respect to Se chemoprevention so far has proved to be very elusive. Our effort on carcinogen metabolism, glutathione pertubation, and immune responses has not been rewarded with promising clues. On the other hand, there is little doubt that Se has a direct inhibitory effect on cell proliferation and DNA synthesis in the in vitro system (Ip and Medina 1987). Medina and co-workers are actively pursuing the role of selenoprotein(s) other than glutathione peroxidase in regulating the growth of mammary epithelial cells in culture (Morrison and Medina 1988; Morrison et al. 1988). There is of course the unresolved issue of whether the in vitro mechanism is directly applicable to the in vivo situation. However, it is important to note that the characteristics of dose dependence and reversibility seen in Se inhibition of tumorigenesis are also observed in Se inhibition of cell growth in vitro.

I have provided in this lecture a collection of circumstantial evidence which suggests that Se metabolism via the reduction and methylation pathway is essential for its anticarcinogenic activity. This hypothesis does not preclude the incorporation of Se in the form of selenocysteine for the expression of specific selenoproteins. Our current research is focused on investigating whether Se chemoprevention is preserved if the inorganic Se pool (H_2Se) is bypassed. Although significant advances have been made in the understanding of Se metabolism during the last 2 decades, many unanswered questions and gaps in our knowledge still remain especially under conditions of high Se intake. Thus,

Se metabolism represents a key area of future investigation which hopefully will lead to valuable information in designing agents and strategies for chemoprevention.

Acknowledgments. This work was supported by USPHS grants CA27706 and CA45164 from the National Cancer Institute, National Institutes of Health. The author is grateful for the input and comments of Drs. Howard Ganther and Daniel Medina in the preparation of this manuscript.

References

Bertram JS, Kolonel LN, Meyskens FL Jr (1987) Rationale and strategies for chemoprevention of cancer in humans. Cancer Res 47:3012–3031

Foster SJ, Kraus RJ, Ganther HE (1986) Formation of dimethyl selenide and trimethylselenonium from selenobetaine in the rat. Arch Biochem Biophys 247:12–19

Ganther He, Levander OA, Baumann CA (1966) Dietary control of selenium volatilization in the rat. J Nutr 88:55–60

Hill CH (1975) Interrelationships of selenium with other trace elements. Fed Proc 34: 2096–2100

Horvath PM, Ip C (1983) Synergistic effect of vitamin E and selenium in the chemoprevention of mammary carcinogenesis in rats. Cancer Res 43:5335–5341

Hsieh HS, Ganther HE (1977) Biosynthesis of dimethyl selenide from sodium selenite in rat liver and kidney cell-free systems. Biochim Biophys Acta 497:205–217

Ip C (1981 a) Factors influencing the anticarcinogenic efficacy of selenium in dimethylbenz(a)-anthracene-induced mammary tumorigenesis in rats. Cancer Res 41:2683–2686

Ip C (1981 b) Prophylaxis of mammary neoplasia by selenium supplementation in the initiation and promotion phases of chemical carcinogenesis. Cancer Res 41:4386–4390

Ip C (1985) Attenuation of the anticarcinogenic action of selenium by vitamin E deficiency. Cancer Lett 25:325–331

Ip C (1986) The chemopreventive role of selenium in carcinogenesis. J Am Coll Toxicol 5:7–20

Ip C (1987) Susceptibility of mammary carcinogenesis in response to dietary selenium levels: modification by fat and vitamin intake. In: Combs GF Jr, Spallholz JE, Levander OA, Oldfield JE (eds) Selenium in biology and medicine. 3rd Int Symp, part B. Van Nostrand Reinhold, New York, pp 1009–1022

Ip C (1988 a) Feasibility of using lower doses of chemopreventive agenst in a combination regimen for cancer protection. Cancer Lett 39:239–246

Ip C (1988 b) Differential effect of dietary methionine on the biopotency of selenomethionine and selenite in cancer chemoprevention. J Natl Cancer Inst 80:258–262

Ip C, Daniel FB (1985) Effects of selenium on 7,12-dimethylbenz(a)anthracene-induced mammary carcinogenesis and DNA adduct formation. Cancer Res 45:61–65

Ip C, Ganther H (1988) Efficacy of trimethylselenonium versus selenite in cancer chemoprevention and its modulation by arsenite. Carcinogenesis 9:1481–1484

Ip C, Medina D (1987) Current concept of selenium and mammary tumorigenesis. In: Medina D, Kidwell W, Heppner, G, Anderson EP (eds) Cellular and molecular biology of breast cancer. Plenum, New York London, pp 479–494

Ip C, Sinha DK (1981) Enhancement of mammary tumorigenesis by dietary selenium deficiency in rats with a high polyunsaturated fat intake. Cancer Res 41:31–34

Ip C, White G (1987) Mammary cancer chemoprevention by inorganic and organic selenium: single agent treatment or in combination with vitamin E and their effects on in vitro immune functions. Carcinogenesis 8:1763–1766

LeBoeuf RA, Hoekstra WG (1983) Adaptive changes in hepatic glutathione metabolism in response to excess selenium in rats. J Nutr 113:845–854

LeBoeuf RA, Hoekstra WG (1985) Changes in cellular glutathione levels: possible relation to selenium-mediated anticarcinogenesis. Fed Proc 44:2563–2567
Morrison DG, Medina D (1988) Intracellular 58K selenoprotein levels correlate with inhibition of DNA synthesis in MMEC lines. Proc Am Assoc Cancer Res 29:19 (Abstr No 74)
Morrison DG, Berdan RC, Pauly DF, Turner DS, Oborn CJ, Medina D (1988) Selenium distribution in mammary epithelial cells reveals its possible mechanism of inhibition of cell growth. Anticancer Res 8:51–64
Obermeyer BD, Palmer IS, Olson OE, Halverson AW (1971) Toxicity of trimethylselenonium chloride in the rat with and without arsenite. Toxicol Appl Pharmacol 20:135–146

Low Serum Selenium Level and Elevated Risk of Lung Cancer

P. Knekt[1,6], G. Alfthan[2], A. Aromaa[1], M. Hakama[3,4], T. Hakulinen[3], J. Maatela[1], R. Peto[5], E. Saxén[3], and L. Teppo[3]

1 Introduction

There is some evidence that selenium has an anticarcinogenic effect. One proposed mechanism of action is its antioxidant property (Chow 1979). Experimental studies in animals and ecological studies as well as case-control studies based on selenium measurements after diagnosis support the hypothesis that low intake of selenium is associated with an increased risk of cancer (Willett and MacMahon 1984). Several studies based on prediagnostic serum selenium samples have also reported an inverse association between the level of selenium and later occurrence of cancer (Willett et al. 1983; Salonen et al. 1984, 1985; Fex et al. 1987; Kok et al. 1987). However, because of the relatively small number of cancer cases in the longitudinal studies performed so far, relatively little is known about the site-specific strength of the association.

In the present study we used a longitudinal design to test the hypothesis that a low selenium intake, as reflected by a low serum selenium concentration, is a risk factor for lung cancer in man.

2 Subjects and Methods

In 1968–1972 althogether 21172 men, aged 15–99 years, participated in the Social Insurance Institution's Mobile Clinic Health Examination Survey, which was carried out in 25 population groups from different parts of Finland (Knekt et al. 1988). A questionnaire supplied information about occupation, history of diseases, and smoking habits. Body height and weight were measured and the body mass index (weight/height^2) was computed. Hematocrit was estimated and the serum cholesterol concentration was determined. During the median follow-up of 11 years between the date of baseline examination and December 31, 1980, 189 new lung cancer cases were diagnosed and notified to the national Finnish Cancer Registry.

[1] Social Insurance Institution, 00381 Helsinki, Finland
[2] National Public Health Institute, Helsinki, Finland
[3] Finnish Cancer Registry, Helsinki, Finland
[4] Department of Public Health, University of Tampere, Tampere, Finland
[5] University of Oxford, Oxford, England
[6] To whom correspondence should be addressed

The serum samples were kept frozen at −20 °C until they were thawed for selenium, retinol, beta-carotene, and alpha-tocopherol analyses in 1983. To cut down the costs of the laboratory determinations, a case-control design nested in the longitudinal study was adopted. Two controls per case were selected by individual matching for sex, place of residence, and age. Matching for residence controlled both for the time of the baseline examination and for the duration that the serum samples were stored. The final data comprised 342 controls. The serum selenium concentrations were determined by an electrothermal atomic absorption spectrometric method (Alfthan and Kumpulainen 1982). The serum retinol, serum beta-carotene, and serum alpha-tocopherol levels were determined simultaneously, using high pressure liquid chromatography.

The association between the serum selenium level and the occurrence of lung cancer was estimated with the conditional logistic model (Breslow and Day 1980). Confounding factors were adjusted for by including them in the model. Relative odds were computed for quintiles of adjusted serum selenium levels. Statistical significances were tested with the likelihood ratio test based on the model.

3 Results

There was an inverse association between the serum selenium level and the incidence of lung cancer. The mean level of serum selenium was 57 $\mu g\,l^{-1}$ among the lung cancer cases and 61 $\mu g\,l^{-1}$ among the controls. The difference between the means was statistically significant ($p<0.001$). The smoking-adjusted relative odds of lung cancer between the highest and the lowest quintile of serum selenium was 3.3 (Fig. 1) and there was a significant inverse gradient between the serum selenium level and the incidence of lung cancer.

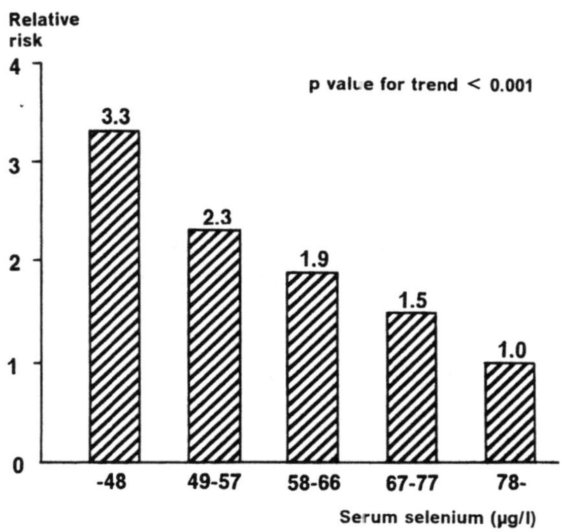

Fig. 1. Smoking-adjusted relative odds of lung cancer according to quintile of serum selenium

Table 1. Relative odds of lung cancer in the lowest quintile of serum selenium in comparison to the other quintiles after various adjustments

Factors adjusted for	Relative risk	95% Confidence interval
Smoking	1.8	1.1 – 2.8
Smoking, vitamin A, vitamin E	1.7	1.0 – 2.8
Smoking, cholesterol, body mass index, hematocrit, occupation	1.7	0.9 – 3.2

The smoking-adjusted relative odds of lung cancer between the lowest quintile ($<49\ \mu g\ l^{-1}$) and the four highest quintiles of serum selenium was 1.8 [95% confidence interval (Cl) = 1.1 – 2.8; Table 1]. The association was not notably altered by further adjustment for serum retinol, serum beta-carotene, serum alpha-tocopherol, serum cholesterol, body mass index, hematocrit, and occupation.

To study the effect of possible early undiagnosed cancer on the serum selenium level, we studied separately the cases observed during the first 4 years of follow-up and cases observed later. The relative odds of the two time intervals did not substantially differ, being 1.9 (Cl = 0.8 – 4.2) for the former and 1.7 (Cl = 0.8 – 3.3) for the latter.

4 Discussion

Men initially free of cancer who had a low level of serum selenium had an excess risk of lung cancer in the present study. This finding is in agreement with the observations reported in the few previous studies based on Finnish populations, in which the serum selenium levels have been very low (Salonen et al. 1984, 1985; Virtamo et al. 1987). In populations with higher serum selenium levels, a similar association between the serum selenium level and the incidence of cancer has been demonstrated in some studies (Willett et al. 1983; Kok et al. 1987) but not in all (Menkes et al. 1986; Nomura et al. 1987). We also found a significant gradient between the serum selenium level and the incidence of cancer, a finding which is in agreement with the results of a previous study concerning all sites of cancer (Salonen et al. 1984) and which suggests a dose-response relationship at low levels of serum selenium. A similar trend has not been observed at high levels (Willett et al. 1983).

As adjustment for several known risk factors and factors reflecting health behaviour and health status did not notably alter the findings, a low serum selenium level appears to be an independent risk indicator for lung cancer. Preclinical disease may depress the serum selenium levels. In the present study, however, the inverse association persisted after the exclusion of cancer cases occurring during the first years of follow-up. This finding suggests that the association cannot mainly be due to preclinical cancer.

In conclusion, lung cancer risk seems to be increased in individuals with a low level of serum selenium. This is in accordance with the hypothesis that a very low serum selenium intake is causally related to an increased lung cancer risk.

References

Alfthan G, Kumpulainen J (1982) Determination of selenium in small volumes of blood plasma and serum by electrothermal atomic absorption spectrometry. Anal Chim Acta 140:221–227

Breslow NE, Day NE (1980) Statistical methods in cancer research. Vol 1. The analysis of case-control studies. IARC Sci Publ no 32, Lyon

Chow CK (1979) Nutritional influences on cellular antioxidant defense systems. Am J Clin Nutr 32:1066–1081

Fex G, Pettersson B, Åkesson B (1987) Low plasma selenium as a risk factor for cancer death in middle-aged men. Nutr Cancer 10:221–229

Knekt P, Aromaa A, Maatela J, Aaran R-K, Nikkari T, Hakama M, Hakulinen T, Peto R, Saxén E, Teppo L (1988) Serum vitamin E and risk of cancer among Finnish men during a 10-year follow-up. Am J Epidemiol 127:28–41

Kok FJ, Bruijn AM de, Hofman A, Vermeeren R, Valkenburg HA (1987) Is serum selenium a risk factor for cancer in men only? Am J Epidemiol 125:12–16

Menkes MS, Comstock GW, Vuilleumier JP, Helsing KJ, Rider AA, Brookmeyer R (1986) Serum beta-carotene, vitamins A and E, selenium, and the risk of lung cancer. N Engl J Med 315:1250–1254

Nomura A, Heilbrun LK, Morris JS , Stemmermann GN (1987) Serum selenium and the risk of cancer, by specific sites: case-control analysis of prospective data. J Natl Cancer Inst 79:103–108

Salonen JT, Alfthan G, Huttunen JK, Puska P (1984) Association between serum selenium and the risk of cancer. Am J Epidemiol 120:342–349

Salonen JT, Salonen R, Lappeteläinen R, Mäenpää PH, Alfthan G, Puska P (1985) Risk of cancer in relation to serum concentrations of selenium and vitamins A and E: matched case-control analysis of prospective data. Br Med J 290:417–420

Virtamo J, Valkeila E, Alfthan G, Punsar S, Huttunen JK, Karvonen M (1987) Serum selenium and risk of cancer. A prospective follow-up of nine years. Cancer 60:145–148

Willett WC, MacMahon B (1984) Diet and cancer – an overview (Second of two parts). N Engl J Med 310:697–703

Willett WC, Polk BF, Morris JS, Stampfer MJ, Pressel S, Rosner B, Taylor JO, Schneider K, Hames CG (1983) Prediagnostic serum selenium and risk of cancer. Lancet II:130–134

Study on the Relation of Se, Mn, Fe and Mo to Liver Cancer Mortality

Y.X. WANG[1,3], J.F. QIN[1], Y.L. XU[1], S.M. WU[1], L.B. YAN[1], and J.D. TU[2]

1 Introduction

In cancer prevention work, main efforts had been placed on identifying carcinogens in the hope that they could be removed from the environment. Unfortunately, a complete removal of all causative agents proved to be impossible. Since the beginning of this decade, much attention has transferred to finding factors that can increase resistance in human bodies against cancer. Trace elements represent one such factor.

The Chongming Island is a narrow stretch of land with a length of only 200 km from east to west. It is one of the areas with the highest liver cancer mortality in China, and more noticeably the incidence of liver cancer in the eastern part of the island is considerably higher than that in the western part (Gao et al. 1981). Not much difference is found between the two parts in the natural environment. The only difference found so far is that the western part was geographically formed earlier than the eastern part. Therefore, certain geochemical factors may exist here which influence cancer incidence. Our project was to make use of the Chongming Island as a high prevalence area of liver cancer and its uneven geographical distribution to study the relation between trace elements and high liver cancer incidence. The ultimate purpose of the work was to probe into the essence of carcino mechanism and to look for possible ways to reduce liver cancer mortality.

2 Material and Methods

1. Comparative study of trace elements in scalp hair of normal persons living in areas with different incidences of liver cancer:
We selected two communes (MaQian and Baozhen) with higher liver cancer mortality (above 40/100000 persons) and another two (SanXing and Maozhen) with a lower mortality (below 20/100000 persons) as a pair of groups for the study.

[1] Shanghai Institute of Nuclear Research, Academia Sinica, P.O. Box 8204, Shanghai 201849, Peoples Republic of China
[2] Shanghai Institute of Cancer Research, Shanghai, Peoples Republic of China
[3] To whom correspondence should be addressed

Samples of scalp hair and blood (5 ml) were collected three times respectively in March, July and November. Each time we randomly selected 100 adults (male 75, female 25) aged between 20 to 40 from one commune of each group for sampling. Thus, we obtained 600 samples in total.

2. Case-control investigation matched with sex and age:
All the liver cancer cases diagnosed in 1984 throughout the island were taken as subjects. Each patient was paired with a control subject selected among the local inhabitants. The patient and the control subject shared the same sex, age and similar economical background. Samples of scalp hair and blood (5 ml) were collected simultaneously from each subject; all subjects in this research were peasants from areas with no industrial contamination.

3. Diagnosis criteria of liver cancer cases:
All the diagnoses of the cases for this study conformed with the criteria set up at the Shanghai Conference of China National Liver Cancer Cooperation Team in 1977.

4. Analysis of scalp hair was conducted by use of X-ray fluorescence analysis to determine Fe, Ni, Ti, Zn, Cu, Sr, Pb and Mn. Se and Mo were determined with differential pulse polarography catalytic wave.

3 Results

The data of trace element content in scalp hair collected from areas with different liver cancer mortality are shown in Table 1.

1. It is noteworthy that the selenium content in the scalp hair from the normal persons in the four communes of the Chongming Island is in the range of 0.222–0.378 ppm (see Table 1). The average value is about 0.309 ppm. This level is relatively low compared with other places in China (see Fig. 1) (Tan et al. in press) which might indicate a Se deficiency on the island.

2. Only Fe, Mo and Mn showed obvious differences indicating that the content of these elements was lower in the high incidence part of the island than in the low incidence part in two of the three samplings (see Table 1). The other sampling, however, failed to give a definite result; but it suggested a similar tendency. Therefore, we may claim to have found a reverse correlation between the content of Fe, Mn and Mo in scalp hair and the incidence of liver cancer.

3. Results of the case-control study on trace elements in scalp hair between subjects with liver cancer and normal subjects are listed in Table 2. Fe content in the patients' scalp hair is clearly lower than in the controls.

This is in conformity with the results of the comparative study between high and low incidence areas. Mn content also tends to be lower in the hair of liver cancer patients.

4 Discussion

The cause of the high liver cancer incidence in the Chongming Island is very complex and no definite conclusions can be made at the present time. The risk factors

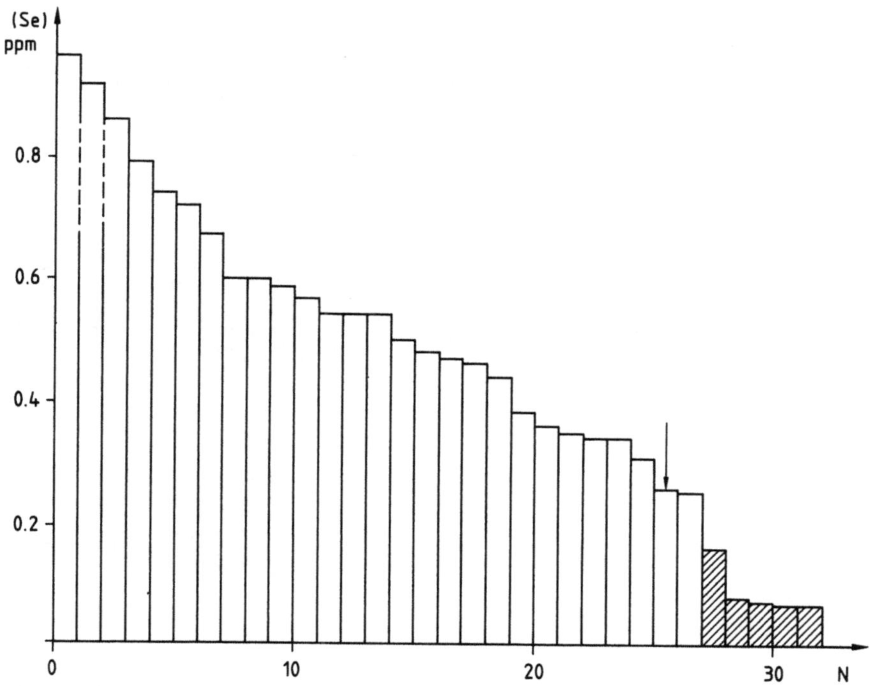

Fig. 1. Selenium content in human hair from different areas of China. The *arrow* indicates the Se content in human hair from subjects from Chongming County

for liver cancer were reported by a prospective cohort study (Tu et al. 1987). The carrier status of HBV was a risk factor of liver cancer occurrence. Drinking stagnant water was another major risk factor. Cigarette smoking might be a secondary and dependent risk factor. However, an interesting fact found by Tu et al. deserves our full attention: the incidence of hepatitis B was lowest in the commune which had the highest liver cancer incidence. In addition, the incidence of liver cancer did not vary among noncarriers of HBV. Variation mainly existed among the carriers. Moreover, liver cancer risk increased greatly with the co-existence of an HBV carrier status and drinking stagnant water.

Stagnant water refers to the water taken from field wells or house ditches, both of which are permeated by rainwater through the soil. The nature of this kind of water depends heavily upon the soil properties. Since the incidence of liver cancer bears a large difference between the eastern part of the island and the western part and, the latter having been formed earlier than the former, we have good reason to expect the existence of a geochemical factor which is directly relevant to the high incidence.

Schrauzer et al. (1977) pointed out that the amount of Se in the diet is reversely correlated with cancer in the main tissues. Later, Schrauzer et al. (1978) elucidated the anticarcinogenic properties of Se. However, he did not associate Se with liver cancer. Chinese researchers (Li et al. 1986; Yu 1985) first reported the finding of a reverse correlation of liver cancer with Se intake and the Se blood serum level.

Table 1. Trace element concentration of scalp hair[a]

Location		San Xing	Ma Qiao	Mao Zhen	Ma Qiao	Mao Zhen	Bao Zhen
Mortality of liver cancer (1/100000)		18	58	17	58	17	41
Season		March	March	July	July	November	November
Sr	Male	4.45 ± 2.24	4.57 ± 3.28	6.26 ± 2.85	4.95 ± 1.86	3.99 ± 2.56	3.88 ± 2.4
	Female	6.70 ± 2.69	7.59 ± 3.5	21.2 ± 12.6	10.66 ± 6.01	8.83 ± 5.92	9.44 ± 5.08
	T	−3.89*	−3.57*	−9.28*	−6.98*	−5.35*	−7.8*
Pb	Male	2.62 ± 1.21	2.55 ± 2.10	5.32 ± 5.71	3.89 ± 3.48	6.51 ± 9.68	7.07 ± 7.34
	Female	2.62 ± 1.31	1.83 ± 0.79	3.43 ± 1.36	2.84 ± 1.33	3.30 ± 2.52	4.20 ± 3.61
	T	−0.017	1.47	1.46	1.72	1.86	2.08*
Zn	Male	181 ± 19	187 ± 33	178 ± 29	178 ± 24	173 ± 20	168 ± 25
	Female	166 ± 24	184 ± 44	169 ± 30	162 ± 22	157 ± 27	167 ± 22
	T	3.07*	0.316	1.27	3.35*	3.21*	0.20
Cu	Male	13.6 ± 4.1	12.6 ± 5.2	12.1 ± 3.5	12.3 ± 3.2	11.9 ± 3.1	12.0 ± 2.7
	Female	16.7 ± 5.2	14.4 ± 3.9	15.7 ± 6.0	15.2 ± 6.6	13.2 ± 4.2	12.8 ± 2.7
	T	−2.89*	−1.43	−3.49*	−2.96*	−1.68*	−1.21
Ni	Male	0.87 ± 0.71	1.02 ± 0.74	0.70 ± 0.66	0.55 ± 0.28	0.72 ± 0.80	0.63 ± 0.57
	Female	1.33 ± 1.14	1.06 ± 1.00	1.76 ± 2.09	1.00 ± 0.88	0.71 ± 0.53	0.80 ± 1.4
	T	−2.24*	−0.19	−3.71*	−3.78*	0.06	−0.95
Fe	Male	26.7 ± 17.1	14.7 ± 8.7	39.8 ± 26.7	32.1 ± 14.5	15.9 ± 6.64	16.1 ± 6.01
	Female	30.4 ± 17.7	12.1 ± 4.6	38.5 ± 12.2	30.9 ± 18.9	21.9 ± 10.6	17.6 ± 7.87
	T	−0.88	1.28	0.22	0.35	−3.3*	−1.13
Mn	Male	3.42 ± 1.73	3.47 ± 2.06	9.49 ± 4.94	4.83 ± 1.92	7.85 ± 4.64	6.60 ± 3.70
	Female	3.8 ± 1.49	4.52 ± 3.06	9.01 ± 4.44	4.74 ± 3.11	9.76 ± 4.68	7.89 ± 5.35
	T	−0.92	−1.82	0.39	0.18	−1.87	−1.44
Cr	Male	1.86 ± 0.74	1.85 ± 0.72	1.69 ± 0.95	1.57 ± 0.62	1.63 ± 1.06	1.91 ± 2.09
	Female	1.59 ± 0.71	1.95 ± 1.07	1.67 ± 1.41	1.50 ± 0.67	1.31 ± 0.69	1.37 ± 0.81
	T	1.53	−0.49	0.09	0.53	1.58	1.39
Ti	Male	6.76 ± 2.62	6.13 ± 3.11	9.86 ± 4.91	7.30 ± 3.00	3.98 ± 2.09	5.31 ± 2.31
	Female	7.23 ± 2.82	5.88 ± 1.83	9.61 ± 2.31	7.35 ± 3.00	4.60 ± 1.92	5.18 ± 2.17
	T	−0.72	0.33	0.22	−0.088	−1.39	0.26

Ca	Male	1285 ± 393	1355 ± 490	1096 ± 482	1027 ± 246	1258 ± 465	1212 ± 554
	Female	1623 ± 455	1751 ± 452	2785 ± 1797	1608 ± 619	1855 ± 861	1968 ± 706
	T	−3.35*	−3.2*	−7.07*	−6.60*	−4.26*	−5.95*
Mo	Male	0.057 ± 0.022	0.053 ± 0.019	0.053 ± 0.020	0.046 ± 0.017	0.066 ± 0.022	0.066 ± 0.019
	Female	0.057 ± 0.018	0.046 ± 0.012	0.042 ± 0.017	0.038 ± 0.017	0.066 ± 0.018	0.067 ± 0.026
	T	0.06	1.35	2.16*	2.05	−0.04	−0.30
Se	Male	0.288 ± 0.202	0.372 ± 0.258	0.347 ± 0.161	0.249 ± 0.174	0.294 ± 0.133	0.378 ± 0.162
	Female	0.343 ± 0.171	0.303 ± 0.207	0.259 ± 0.169	0.222 ± 0.132	0.292 ± 0.169	0.362 ± 0.148
	T	−1.21	1.05	2.16	0.78	0.08	0.48

[a] * Significant difference.

Table 2. Trace element contents in scalp hair from patients with liver cancer and normals

	Sr	Pb	Zn	Cu	Ni	Fe	Mn	Cr	Ti	Ca
Patient with cancer	3.01 ± 2.83	4.25 ± 2.56	163 ± 40	12.2 ± 2.9	1.11 ± 0.93	22.1 ± 11.8	4.44 ± 3.67	3.92 ± 2.75	4.33 ± 2.99	805 ± 344
Control	3.73 ± 3.59	4.04 ± 4.42	161 ± 33	13.0 ± 3.9	1.07 ± 0.60	30.0 ± 16.0	5.19 ± 3.78	4.26 ± 2.68	4.93 ± 2.75	934 ± 551
T Value	1.30	−0.34	−0.36	1.47	−0.37	3.32*	1.19	0.75	1.21	1.65

* Significant difference.

Considering the low selenium level we found that Se deficiency may be an important risk factor for the high incidence on the island, but it is not a direct cause. It was suggested that Se deficiency functioned through decreasing organic resistance against carcinogens.

In Table 1, it can be seen that the distribution of the Fe, Mn and Mo contents is obviously uneven and shows a reverse correlation with liver cancer incidence. Results from the case-control study (see Table 2) also show that the Fe content in scalp hair from liver cancer patients is much lower than that from normal persons. In addition, Mn tends to decline in cancer cases.

Suende county in Guangdong Province is another prevalence area of liver cancer in China. It was reported that (Li et al. 1987) a reverse correlation was found between Mn content in scalp hair and the mortality of liver cancer. Mn and Fe content in the patients' scalp hair was considerably lower than in the controls of the same county. These points agree perfectly with our observations.

In conclusion, the group of HBV carriers living in an area with low Se, Mn, Fe and Mo, and drinking stagnant water was the highest risk group for liver cancer. As multivalent elements, Se, Mo, Mn and Fe, are components of various oxidation-reduction enzymes, playing an important role in the metabolism of radicals in vivo. Radical damage to cells and genetic substances is an essential cause of carcinomas. Therefore, not only Se but Mo, Mn and Fe as well should be taken into consideration in cancer prevention research.

References

Gao RN, Tu JT, Gao XT (1981) Etiology study of primary liver cancer in prevalence area Chongming County of Shanghai. Cancer 4:4–7

Li WG, Xie JH, Yu SY, Zhu JY, Gong SL (1986) Relationship between the PLC mortality and Se level. Trace Element 3:58–61

Li ZX, Li GC, Tan ZS (1987) Determination of trace elements of scalp hair from liver cancer patients and normals in Suende County. In: Feng ZL (ed) Modern trace element research. China Environ Sci Press, p 103

Schrauzer GN, White DA, Schneider CJ (1977) Cancer mortality correlation studies – III. Statistical associations with dietary selenium intakes. Bioinorg Chem 7:23–34

Schrauzer GN, White DA, Schneider CJ (1978) Selenium and cancer: effects of selenium and of the diet on the genesis of spontaneous mammary tumors in virgin inbred female C_3H/St mice. Bioinorg Chem 8:387–396

Tan MG, Wang YX (in press) Evaluation of the literature values of 14 elements' contents in human hair. Zhonghua J Indust Hyg Occup Dis

Tu JT, Gao RN, Zhang DH, Gu BC, Xu GX (1987) Risk factors of primary liver cancer in the high prevalence area Chongming County: results from a five-year follow-up. In: Wagner G (ed) Cancer of the liver, esophagus, and nasopharynx. Springer, Berlin Heidelberg New York Tokyo, p 19

Yu SY (1985) Cancer mortality incidence and its relation to selenium levels in China. Biol Trace Element Res 7:21–29

Determination of Selenium Contents in Sera, Hair and Glutathione Peroxidase Activities in Whole Blood of Nasopharyngeal Carcinoma (NPC) Patients

JIA-CHEN HUANG[1,2], RONG-PU YANG[1], MEI-LIAN ZHENG[1], GONG-KAN FENG[1], ZHANG-EN MEI[1], and KUO-SHAN ZHENG[1]

Abstract

The selenium contents in sera from 50 NPC patients (29 males, 21 females) and 56 healthy persons (29 males, 27 females) in an NPC high risk area, Sihui county, Guangdong Province were analyzed by fluorimetry with 2,3-diaminonaphthalene. The results showed that the selenium level in the sera of NPC patients (0.084±0.016 ppm) is lower than that of healthy persons (0.096±0.029 ppm). The difference is significant, $P<0.05$. The selenium contents in hair from 54 NPC patients (31 males, 23 females) and 53 healthy persons (33 males, 20 females) in this county were analyzed. The selenium level in the hair of NPC patients (0.443±0.070 ppm) is lower than that of healthy persons (0.469±0.066 ppm). The difference is significant, $P<0.05$.

Furthermore, the difference in the selenium level of hair from healthy persons between the NPC high risk area, Sihui county (0.469±0.066 ppm) and the low risk area, Wuhua county (0.556±0.313 ppm), is not significant. No difference was found in the selenium level of rice between these two counties.

The activities of glutathione peroxidase, a selenium-containing enzyme, were analyzed from whole blood of 43 NPC patients (37 males, 6 females) and 37 healthy persons (17 males, 20 females). The glutathione peroxidase activities of

Table 1. Selenium contents in sera from NPC patients and healthy persons in Sihui county

	NPC patients			Healthy persons		
	Male	Female	Total	Male	Female	Total
Cases	29	21	50	29	27	56
Range (ppm)	0.057 −0.124	0.042 −0.124	0.042 −0.124	0.060 −0.260	0.069 −0.137	0.060 −0.260
$\bar{X}\pm SD$ (ppm)	0.090 ±0.014	0.067 ±0.015	0.084 ±0.016	0.098 ±0.037	0.094 ±0.018	0.096 ±0.029

$P<0.05$.

[1] Cancer Institute, Sun Yat-sen University of Medical Sciences, Guangzhou, 510060, Peoples Republic of China
[2] To whom correspondence should be addressed

Table 2. Selenium contents in hair from NPC patients and healthy persons in Sihui county

	NPC patients			Healthy persons		
	Male	Female	Total	Male	Female	Total
Cases	31	23	54	33	20	53
Range (ppm)	0.342 −0.565	0.292 −0.539	0.292 −0.565	0.342 −0.628	0.341 −0.580	0.341 −0.628
$\bar{X} \pm SD$ (ppm)	0.463 ±0.053	0.416 ±0.081	0.443 ±0.070	0.473 ±0.064	0.463 ±0.072	0.469 ±0.066

$P<0.05$

Table 3. Selenium contents in hair from healthy persons in the NPC high area and the low risk area

	NPC high risk area			NPC low risk area		
	Male	Female	Total	Male	Female	Total
Cases	33	20	53	21	20	41
Range (ppm)	0.342 −0.628	0.341 −0.580	0.341 −0.628	0.353 −0.640	0.345 −1.902	0.345 −1.902
$\bar{X} \pm SD$ (ppm)	0.473 ±0.064	0.463 ±0.072	0.469 ±0.066	0.472 ±0.074	0.644 ±0.430	0.556 ±0.313

$P>0.05$

Table 4. Selenium contents of rice from the NPC risk area and the low risk area

	NPC high risk area	NPC low risk area
No. of samples	99	100
Range (ppm)	0.008 − 0.081	0.008 − 0.190
$\bar{X} \pm SD$ (ppm)	0.034 ± 0.018	0.035 ± 0.024

$P>0.05$

whole blood in NPC patients (71.4±9.3) are significantly lower than those in the normal control group (86.9±10.7); $P<0.05$.

The data above mentioned showed that the selenium level in sera, hair and the glutathione peroxidase activities in whole blood of NPC patients is lower than that of healthy persons.

In order to explore the correlation of trace element selenium and NPC, the selenium contents in sera and hair from NPC patients and healthy persons in an NPC high risk area, Sihui county, Guangdong Province, were analyzed by fluorimetry with 2,3-diaminonaphthalene. As shown in Tables 1 and 2, the results showed that the selenium level in sera and the hair of NPC patients was lower than that of healthy persons; the difference is significant.

Table 5. The activities of glutathione peroxidase in whole blood from NPC patients and healthy persons

	NPC patients			Healthy persons		
	Male	Female	Total	Male	Female	Total
Cases	37	6	43	17	20	37
GSH-Px $\bar{X} \pm SD$ (Eu[a])	70.9±9.1	74.7±10.5	71.4±9.3	87.4±8.9	86.4±12.1	86.9±10.7

$P<0.05$
[a] Enzyme unit.

The difference in the selenium level of hair from healthy persons between the NPC high risk area, Sihui county and the low risk area, Wuhua county, is not significant (Table 3). No difference was found in the selenium level of rice between these two counties (Table 4).

The activities of glutathione peroxidase (GSH-Px), a selenium-containing enzyme, were determined with whole blood from NPC patients and healthy persons. The results showed that the glutathione peroxidase activities of whole blood in NPC patients are significantly lower than those in the normal control group (Table 5).

Summary

The selenium level in sera, hair and the glutathione peroxidase activities in whole blood of NPC patients is lower than that of healthy persons. These results suggest that there may be some association between a lower selenium level and NPC.

Subject Index

MIX
Papier aus verantwortungsvollen Quellen
Paper from responsible sources
FSC® C105338

If you have any concerns about our products,
you can contact us on
ProductSafety@springernature.com

In case Publisher is established outside the EU,
the EU authorized representative is:
Springer Nature Customer Service Center GmbH
Europaplatz 3, 69115 Heidelberg, Germany

Printed by Libri Plureos GmbH
in Hamburg, Germany